Introduction to
Engineering
Materials

Second Edition

MATERIALS ENGINEERING

Introduction to
Engineering
Materials

Second Edition

George Murray
Charles V. White
Wolfgang Weise

CRC Press
Taylor & Francis Group
Boca Raton London New York

CRC Press is an imprint of the
Taylor & Francis Group, an **informa** business

CRC Press
Taylor & Francis Group
6000 Broken Sound Parkway NW, Suite 300
Boca Raton, FL 33487-2742

First issued in paperback 2019

© 2008 by Taylor & Francis Group, LLC
CRC Press is an imprint of Taylor & Francis Group, an Informa business

No claim to original U.S. Government works

ISBN-13: 978-1-57444-683-8 (hbk)
ISBN-13: 978-0-367-38866-9 (pbk)

Library of Congress Cataloging-in-Publication Data

Introduction to engineering materials / George Murray ... [et al.]. -- 2nd ed.
 p. cm. -- (Materials engineering)
 Rev. ed. of: Introduction to engineering materials / George Murray.
 Includes bibliographical references and index.
 ISBN-13: 978-1-57444-683-8 (alk. paper)
 1. Materials. I. Murray, G. T., 1927- II. Murray, G. T., 1927- Introduction to engineering materials. III. Title. IV. Series.

TA403.M84 2007
620.1'1--dc22   2006032010

Visit the Taylor & Francis Web site at
http://www.taylorandfrancis.com

and the CRC Press Web site at
http://www.crcpress.com

Dedication

To my wife Tonny
GM
To Barbara
CW
To Stephanie
WW

Table of Contents

Authors

George T. Murray, Sc.D., is professor emeritus at the Department of Materials Engineering, California Polytechnic State University, San Luis Obispo. In addition to his teaching materials engineering for 15 years, he spent 28 years in the nuclear, aerospace, and electronic materials fields. He is the author of *Introduction to Engineering Materials* and editor and coauthor of the *Handbook of Materials Selection for Engineering Applications,* both published by Marcel Dekker, Inc. His interest in materials engineering is further reflected in the more than 35 technical papers published in recognized journals. Dr. Murray is a member of the Minerals, Metals and Materials Society of AIME and American Society for Metals (ASM) International. He received a B.S. degree in 1949 from the University of Kentucky, Lexington, an M.S. degree in 1951 from the University of Tennessee, Knoxville, and an Sc.D. degree from Columbia University, New York, all in metallurgical engineering.

Charles V. White, Ph.D., holds degrees in metallurgical engineering from the University of Illinois, Urbana, and the University of Wisconsin, Madison, and a Ph.D. from the University of Michigan, Ann Arbor. He has worked in the casting and forging industry for more than 40 years and as a consultant to the U.S. Navy where he has worked on the development of ductile iron products for military applications. He is involved in manufacturing research and consulting activities for many industries throughout the United States and Canada. Dr. White is a professor in the Industrial and Manufacturing Engineering Department at Kettering University (formerly General Motors Institute) in Flint, Michigan. He teaches in the area of forging and casting processing, failure analysis, and ethics in engineering practice. He is also active in professional registration activities and is a visiting professor at the University of Applied Sciences in Esslingen, Germany.

Wolfgang Weise, Ph.D., studied material science at the University of Erlangen–Nurnberg in Germany. He carried out his Ph.D. work at the Institute for Advanced Materials in Petten, The Netherlands, which is part of the Joint Research Center for the Commission of the European Communities. His thesis work was approved by the Ecole Polytechnique Federale de Lausanne Switzerland. For more than 10 years at Degussa AG, he worked on materials development in the Research and Development Department with a focus on metal matrix composites, joining of metals and ceramics, powder metallurgy, and the manufacturing of metal products by rolling, drawing, and sintering technologies. In 1998, he joined the faculty of the Mechanical Engineering Department at the University of Applied Sciences in Esslingen, Germany, as professor of materials technology. He is currently dean of the graduate school, which is responsible for the international exchange masters program. Dr. Weise continues to have an active roll in the industry as a consultant and as a teacher.

1 Classification of Materials

1.1 INTRODUCTION

Advancements in technology in most industries have been associated with the development of new materials and processes as well as advances in the state of the art of existing materials and combinations thereof. It is estimated that currently about 85,000 materials are available for industrial applications. In considering which materials to use for a particular structure or device, the selection process is further complicated by the wide variation in properties of materials with the manner in which the material is processed, for example, the heat treatment time, temperature, and cooling rate used for certain alloys. Some type of materials classification is an essential part of the selection process and an important element of engineering education.

In the following sections, materials will be classified in the broad categories of metals, polymers, ceramics, composites, and semiconductors. In subsequent chapters, these materials and their associated processing methods will be presented in more detail.

In the design and material selection procedures, items such as the recycling potential of the material and environmental problems must be considered.

1.1.1 RECYCLING

Recycling is generally considered to be a part of solid waste management strategies. Any strategy devised must be incorporated in the material selection and design steps.

Polymers are of much concern in recycling because many are nonbiodegradable and consist of about 20% of municipal solid waste. The thermoplastic polymers, which are easily formed and are abundant in packaging materials, have been given the most attention in terms of recycling technology. Representing about 90% of all plastics sold, they consist primarily of polyethylene, polyethylene terephthalate (PET), polystyrene, polypropylene, and polyvinyl chloride (PVC). For plastic containers, which make up a large part of their usage, codes have been molded into the container so that they can be identified in the separation process, remelted, and reformed into new containers. When these polymers have been comingled because of separation difficulties, they can still be added to virgin material and manufactured into useful products.

The thermosetting polymers, which include epoxies, phenol-formaldehyde, polyurethane, certain polyesters, and the polyimides, decompose on melting, and thus waste usage is limited to shredding and mechanically mixing with other materials to form some kind of useful particulate composite. Elastomers (rubbers) have the same nonmelting characteristics. After shredding they have been fabricated into floor mats, gaskets, sandals, and so on. Finely ground rubber can be mixed with other polymers to fabricate useful products. Scrap tires are reused in asphalt paving as part of the asphalt binder. In summary, if at all feasible, a thermoplastic polymer should be chosen when selecting a polymer for a design.

Metal recycling is much simpler and more economical than polymer recycling. In 1997, 64% of aluminum beverage cans were recycled in a simple melting operation. The energy required for melting aluminum cans is about 5% of that required for extracting aluminum from its ore. The percentage of aluminum cans recycled is increasing at a significant rate.

About 65% of steel, produced from a wide range of sources including automobiles, cans, appliances, construction, and railroads, is recycled by melting scrap, which in most cases is added to virgin metal or pig iron. In most situations, recycling of steel is economically feasible, exceptions being some appliances and other products for which collection and separation costs must be considered.

Clean copper scrap is marketable at about $0.80/lb. Brass runs about half of that, depending on the copper content. The prices for zinc, magnesium, and other metals vary significantly, but they are often recycled. Clean titanium scrap is valuable, but the supply is scarce.

In summary, the recycling of metals is an important factor in the design process. The life cycle of a product must always be considered.

Glass bottles represent about 10% of solid waste. Glass products must be separated by color, for example, clear, green, and brown, except when crushed glass (cullet) is used as a road base or some similar application. Some cullet can be remelted and added to virgin stock to make green or brown bottles. Windowpanes, light bulbs, mirrors, fiberglass, glassware, and so on are considered contaminants in container glass recycling.

1.2 METALS

Of all structural and device materials in use today (wood and concrete are excluded here), metals are probably the most widely used by either weight or volume percent, although the trend toward increasing polymer usage is evident (Figure 1.1). Metals are strong, hard, possess high electrical and thermal conductivities, and can be formed into many complex shapes by plastic deformation, machining, casting, and powder metallurgy methods.

1.2.1 IRON AND STEEL

Iron-based alloys (that is, those in which iron is the major constituent) include the cast irons, a number of steels, and a few iron-based alloys that are not called steels. Some wrought iron, which consists of iron silicate fibers in an iron matrix, is used today in the form of pipe, grills, and decorative objects.

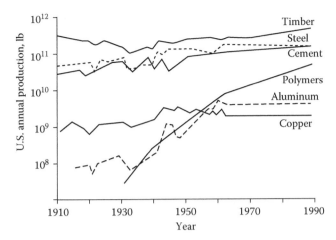

FIGURE 1.1 Trends in material usage in the United States. (From W. F. Smith, *Principles of Materials Science and Engineering,* 2nd ed., McGraw-Hill, New York, 1990. With permission.)

Tools made of iron first appeared around 1300 B.C. in Palestine, where an iron furnace has been found. Smelting of iron to extract it from its ore is believed to have begun about this time. Both steels and cast irons are basically alloys of iron and carbon, with the former containing up to about 2 wt % carbon and the latter about 2 to 4 wt % carbon. Why are the higher-carbon alloys called "cast irons?" It is simply because they are too brittle to permit forming to the desired shape and therefore must be cast to shape. Cast iron first appeared in China around 200 B.C. It is widely used in large, intricately shaped structures that cannot be machined or forged to shape. It comprises approximately 95% of the weight of a typical automobile engine, in the form of the engine block, camshaft, piston rings, lifters, and manifolds. There are several types of cast iron from which to choose, referred to as either white, gray, ductile, or malleable iron. These are examined in more detail in later chapters.

Steel, because of its strength, formability, and abundance, and therefore lower cost, is still, and will be for some time to come, the dominant metal for structural applications. Numerous steel alloys have emerged over the years, each class of such alloys having its own niche among the myriad of modern designs. These classes are listed and discussed briefly below, but first the steel numbering system needs to be examined.

1.2.1.1 Steel Numbering System

Many numbering systems are used to designate the composition of steels. The most common is that used by the Society of Automobile Engineers (SAE). Examples are given in the following sections. The American Society for Testing and Materials (ASTM) has formulated a system that is becoming more widely used, especially for procurement specifications. An ASTM specification may include more than one SAE steel since a given ASTM specification number frequently designates a range of properties plus the finishing process (for example, hot-rolled plate, cold-rolled strip, etc.). There also exist many foreign numbering systems. The Germans use a DIN number, the D standing for Deutsch; the Japanese use a JIS number; and the British use BS (British steel) number. Fortunately, a unified number system (UNS) has been developed by ASTM, SAE, other technical societies, and the U.S. government. These numbering systems have been compared in handbooks. *Metals and Alloys in the UNS System* (5th Edition, 1989) is updated every few years by ASTM and SAE. The *Metals Handbook*, Vol. 1, 10th Edition (American Society for Materials [ASM], International, Metals Park, Ohio, 1990), is also an excellent source, not only for clarifying the numbering system, but also for the properties, processing, and selection of all ferrous alloys.

1.2.1.2 Plain Carbon Steels

As the term *plain carbon steels* implies, these are alloys of iron and carbon. Some small quantities of manganese and silicon may be present, but carbon is the alloying element that dictates the properties. These were the first steels developed, are the least expensive, and have the widest range of applications. According to the SAE system, plain carbon steels are identified by a four-digit number, the first two digits being 1 and 0 and the second two digits indicating the wt % carbon content. For example, a 1045 steel is a plain carbon alloy containing 0.45 wt % carbon. The corresponding UNS number is GI0450.

1.2.1.3 Alloy Steels

The alloy steels contain small amounts, on the order of 2 to 5 wt % alloying elements in addition to carbon. Sometimes these steels are called low-alloy steels to distinguish them from high-strength steels such as maraging steels, which contain about 30 wt % alloying elements. The most common alloying elements in the low-alloy steels are nickel, chromium, and molybdenum. These alloys permit the hardening of the steel to greater depths when quenched from elevated temperatures. Whereas plain carbon steels can achieve only about 1/8 in depth of hardening, the alloy steels can be hardened to depths of 2 in. This means that a shaft of 4-in. diameter could be hardened throughout the entire cross section. As might be expected, these steels are used in large-diameter shafts such as crankshafts, axles, rolls, and large bolts. In the heat-treated condition their strengths are usually in the vicinity of 1035 to 1380 MPa (Mea pascals) (150 to 200 ksi [thousand pounds per square inch]). Plain carbon steels can also attain this strength level, but only in thin sections.

SAE numbers for alloy steels are also four-digit numbers, with the first two digits specifying the alloying elements and their percentage and the last two digits their carbon content. One cannot look at this number and immediately know the alloy content unless it has been memorized. To find the alloy content, one must refer to handbooks. The last two digits indicate the carbon content, just as in the case of the plain carbon steels. Page 152 of the *Metals Handbook,* Vol. 1, 10th Edition (ASM International, Metals Park, Ohio), shows that an SAE 4340 steel contains 0.40 wt % carbon, 1.65 to 2.00 wt % nickel, 0.70 to 0.90 wt % chromium, 0.20 to 0.30 wt % molybdenum, 0.65 to 0.85 wt % manganese, and small amounts of silicon, phosphorus, and sulfur. The corresponding UNS number is G43400.

1.2.1.4   High-Strength Low-Alloy Steels

The term *high-strength low-alloy steels* is a misnomer since the name implies that such a steel is of a higher strength than an alloy steel, but this is not the case. They are stronger than the plain carbon steels and have a lower alloy content than the alloy steels. They are primarily structural steels used in large buildings, bridges, ships, oil and gas pipelines, automobiles, and pressure vessels. They are sometimes called microalloyed steels, since they contain minute quantities of niobium, titanium, vanadium, and molybdenum. Their exact compositions and processing methods are often proprietary information. They contain less than 0.2 wt % carbon, such a low carbon content being necessary for welding operations. The SAE numbering system described above is not used for these steels. They were identified by trade names as they were developed, but the trend today is to use ASTM designations or UNS numbers. ASTM specification A572 covers an important group of these steels. Their strength levels are on the order of 552 MPa (80 ksi). An example of the application of plain carbon and low-alloy steels, where cost is a major factor, is shown in Figure 1.2 for an aircraft cargo loader. The majority of the heavy plates, rectangular bars, channels, and angles employ inexpensive 1022 steel in the hot-rolled (hot-finished) condition. The bridge scissors require a higher strength, and thus a slightly more expensive cold-drawn 1018 steel is selected. For the sections requiring higher strength, a plain carbon 1018 quenched and tempered steel is appropriate.

1.2.1.5   Specialty Steels

The three categories of steel listed above probably account for more than 90% of steel use. The remainder includes a group of steels that we call specialty steels. The most prominent steels in this group are the stainless steels. Another class of specialty steels is the tool steels.

Stainless steels require a minimum of 10.5% chromium to be stainless (that is, to be very highly resistant to corrosive environments). They achieve their stainless properties through the formation of a thin, adherent chromium oxide surface layer. Other elements are added for a variety of reasons, which will be considered in more detail in later chapters. The stainless steels are classified as martensitic, ferritic, austenitic, and precipitation-hardening stainless steels. Their numbering system is somewhat confusing and fits into no particular pattern as

FIGURE 1.2 Cargo loader for wide-body aircraft. (Courtesy of the Lantis Corp.)

do the plain carbon and alloy steels. They can also be identified by a UNS number, but this method has not yet caught on.

Austenitic stainless, the most common and widely used stainless steel, falls into the 300 series. Type 304 is best known because it is used for most household applications. The ferritic and martensitic groups both have a 400 series of numbers.

Of the martensitic steels, the 410 and 440 stainless are the most popular. The precipitation-hardened stainless steels are numbered according to the percent of chromium and nickel content. PH 17-4 is the best known steel in this group, probably because it is used for golf club heads and other consumer products. It contains 17 wt % chromium and 4 wt % nickel. Other such steels, PH 15-5 and 17-7, are similarly numbered according to chromium and nickel content. A relatively recent addition to the family of stainless steels is the duplex stainless steels, which are a mixture of the ferritic and austenitic types. How these names emerged during the development of stainless steel metallurgy will be discussed later. The composition of these stainless steels, using both the numbering system above and the UNS system, can be found on page 843 of Vol. 1 of the *Metals Handbook,* 10th Edition.

Tool steels, which are used for cutting and forming operations such as shears and dies, are a subject by themselves. Another long-standing reference of a type similar to the ASM *Handbook* deals only with tool steels. The tool steels are normally divided into four subgroups: air-hardened, oil-hardened, shock-resistant, and high-speed steels. The first two groups are so named because of the quenching media used in heat treatment. The shock-resistant tool steels are used for chisels, punches, driver bits, and other applications in which the tool is subjected to impact loading. The high-speed group consists of those that experience very high temperatures during use, such as machine tools and hot-working dies. A significant letter preceding the number designates the high-speed steels. The molybdenum-containing high-speed tool steels have the letter M preceding their number, the tungsten-containing high-speed steels have the letter T preceding their number, and those used at very high temperatures for hot forming of metals have the letter H preceding their number, which is independent of their major alloying element. The letters preceding the numbers for the air-hardened, oil-hardened, and shock-resistant steels are A, 0, and S, respectively. These designations, together with their corresponding UNS number, can be found on page 758 of the *Metals Handbook.*

1.2.2 Nonferrous Alloys

The principal nonferrous alloys are those in which the base metal (that is, the major constituent) consists of either aluminum, copper, nickel, magnesium, titanium, or zinc. Just as there is a *Metals Handbook* for iron and steel, there is also one for the nonferrous alloys. The 1990 *Metals Handbook,* Vol. 2, 10th Edition by ASM International includes properties, processing, selection, and numbering systems for all prominent nonferrous alloys. The numbering system for nonferrous alloys is just as confusing as that for ferrous alloys. There is the old numbering system, mostly developed by alloy manufacturers, plus the UNS numbers and ASTM specifications that may include more than one alloy.

1.2.2.1 Aluminum Alloys

Aluminum is produced by the electrolytic reduction of aluminum oxide. In 1900, only 6384 tons of aluminum was produced, compared to about 20 million tons in 1997. The advent of the aircraft industry early in the twentieth century propelled aluminum to the top of the list of nonferrous metal use. Subsequently, many other uses for aluminum were developed. About 1.5 million tons were used in 1997 for beverage cans alone, of which approximately 64% was recycled. The percentage of recycled aluminum is expected to grow rapidly with the introduction of recent conservation and environmental regulations. The energy required for recycling aluminum is about 5% of that required to extract aluminum from aluminum oxide, a tremendous savings in dollars and energy.

The attractiveness of aluminum as a structural metal resides in its light weight. Its density is only 2.7 g/cm³, compared to 7.87 g/cm³ for iron. Aluminum alloys can be heat-treated to attain yield strengths in the vicinity of 550 MPa (80 ksi), compared to high-strength steels of about 1550 MPa (225 ksi). On a strength-to-density ratio (sometimes referred to as *specific strength*), aluminum alloys come out ahead, being approximately 204, while the higher-strength steels have a ratio of 197. These high-strength steels are most often used in gears, shafts, axles, and the like. More appropriately, we should compare the strength-to-density ratios of aluminum alloys to those for structural steels. The latter have a strength-to-density ratio of about 70. In this case, aluminum alloys come out on top by a factor of about 3. It is obvious why aluminum became the structural material for aircraft fuselages, wings, and support structures.

A bonus also arises when comparing the corrosion resistance of aluminum to steels (other than stainless steels, which, aside from weight problems, are also quite expensive). The most common aluminum alloys are those containing quantities of 3 to 4% copper and zinc. These alloys are in the 2000 and 7000 series in the four-digit numbering system. Aluminum alloys produced in wrought form, such as sheet, rod, wire, and extruded shapes, are classified according to their content of major alloying elements. The first digit indicates the alloy group, while the other digits refer to modifications of the original alloy. The second digit originally had something to do with aluminum purity, but since all alloys are now made from essentially the same commercial-purity aluminum, this designation has lost its significance. The numbers are listed according to alloying element in Table 1.1. These alloys can also be specified by a UNS number.

TABLE 1.1
Wrought Aluminum Alloy
Number System

Aluminum	
(99.0% minimum aluminum)	**1xxx**
Alloys by Major Element	
Copper	2xxx
Manganese	3xxx
Silicon	4xxx
Magnesium	5xxx
Magnesium and silicon	6xxx
Zinc	7xxx
Other elements	8xxx

TABLE 1.2
Cast Aluminum Alloy Numbering System

Aluminum (99.0% minimum aluminum)	1xx.x
Alloys by Major Element	
Copper	2xx.x
Silicon with copper and magnesium	3xx.x
Silicon	4xx.x
Magnesium	5xx.x
Zinc	7xx.x
Tin	8xx.x
Unused series	6xx.x

In recent years, 2024 (UNS A92024) and 7075 (UNS A97075) have been the most widely used wrought alloys where specific strength is important. Alloy 3003, which contains 1.2 wt % manganese, is used for beverage cans. Alloy 1350, which is 99.5% aluminum, makes up most aluminum electrical connectors. Electronic interconnects in integrated circuits are made from an extra-high-purity aluminum with a few percent copper added to reduce electromigration. Alloy 6061 is used for car bodies.

Aluminum casting alloys now have a four-digit series numbering system similar to that for wrought alloys. Their numbering system is listed in Table 1.2. Aluminum casting alloys were for many years designated by a three-digit number, sometimes with a letter preceding the number. Alloy 242, for example, is a sand-cast aluminum–copper alloy that now has the number 2420 (UNS A02420).

1.2.2.2 Copper Alloys

Copper ores found in the United States contain about 1% copper in the form of copper sulfide. Copper sulfide concentrates are smelted to yield a matte that is a mixture of copper and iron sulfides. The copper-containing mattes are melted again in a converter. Air is blown through the mattes to oxidize the remaining sulfur. The remaining copper is called *blister copper* and is about 99% pure. The blister copper is further fire-refined to remove other impurities, leaving a *tough-pitch* copper of around 99.5% purity. This is suitable for many alloys, but for some a purity of 99.9% is obtained by electrolytic refinement of the tough-pitch copper.

Copper–zinc brasses are the most frequently used copper alloys. The zinc content can vary from 5 to 40 wt % and range in yield strengths from 100 to about 345 MPa (14.5 to 50 ksi), depending on the zinc content and the amount of cold-work. Known as cartridge brass, 70% Cu–30% Zn is also used for radiator cores, plumbing accessories, nuts, bolts, fasteners, heat exchangers, lamp fixtures, and a host of small parts.

Copper–tin alloys are the original bronzes for which the Bronze Age was named. They are sometimes referred to as tin bronzes, containing up to 10 wt % tin, but are also known as phosphor bronzes because of the addition of phosphorus to remove oxygen. These bronzes can be cold-worked to increase strength and in this form are stronger but more expensive than the brasses. They are used in the as-cast condition for gears, bells, and bearings. Wrought tin bronzes have strengths up to 840 MPa (120 ksi), while the as-cast parts are much weaker, being on the order of 140 MPa (20 ksi).

Other copper alloys of significance include the copper–beryllium alloys, which can be heat-treated to exceptionally high strength levels, on the order of 1070 MPa (155 ksi). These alloys are the strongest of the copper-based alloys and require only about 2 wt % beryllium to achieve this strength level. They are used for springs, electrical switch contacts, diaphragms, and fasteners in the cold-worked state and to some extent as molds, bushings, gears, and valves in the cast state. Copper–aluminum alloys, usually containing about 6 to 8 wt % aluminum, are known as aluminum bronzes and are also relatively high-strength copper alloys achieving about 517 MPa (75 ksi) strength levels in the wrought cold-worked state and around 370 MPa (54 ksi) in the cast state. They also have good corrosion resistance, owing to the aluminum content, and are frequently used for gears, bearings, bushings, valves, and parts that may be exposed to marine environments. Copper alloys now have a UNS, which was derived from the Copper Development Association numbers, but as with other alloys they can also be described by an ASTM specification.

1.2.2.3 Other Nonferrous Alloys

Nickel-based alloys, although in fifth place on the nonferrous metals production list, are far more important than this statistic suggests. Their claim to fame is their good high-temperature strength and corrosion resistance. Both nickel- and cobalt-based alloys are used almost exclusively for the components in the hottest regions of jet engines (that is, the turbine blades and compressor parts). For this reason, they have been termed *superalloys*. Although cobalt-based alloys were the front-runners here, nickel alloys have taken the lead, owing primarily to the scarcity and uncertainty of the cobalt supply. All of these alloys contain up to 25 wt % of chromium for high-temperature oxidation resistance. They are known by trade names such as Inconel, Incalloy, Waspalloy, Udimet, Hastelloy, and Nimonic as well as by UNS numbers. Inconel 625 is designated N06625 in the UNS system.

The room-temperature use of nickel alloys includes the use of some of the Inconels, but most are nickel–copper alloys, known as Monels. These contain about 30% copper and have a room-temperature strength on the order of 262 MPa (38 ksi), although one alloy can be heat-treated to strengths nearly twice this level. They have good corrosion resistance and find frequent application as valves, pumps, springs, and so on, in the marine and petroleum industries. To some extent they are in competition with the aluminum bronzes (copper–aluminum), and in most cases would win out if it were not for the high cost of nickel, which is nearly three times that of copper.

When it first emerged in the 1940s, titanium was called the *wonder metal* because it appeared to have all the desirable characteristics of strength, low density, corrosion resistance, and formability. It has not quite lived up to the forecasts but is, nevertheless, still in the running to replace certain other metals. The density of titanium, 4.5 g/cm^3, falls between that of aluminum and iron. Alloys such as the common Ti-6Al-4V have been heat-treated to strength levels of about 1170 MPa (170 ksi), and on a strength-to-density ratio beat both aluminum and iron alloys. Because of this characteristic, and because it retains a relatively high strength at temperatures reaching 500°C (930°F), titanium is the obvious metal for the fuselage structure of supersonic aircraft. It cannot be used at temperatures above 500°C because of its high reactivity with gases. Thus, it cannot be used in jet engines unless a suitable protection coating is developed. Also because of the high reactivity, precautions must be taken during melting and hot-forming operations. Such difficulties, along with the high cost of extracting it from its ore, have priced titanium out of reach for many applications. It is being used extensively in sports equipment and medical body implants where material cost is not always the deciding factor, the strength-to-weight ratio being more important. The numbering system for titanium alloys created by manufacturers is based on the percent alloying element. The popular alloy titanium–6% aluminum–4% vanadium is known as Ti-6-4. The corresponding UNS number is R56400.

Magnesium alloys are in competition with aluminum alloys because of their low density (1.74 g/cm). They are used in the aircraft industry and are competing with aluminum for several applications in the automotive industry. Their disadvantages include their brittleness, requiring hot- rather then cold-forming procedures, and their relatively high cost — about 2.5 times that of aluminum. Their strengths are on the order of 200 MPa (29 ksi) and are not competitive with aluminum on this basis. They are used in cast, extruded, and forged conditions. Magnesium alloys are identified by a UNS number in addition to their original numbering method, which included both letters and numbers. An AZ92A alloy contains 9 wt % aluminum and 2 wt % zinc and has a UNS number of M11920.

Zinc is used more as an alloying element and in galvanizing steel than as a base metal. A few zinc alloys containing a small percentage of aluminum are used for die castings, a process whereby the molten alloy is squeezed into a die. Zinc alloys also have a UNS numbering system in addition to their manufacturer's number and ASTM specifications.

Tin, although not used in large quantities, is a major constituent in solders, some bearing alloys, and pewter. It has been used for many years as a coating for other metals. It is gaining more importance today in solders because of the ongoing effort to eliminate lead from solders.

The precious metals, especially gold, platinum, and silver, have found extensive usage in the microelectronics industry. Other significant applications are still found in jewelry and dental applications.

1.3 CERAMICS

Ceramics consist of the combination of one or more metals with a nonmetal. They are often classified according to the nonmetallic element, for example, as oxides, carbides, nitrides, and hydrides, depending on whether the metal is combined with oxygen, carbon, nitrogen, or hydrogen, respectively. The halides, such as the chlorides, bromides, iodides, and fluorides, are today considered to be a special type of ceramic not discussed in this book. In addition the metal–nonmetal semiconductors will be considered under that heading.

The largest group of ceramics found in nature is the silicates. Silicon and oxygen are the most abundant elements in the Earth's crust. Silicon dioxide, known as silica, is crystalline in natural form, and a common source is the tiny crystals found in beach sand. When silica is heated above 1700°C it becomes a liquid and if slowly cooled, it forms crystals where the atoms take up specific positions in the crystal lattice. If cooled at more accelerated rates the atoms do not have time to take up these positions, and it becomes a glassy amorphous solid. Silica is the main ingredient in glass.

During the past several decades high-performance ceramics have emerged as prominent materials in electronic packages, fiber optics, advanced cutting tools that require hot hardness, bearings for their wear resistance, certain automotive parts (for example, spark plugs, fuel injector parts, diesel precombustion chambers, and exhaust port liners), aircraft and missile parts, and medical device applications such as body implants (for example, hip replacements). The more widely used high-performance ceramics include alumina, magnesia, silicates, silicon carbide, and silicon nitride. Often the primary selection criteria for a ceramic material for a specified application is based on one or on a combination of properties involving thermal properties, wear resistance, electrical insulation, and strength or hardness at elevated temperatures. Table 1.3 illustrates these applications for some popular high-performance ceramic materials. (The traditional ceramics such as pottery are not subjects of this book.)

TABLE 1.3
Ceramic Materials and Their Applications

Material	Significant Property	Applications
AlN	Thermal expansion coefficient	Microelectronic packages
Si_3N_4	Wear, tribological properties	Ball bearings against steel
Al_2O_3 with SiC whiskers	With hot hardness	Cutting tools
Si_3N_4 hot pressed	Wear, thermal shock resistance	Cutting tools for cast iron
Stabilized zirconia	Wear resistance	Drawing dies, paper slitters
Si_3N_4, SiC	High temperature strength, impact resistance	Turbocharger rotors
Aluminum titinate	Thermal shock resistance	Exhaust port liner

1.4 POLYMERS

Polymers can generally be classified into three categories: thermoplastic polymers, called *thermoplasts*; thermosetting polymers, called *thermosets*; and *elastomers*, better known as rubbers. Thermoplasts are long-chain molecules that can be easily formed by heat and pressure at temperatures above a critical temperature referred to as the *glass temperature*. This term was first applied to glass to represent the temperature at which glass became plastic and easily formed. The glass temperature for many polymers is below room temperature, and hence these polymers are brittle at room temperature.

Polyethylene is the most common thermoplast and consists of a long chain of hydrogen atoms attached to each side of the carbon atoms. It has a nominal glass transition of −100°C and is easily formed at room temperature. PVC is another popular member of this group and is obtained by replacing one of the hydrogen atoms of polyethylene with a chlorine atom. It has a glass transition temperature of 87°C. The most widely used thermoplasts and some of their applications are listed in Table 1.4.

Thermosets are polymers that take on a permanent shape or set when heated, although some will set at room temperature. An example of the latter are epoxies that result from combining an epoxy polymer with a curing agent or catalyst at room temperature. Thermosets consist of a three-dimensional network of atoms rather than being a long-chain molecule. They decompose on heating and thus cannot be reformed or recycled by melting. Some common thermosets and their applications are listed in Table 1.5.

Elastomers are polymeric materials whose dimensions can be changed drastically by applying a relatively small force, but they return to their original shape when the force is released. The molecules are extensively kinked such that when

TABLE 1.4
Common Thermoplastic Polymers and Their Applications

Polymer	Applications
Polyethylene	Sheet forms, tubing, tanks, bottles, irrigation pipe, laboratory instruments
Polyvinyl chloride (PVC)	Pipe, conduits and fittings, cable insulation, down spouts, footware, houseware, chemical equipment
Polystyrene	In clear form used for plastic containers; in foam form used as insulating materials, packaging
Acrylonitrile-butadiene-styrene (ABS)	Pipes, tanks, appliances, artificial rubbers, recreation; competes with PVC
Polycarbonates	Helmets, windshields, electrical insulators, tubs; relatively high cost; can be used to 140°C (285°F)
Polypropylene	Piping for chemicals, duct systems, housewares, housings
Polyimides	Higher temperature usage — to 200°C

TABLE 1.5
Application of Reinforced Thermosets

Epoxies	Electrical molding, adhesives, printed wiring boards, aerospace
Unsaturated polyesters	Appliance housings, automotive body panels, boats, shower stalls, fans, pipes, tanks, electrical
Phenolics	Motor housings, telephones, electrical fixtures, particle board, brake and clutch linings, electrical, decorative
Polyurethanes	Automotive parts such as steering wheels, head rests, and instrument and door panels
Bismaleimides	Elevated temperature applications, electronics, aerospace
Polyimides (general)	Excellent dielectric properties, wear- and heat-resistant applications
Vinylesters	Corrosion-resistant uses, electrical equipment, chemical process equipment

Source: G.T. Murray, Ed., *Handbook of Materials Selection for Engineering Applications*, Marcel Dekker, New York, 1997. (With permission.)

a force is applied, they unkink, straighten, or uncoil and can be extended up to 1000% with minimal force and returned to their kinked shape when the force is released. In general, they must be cooled below room temperature to be made brittle (that is, their glass temperature is below room temperature). Natural rubber, polyisoprene, as obtained from rubber trees is a sticky, gum-like material that in addition to the isoprene molecule contains small amounts of liquids, proteins, and inorganic salts. To be useful it must be further processed. In 1839, Goodyear developed the vulcanization process whereby the addition of sulfur atoms formed cross-links between the long-chain molecules. These cross-links pull the molecules back to their original positions when the force is released. Some widely used elastomers and their applications are listed in Table 1.6.

Polymers are usually designated by their name or an abbreviation. Polyvinylchloride is known as PVC. Acrylonitrile-butadiene-styrene, a long-chain copolymer, is known as ABS. The Society of Plastic Engineers (SPE) has established identification numbers that identify the material, its form, and additives that are introduced for different applications. For example, flexible PVC has an SPE number of 29, and rigid PVC has and SPE number of 30.

1.5   COMPOSITES

The term *composites* as used here describes a mixture of two or more materials, each being present in significant quantities and each imparting a unique property to the mixture. These materials may be in the shape of fibers, sheets, or particulates usually embedded in a metal, polymer, or, in a few cases, a carbon or ceramic matrix. The object is to use the most desirable characteristic of each material to achieve properties in composite form that exceed those of the individual components alone. Combining a strong but brittle ceramic fiber in a ductile and weaker

TABLE 1.6
Commercial Elastomers and Their Applications

Material	ASTM Class	Applications
Natural rubber		Styrener
Styrene–butadiene rubber	R	Synthetic rubber; used in place of natural rubber in tires, belts, and mechanical goods
Polybutadiene	R	Synthetic rubber
Polychloroprene	R	Synthetic rubber; gaskets, V-belts, cable coatings
Ethylene–propylene rubbers	M	Saturated carbon chain elastomer; copolymer-resistant to ozone and ultraviolet radiation, good chemical resistance to acids but not to hydrocarbons; cable coating, hoses, roofing, automotive parts
Poly(propylene oxide) rubbers	O	Polyester elastomer
Silicone elastomers	Q	Useful from −101 to 316°C; tubing, gaskets, molded products
Polyurethane polyester	Y	Thermoplastic elastomers; combination of a rubbery and a hard phase
Polyester polyether	Y	Thermoplastic elastomers; footwear, injected-molded parts
Elastomeric alloys	Y	Extruded, molded, blow-molded, and calendered goods (thermoplastic)

Source: G.T. Murray, Ed., *Materials Selection for Engineering Applications*, Marcel Dekker, New York, 1997. (With permission.)

metal matrix results in a composite whose strength lies somewhere between that of the ceramic fiber and that of the metal matrix but at the same time is not as brittle as the ceramic alone.

Fiberglass is an example of the fiber type of composite, plywood the laminar kind, and concrete the particulate composite in which gravel is mixed with cement. Wood is a natural composite consisting of cellulose, lignin, and other organic compounds. The possibilities of material combinations are extensive. Some advanced composite types in use today are polymer fibers in a polymer matrix, such as Kevlar fibers, introduced in 1972 by Dupont; ceramic fibers in a metal matrix; ceramic particles in a metal matrix; metal fibers in a polymer matrix; and carbon fibers in several matrices, including a carbon matrix. Some typical composite materials and their applications are listed in Table 1.7 for the metal matrix and in Table 1.8 for the polymer matrix type. The ceramic matrix and carbon–carbon composites are less developed and accordingly have fewer applications than the ones listed in Tables 1.7 and 1.8. Nevertheless, they have found applications in rocket nozzles, nose cones, heat shields of reentry vehicles, and aircraft brakes.

Graphite–epoxy composites were used in conjunction with a honeycomb structure to attain the light weight and strength necessary to enable *Voyager* (Figure 1.3) to achieve its global circuit without refueling. Scaled Composites

TABLE 1.7
Manufacturers of Metal Matrix Composites (MMCs)

Manufacturer	Type, Method, and Potential Applications
Duralcan USA (Division of Alcan)	Al alloy-particulate SiC for gravity casting and aluminum–alumina for wrought materials. Driveshaft tubing, automotive front disc-brake rotor, golf club heads, bicycle frame tubes, and miscellaneous other parts. Both wrought and cast products are produced by stirring particulate reinforcement into liquid metal.
Lanxide Corporation	Pressureless molten metal infiltration of a ceramic particle bed or perform. Includes a variety of aluminum alloys with SiC and alumina particles. Net and near-net shapes by investment, die, and sand casting methods. Squeeze casting may also be applied to some products. Automotive brake calipers, disc brake rotors, connecting rods, and electronic applications.
Dynamet Technology, Inc.	Cold and hot isostatic processing of titanium alloys with TiC, TiAl, and TiB_2 particulate. Medical implants, automotive, and sporting goods.
Alcoa	Pressure infiltration of ceramic performs by aluminum to produce composites of about 70% volume reinforcement for net shape components for electronics and thermal management systems. A second family of products is produced by the blending of particulate ceramic particles (15–40%) with aluminum powders followed by powder metallurgy consolidation. Aerospace and automotive uses.
Thermal Ceramics (Babcock & Wilcox)	Discontinuous fiber MMC. Infiltration of alumina–silica (Kaowool) or alumina fiber performs with liquid Al by squeeze casting.
Ametek, Inc.	Copper with 100 and 20 μm Mo particles. Used for heat sinks, substrates, and thermal spreaders.
SCM Metal Products, Inc.	Dispersion-strengthened copper with alumina and with 10% niobium additions. Internal oxidation process in which a dilute Cu–Al alloy is preferentially oxidized within a copper matrix. Mill products are produced using HIP and extrusion. Used for resistance-welded electrodes and components requiring high conductivity and strength.
DWA Aluminum Composites, Inc.	Powder metallurgy–processed MMCs with a range of ceramic particulates for automotive, aerospace, and specialized high-performance applications.
PCC Composites, Inc.	Pressure infiltration by a proprietary process of ceramic particle performs to produce MMCs with high reinforcement volumes for thermal management and electronic applications.
Textron Specialty Materials	SiC continuous fibers are woven to produce fabric or drum wound for plasma spraying. Used to reinforce Al alloys by hot molding and titanium alloys by hot pressing with titanium foils. Aerospace and defense applications.
3M Company	Continuous Al matrix composite reinforced with alumina fibers and titanium matrix composite with SiC fibers. Applications same as for Textron material.

Source: G.T. Murray, Ed., *Materials Selection for Engineering Applications*, Marcel Dekker, New York, 1997. (With permission.)

TABLE 1.8
Comparison of Common Polymeric Matrix Materials

Polymer	Processing Temperature (°C)	Mold Shrinkage (%)	Continuous Use Temperature (°C)	Chemical Resistance	Cost	Principal Application Market	Comments
Epoxies	120–177	1–5	80–180	Fair	Medium	Aerospace	High moisture absorption
Polyesters	60–150	5–12	70–130	Fair	Low	Automotive	
Vinyl esters	100–150	5–12	80–150	Good	Medium	Automotive	Higher fatigue resistance than polyesters
Phenolics	90–125	0.5–1	120–260	Fair	Low	Aerospace	Excellent fire resistance
Polyurethanes	60–120	2	80	Good	Low	Automotive	
Polyetherketones (PEEK)	350–430	1.1	120–250	Good	Very high	Aerospace	
Polyphenylene sulfides (PPS)	315–430	0.6–0.8	100–200	Good	High	Aerospace	
Nylons	260–330	0.7–1.8	140	Good	Medium	Automotive	High moisture absorption
Polypropylene	200–290	1–2.5	110	Good	Low	Automotive	

Source: G.T. Murray, Ed., Materials Selection for Engineering Applications, Marcel Dekker, New York, 1997. (With permission.)

FIGURE 1.3 *Voyager* attained its light weight through the use of graphite–epoxy composites in conjunction with a honeycomb structure. (Courtesy of Voyager Enterprises, Inc.)

Corporation, the manufacturer of *Voyager*, has also designed and is constructing the *Global Flyer*, which is expected to be the first solo-piloted aircraft to fly nonstop around the world in 80 hours. Scaled Composites Corporation has recently developed a spacecraft that is ejected into space from a plane and lands similarly as any other type of aircraft. On October 4, 2004, their *SpaceShipOne* was the first to fly three people into suborbital space twice within 5 days and return, thereby winning a coveted $10 million prize. The fuselage and wings of the new Boeing 7E7 are constructed of composite materials, primarily graphite–epoxy. All of these achievements have been made possible by the generous use of composite materials.

1.6 SEMICONDUCTORS

Semiconductors are a more recent addition to the classes of materials that have become important during the past 50 years. Semiconductor materials have an electrical conductivity between conductors, which are usually metals, and insulators, which are primarily ceramics, although some polymers could be considered as insulators. A few polymers are also considered to be conductors, and a number of organic materials possess semiconducting properties. Semiconductors can be pure materials such as germanium, silicon, boron, gray tin, carbon (in diamond form), selenium, tellurium, or compounds such as gallium arsenide, gallium phosphide, and a multitude of other semiconducting compounds. Although the first transistor was made of germanium, silicon is the dominant material used for chip substrates in the microelectronics industry, for example, in integrated circuits and many other devices.

The binary compounds can be grouped according to their respective valencies and denoted as III-V, IV-VI, and II-VI semiconductors. There are also a number of ternary semiconductng compounds. Gallium arsenide is the most widely used

TABLE 1.9
Applications of Compound Semiconductors

Compound	Applications
GaAs	Integrated circuit transistors; high speed, light emitting diodes, lasers, photovoltaic cells
AlGaAs	Near-infrared photoluminescence, bipolar transistor heterojunction
GaP	Light emitting diodes, high-temperature circuits, magneto-optical devices
GaN, AlGaN, InGaN	Materials emitting in the ultraviolet region, blue and green diodes, photoluminescence
ZnO	Varistors, optoelectronics
SiC	Field effect transistor, light emitting diodes
ZnSe	High-power laser windows
InP	Optical band-gap thermometry, lasers
InAs	Infrared detectors, photon detectors
AlN	Ultraviolet source, microwave packages
BN	Light emitting junctions
CdS	Metal-insulator-conductor

semiconducting compound. In addition to its use in integrated circuits, it is used in lasers, light emitting diodes, and photovoltaic cells. In integrated circuits it shows faster switching speeds, passes signals more rapidly, consumes less power at high speeds, and operates over a wider temperature range than do similar silicon circuits. Because of its high cost, it will probably never be a significant threat to silicon circuits. Many, if not most, of the compound semiconductors are used in the broad field of optoelectronic devices, which include the light emitting diodes, lasers, infrared and photodetectors, and photoluminescence devices. Some of the more common compound semiconductors and their associated applications are listed in Table 1.9.

Pure semiconductors, known as intrinsic semiconductors, have too large a band-gap for practical device construction and too much energy is required to move the electrons from the filled to the conduction band. Consequently, impurities (dopants) are added in small quantities (parts per million), which causes a marked increase in conductivity. By controlling the concentration of dopant, the characteristics of the device can be tailored to obtain the desired performance.

1.7 MORE RECENT DEVELOPMENTS

1.7.1 NANOTECHNOLOGY

The terms *nanotechnology, nanoscience,* and *nanosystems* have been described and defined in different ways. The prefix *nano* is a unit of measurement that involves the number 10^{-9}. The more familiar prefixes 10^{-3} and 10^{-6} are referred to as *milli* and *micro* measures, respectively, and are frequently used in the measure of time, distance, and weight, such as milligram, millisecond, millimeter, microgram, microsecond, and micrometer. Accordingly, the prefix *nano* is used as nanogram, nanometer, and nanosecond. For this discussion, the primary interest is in size and as such the focus will be on the term *nanometer*. Thus, nano-technology is the science and engineering of materials, structures, and devices on a nanoscale. In the more frequently used metric system we are speaking of nanometer structures. A micrometer, referred to as micron, is a distance that can be easily measured with an electron microscope. Nanostructures are somewhat more difficult to visualize and to construct. However, with the more recent developments such as the scanning tunneling and atomic force microscopes, nanometer distances can be measured, and devices constructed to move atoms or molecules. The typical interatomic distance in crystals is the order of 0.5 nanometers.

The first significant applications of nanotechnology will probably involve taking microelectronics to the nanolevel. Existing integrated circuits are fabricated on a micron scale. By going to the nanoscale, integrated circuits could possibly be fabricated with as many as 10^{12} circuits on a single chip. Simple nanoscale circuits, for example, logic gates, and electronic switches have already been constructed in the laboratory. The first nanoelectronic materials and devices will most likely be combined in some fashion with existing chip manufacturing processes. Entrepreneurs speculate that nanoelectronic memory products will enter the market by 2008, and logic products by 2013.

Mechanical properties are also of interest in materials prepared on a nano-scale. It is believed that superstrength materials can be produced by embedding nanocrystalline particles in a noncrystalline matrix. Graphite needles, called carbon nanotubes, have been shown to possess strengths and elastic moduli larger than those of the high-strength steels. Machine tools of higher hardness, high-resolution phosphors for cathode ray tubes, and high-power magnets are examples of other potential applications.

Nanostructural materials have been made by chemical vapor deposition, electrochemical methods, crystallization, or precipitation processes and by synthesis using inert gas evaporation techniques.

Another nanotechnology-related area that is getting much attention is micro-electromechanical systems (MEMS) research, which involves the integration of mechanical elements, sensors, actuators, and electronics on a common silicon substrate. The electronics part includes a variety of integrated circuits, while the mechanical components are fabricated using micromachining processes that selectively etch away parts of the silicon wafer or add new parts to form electro-mechanical and mechanical devices. This will allow microsystem sensors to gather information from the environment through the measurement of mechanical, thermal, biological, chemical, magnetic, and optical phenomena. The electronics then process the information derived from the sensors and direct actuators to respond by moving, positioning, and regulating, thereby controlling the environment for some desired purpose. A typical machine part that has been fabricated in the laboratory is shown in Figure 1.4.

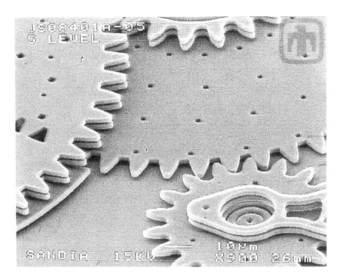

FIGURE 1.4 Micromachined mechanical elements. (Courtesy of Sandia National Laboratories, SUMMIT, www.mems.Sandia.gov.)

1.7.2 LIQUID METALS

Liquid metals, also known as metallic glasses, have been so-called because of their noncrystalline structure. These amorphous metals were first made by employing an extremely rapid solidification rate such that the resulting solid did not have time to form as crystals, the atoms not having sufficient time to take up their prescribed lattice positions. More recently it has been found that by introducing other larger atoms, for example, yttrium, amorphous structures can be obtained by cooling at rates approaching those more conventionally employed. These alloys have some very attractive properties.

DEFINITIONS

Biodegradable: Decomposition of materials by naturally occurring processes, e.g., bacterial attack.

Elastomers: Rubber and rubber-like materials. A type of polymer.

Nanotechnology: Technology involving materials or devices with dimension on a nanoscale, 10^{-9}.

Optoelectronics: Devices that use materials whose functions are related to optical properties of materials, e.g., lasers, photoluminescence, detectors.

Recycling: Reuse of discarded materials by separation, cleaning, and reprocessing procedures.

Specific Strength: Strength-to-density ratio.

Thermoplasts: Types of polymers that can be easily formed and recycled by melting.

Thermosets: Types of polymers that have a three-dimensional type network and that decompose on melting.

QUESTIONS AND PROBLEMS

1.1   What are the differences between cast iron and steel? Do we or should we have a category called "cast steels?"

1.2   Convert the following Fahrenheit temperatures to Celcius.
(a) −32
(b) 32
(c) 212
(d) 1500

1.3   What single property of a metal which more than any other property distinguishes a metal from a polymeric or ceramic material?

1.4   What do the following groups of letters mean?
(e) SAE
(f) UNS
(g) ASTM

1.5   What carbon content does an SAE 1010 steel have? What other name is often used for low-carbon steels that contain only iron and carbon?

1.6   What is the composition of a 4340 steel? Where could you find the composition of a 4140 steel?

1.7   What is the major alloying element in a 2024 aluminum alloy? For what aircraft component or part is this alloy frequently used?

1.8   A steel with the designation 17-7 PH is what type of steel? What are the two major alloying elements and the amount of each present?

1.9   What is the approximate copper content in aluminum bronze?

1.10 What is a common thermoplastic polymer? How does it differ from the elastomers and thermoset polymers?

1.11 List some materials that are currently recycled? What is the most common and most economical recycled material?

1.12 What is the most common class of alloys used for structural applications? What are the favorable properties of these alloys?

1.13 How do stainless steels achieve their stainless properties?

1.14 List the three major polymer types and how they differ in processing and properties.

SUGGESTED READING

Callister, Jr., W. D., *Materials Science and Engineering*, Wiley, 1990.

Rittner, M. N., Nanomaterials in nanoelectronics: who's who and what's next, *JOM*, 56(6), 22–26, 2004.

Roberts, G. A. and Gary, R. A., *Tool Steels*, 4th ed., ASM International, Metals Park, OH, 1980.

Schaffer, J. P. et al., *The Science and Design of Engineering Materials*, McGraw-Hill, New York, 1999.

Smith, W. F., *Principles of Materials Science and Engineering*, McGraw-Hill, New York, 1990.

2 Properties and Their Measurement

2.1 INTRODUCTION

For an engineer to use materials effectively in design or in production, he or she must be able to measure and manage the mechanical and physical properties of the materials. In manufacturing, properties measurement is essential in benchmarking incoming materials and assessing the changes occurring during the manufacturing process.

Mechanical behavior deals with the reaction of the body to a load or force, whereas physical behavior deals with electrical, optical, magnetic, and thermal properties. One could also discuss chemical properties, but the only chemical characteristics discussed here are those pertaining to environmental effects, and those measurements will be dealt with in the appropriate chapter. For some applications, mechanical properties will be of prime interest, while for others the physical properties will be of major concern. The mechanical properties of most interest and the ones that will be considered here include hardness, strength and ductility, and creep and fracture behavior.

2.2 HARDNESS

Hardness is the easiest and most commonly measured property. The principal of most modern hardness testing is the resistance of the material to penetration by an object under a known load. This principal has applications in metals, plastics, rubbers, molding sand in the foundry, minerals, and a host of other situations. The first method devised to measure hardness was developed by Freiedrich Mohs in 1832. It was a relative hardness scale based on minerals found in the Earth's crust. The method consisted of a hand abrasion of one mineral against another. Diamond was given a 10 value since nothing would scratch the diamond, and other minerals were ranked in descending order. Table 2.1 shows the scale for a variety of minerals.

For practical reasons, the Mohs measurement system is not appropriate or accurate enough for modern industrial measurements. More quantitative measures have been developed that allow fast and inexpensive measurement of the relative hardness of a material. There is no absolute hardness scale; all measurement is relative.

Hardness measurement is universally accepted as a fast, convenient way of benchmarking a material or assessing a change as a result of processing. Hardness

TABLE 2.1
Mohs Hardness
Scale

Mineral	Hardness
Diamond	10
Corundum	9
Topaz	8
Quartz	7
Feldspar	6
Apatite	5
Fluorite	4
Calcite	3
Gypsum	2
Talc	1

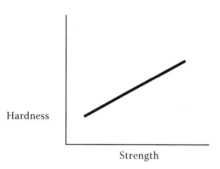

FIGURE 2.1 Hardness vs. strength.

is directly related to strength and for a given family of materials can be used as a predictor of properties (Figure 2.1).

As with the Mohs scale, modern hardness tests are based on resistance to penetration of the subject material by an indenter under a known load. Various methods of assessing the penetration are used in the tests.

2.2.1 BRINELL HARDNESS TEST

The Brinell test is one of the oldest test methods. It consists of pressing a 10-mm hardened steel or carbide ball into the surface of the sample with loads ranging from 500 to 3000 kg. The impression made in the surface is measured with a 10x or 20x microscope with a scale in the lens. The diameter in millimeters is measured and compared to a chart to get a Brinell hardness number. The advantages of the Brinell measurement method are that only minimal surface preparation is needed and the large size of the impression takes into account more material for a better statistical measurement. The disadvantages are that the large size of the impression is often cosmetically or mechanically unacceptable for some products and the readings are taken manually, resulting in some operator variability. Automated and computer-assisted systems can be used to minimize variability, but they are expensive.

2.2.2 ROCKWELL HARDNESS TEST

This test method is the most widely used because of its simplicity and versatility. Developed in 1919 by S. P. Rockwell, it is not as linear as other tests, so some care must be used when reporting data at the extremes of any scale range. This test consists of an indenter load combination ranging from a 1/8-in steel ball to a diamond pyramid indenter forced into the sample with a load from 15 to 150 kg. The test measures the difference in depth of penetration between a reference load placed on the sample by a spring mechanism and the major load. After measuring the differential penetration, the machine directly displays a numerical value for the test. Because of this direct reading feature, these testers can be fully automated in a production-control situation.

Hardness is reported as HR_ and the code letter for the scale used. A 43 Rockwell C would be reported as 43HRC. The Rockwell C test is used for steel and harder material with its 150-kg major load and diamond indenter. Aluminum is commonly measured on the B scale, as are some coppers. Polymers are measured on the E and M scales, with the M scale being preferred for harder polymers such as phenolics.

2.2.3 MICROHARDNESS TESTS

The Knoop (HK) and Vickers (diamond pyramid hardness [DPH]) tests were developed for precise measurement of individual phase components, thin cases, or areas smaller than those measured by either the Rockwell or Brinell tests. The tests are based on the Brinell principle, but a small diamond indenter is used. An impression is made in the sample using loads of 1 kg or less, and the diagonals of the diamond-shaped indenter are measured using an optical microscope. The DPH hardness tests are more linear. For example, a value of 300 DPH is close to three times the hardness of a material that has a measured 100 DPH.

Ceramic materials are relatively brittle and frequently crack if too much load is applied during the test. Nevertheless, their hardnesses are measured using Vickers (the DPH) and the Knoop tests. These are the preferred methods because of their linearity and because cracking can be seen through the microscope.

2.2.4 OTHER TESTING METHODS

Hardness testing is based on the resistance to the penetration of the material to a known load and indenter. This theme is carried out in the foundry in measuring the compaction hardness of a sand mold or in plastics and rubber manufacture using the Durometer test. Alternative methods have used the rebound of a hardened stylus as in the Shores hardness test.

American Society for Testing and Materials (ASTM) specifications have been developed for the common tests: ASTM E10 for the Brinell method, ASTM E18 for the Rockwell test, and ASTM E92 for the DPH methods. Hardness tests consume little time and are very informative. They do not have the precision that other tests do, but one can gain an enormous amount of information quickly. The hardness is related to strength. The resistance to deformation is the reason the hardness value obtained can be related to the ultimate tensile strength of the material. It takes into account any strain-hardening effects. Empirical relationships have been developed and published in the form of charts. By far the most widely used chart is the one that was published by the developers of the Rockwell testers, the Wilson Company. Their chart is reproduced in Table 2.2. One must remember that these comparisons are empirical, but they can give an approximate strength of the material without the time-consuming operations of machining and testing a tensile bar.

A comparison of the various methods, together with the indenter shape, is shown in Figure 2.2, while a digital-readout Rockwell tester is shown in Figure 2.3.

2.3 MECHANICAL PROPERTY MEASUREMENTS

Strength, modulus, and ductility are of prime importance and are most often determined in a uniaxial tension test. Brittle materials are frequently tested in uniaxial compression since it is difficult to grip a brittle specimen for a tensile or pulling load. In general, properties determined in compression would be similar to those determined in tension. Another test that is occasionally used for strength and ductility is the bend test. Sometimes a bend test is easier to perform than a tension or compression test. In a bend test, the outer portion of the specimen is in tension while the inner part is in compression. Ceramic materials are commonly assessed in this manner. The torsion test, in which a twisting force is applied, is used to determine the torsion yield strength and the torsion modulus, which are not equal to the yield strength and modulus values obtained in tension. This is a very important point to remember and is emphasized more strongly later. In selecting a material for a certain design, it is important to know the type of loading to which the material will be subjected. Generally, the yield strength under a torque load (torsion) will be about half of that in tension, and the corresponding torsion modulus of elasticity, sometimes called the shear modulus, will be about 40% of that measured in the uniaxial tension test.

2.3.1 STRESS–STRAIN RELATIONS IN TENSION

All materials deform to some extent when subjected to a force. If the material returns to its original shape, it is said to have deformed *elastically*. If some permanent change remains after removal of the force, it has been *plastically*

TABLE 2.2
Hardness Conversion Numbers from a Variety of Test Methods

C 150 kg Brale (Rockwell)	A 60 kg Brale (Rockwell)	15-N 15 kg N Brale (Rockwell Superficial)	30-N 30 kg N Brale (Rockwell Superficial)	Vickers 10 kg 136° Diamond (Vickers)	Knoop 500 Gr. & Over (Knoop)	Brinell 3000 kg 10 mm Ball (Brinell* Standard Ball)	Tensile Strength (Thousand lbs. per sq. in.)
70	86.5	94.0	86.0	1076	972	—	
69	86.0	93.5	85.0	1004	946	—	
68	85.6	93.2	84.4	940	920	—	
67	85.0	92.9	83.6	900	895	—	Inexact and only for steel
66	84.5	92.5	82.8	865	870	—	
65	83.9	92.2	81.9	832	846	—	
64	83.4	91.8	81.1	800	822	—	
63	82.8	91.4	80.1	772	799	—	
62	82.3	91.1	79.3	746	776	—	
61	81.8	90.7	78.4	720	754	—	—
60	81.2	90.2	77.5	697	732	—	—
59	80.7	89.8	76.6	674	710	634	351
58	80.1	89.3	75.7	653	690	615	338
57	79.6	88.9	74.8	633	670	595	325
56	79.0	88.3	73.9	613	650	577	313
55	78.5	87.9	73.0	595	630	560	301
54	78.0	87.4	72.0	577	612	543	292
53	77.4	86.9	71.2	560	594	525	283
52	76.8	86.4	70.2	544	576	512	273
51	76.3	85.9	69.4	528	558	496	264
50	75.9	85.5	68.5	513	542	481	255
49	75.2	85.0	67.6	498	526	469	246
48	74.7	84.5	66.7	484	510	451	238
47	74.1	83.9	65.8	471	495	442	229
46	73.6	83.5	64.8	458	480	432	221
45	73.1	83.0	64.0	446	466	421	215
44	72.5	82.5	63.1	434	452	409	208
42	71.5	81.5	61.3	412	426	390	194
40	70.4	80.4	59.5	392	402	371	182
38	69.4	79.4	57.7	372	380	353	171
36	68.4	78.3	55.9	354	360	336	161
34	67.4	77.2	54.2	336	342	319	152
32	66.3	76.1	52.1	318	326	301	146
30	65.3	75.0	50.4	302	311	286	138
28	64.3	73.9	48.6	286	297	271	131
26	63.3	72.8	46.8	272	284	258	125
24	62.4	71.6	45.0	260	272	247	119
22	61.5	70.5	43.2	248	261	237	115
20	60.5	69.4	41.5	238	251	226	110

TABLE 2.2 (Continued)
Hardness Conversion Numbers from a Variety of Test Methods

B 100 kg 1/16″ Ball	F 60 kg 1/16″ Ball	30-T 30 kg 1/16″ Ball	E 100 kg 1/8″ Ball	Knoop 500 Gr. & Over	Brinell 3000 kg DPH 10 kg	Tensile Strength
Rockwell	Rockwell	Rockwell Superficial	Rockwell Superficial	Knoop	Brinell	Thousand lbs. per sq. in.
100	—	83.1	—	251	240	116
99	—	82.5	—	246	234	114
98	—	81.8	—	241	228	109
97	—	81.1	—	236	222	104
96	—	80.4	—	231	216	102
95	—	79.8	—	226	210	100
94	—	79.1	—	221	205	98
93	—	78.4	—	216	200	94
92	—	77.8	—	211	195	92
91	—	77.1	—	206	190	90
90	—	76.4	—	201	185	89
89	—	75.8	—	196	180	88
88	—	75.1	—	192	176	86
87	—	74.4	—	188	172	84
86	—	73.8	—	184	169	83
85	—	73.1	—	180	165	82
84	—	72.4	—	176	162	81
83	—	71.8	—	173	159	80
82	—	71.1	—	170	156	77
81	—	70.4	—	167	153	73
80	—	69.7	—	164	150	72
79	—	69.1	—	161	147	70
78	—	68.4	—	158	144	69
77	—	67.7	—	155	141	68
76	—	67.1	—	152	139	67
75	99.6	66.4	—	150	137	66
74	99.1	65.7	—	147	135	65
72	98.0	64.4	—	143	130	63
70	96.8	63.1	99.5	139	125	61
68	95.6	61.7	98.0	135	121	59
66	94.5	60.4	97.0	131	117	57
64	93.4	59.0	95.5	127	114	
62	92.2	57.7	94.5	124	110	

TABLE 2.2 (Continued)
Hardness Conversion Numbers from a Variety of Test Methods

B 100 kg 1/16" Ball	F 60 kg 1/16" Ball	30-T 30 kg 1/16" Ball	E 100 kg 1/8" Ball	Knoop 500 Gr. & Over	Brinell 3000 kg DPH 10 kg	Tensile Strength
Rockwell	Rockwell	Rockwell Superficial	Rockwell Superficial	Knoop	Brinell	Thousand lbs. per sq. in.
60	91.1	56.4	93.0	120	107	
58	90.0	55.0	92.0	117	104	
56	88.8	53.7	90.5	114	101	
54	87.7	52.4	89.5	111	**87	
52	86.5	51.0	88.0	109	**85	
50	85.4	49.7	87.0	107	**83	
48	84.3	48.3	85.5	105	**81	
46	83.1	47.3	84.5	103	**79	
44	82.0	45.7	83.5	101	**78	
42	80.8	44.3	82.0	99	**76	
40	79.7	43.0	81.0	97	**74	
38	78.6	41.6	79.5	95	**73	
36	77.4	40.3	78.5	93	**71	
34	76.3	39.0	77.0	91	**70	
32	75.2	37.6	76.0	89	**68	
30	74.0	36.3	75.0	87	**67	
28	73.0	34.5	73.5	85	**66	
24	70.5	32.0	71.0	82	**64	
20	68.5	29.0	68.5	79	**62	
16	66.0	26.0	66.5	76	**60	
12	64.0	23.5	64.0	73	**58	
8	61.5	20.5	61.5	71	**56	
4	59.5	18.0	59.0	69	**55	
0	57.0	15.0	57.0	67	**53	

Even for steel, tensile strength relation to hardness is inexact unless determined for specific material.

* Above Brinell 451 HB tests were made with 10-mm carbide ball.
** Below Brinell 101 tests were made with only 500-kg load and 10-mm ball.

Source: (Courtesy of Page-Wilson Corporation.)

deformed. In the latter case it is also elastically deformed since, for engineering materials, some elastic deformation precedes the plastic deformation. The elastic part, however, disappears when the force is removed (see Figure 2.4). It is common engineering practice to use the term stress (σ), which is defined as the applied force per unit area of the specimen, and the resulting deformation as strain (ε), which can be defined as a change in volume per unit volume, or as is more often done for convenience, as a change in length per unit length or a change

Shape of Indentation

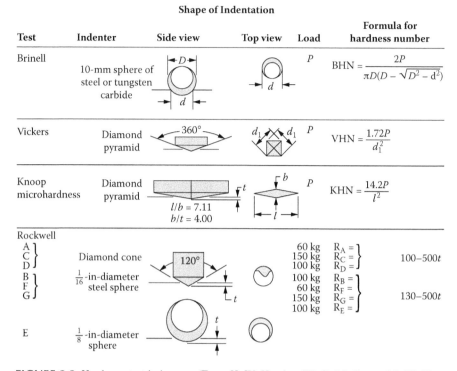

Test	Indenter	Side view	Top view	Load	Formula for hardness number
Brinell	10-mm sphere of steel or tungsten carbide	D / d	d	P	$BHN = \dfrac{2P}{\pi D(D - \sqrt{D^2 - d^2})}$
Vickers	Diamond pyramid	$360°$	d_1 d_1	P	$VHN = \dfrac{1.72P}{d_1^2}$
Knoop microhardness	Diamond pyramid	$l/b = 7.11$ $b/t = 4.00$ t	b l	P	$KHN = \dfrac{14.2P}{l^2}$

Rockwell

				Load		
A C D	Diamond cone	$120°$		60 kg 150 kg 100 kg	$R_A =$ $R_C =$ $R_D =$	100–500t
B F G	$\frac{1}{16}$-in-diameter steel sphere			100 kg 60 kg 150 kg	$R_B =$ $R_F =$ $R_G =$	130–500t
E	$\frac{1}{8}$-in-diameter sphere	t		100 kg	$R_E =$	

FIGURE 2.2 Hardness test indenters. (From H. W. Hayden, W. G. Moffat, and J. Wulff, *The Structure and Properties of Materials*, Wiley, New York, 1965. With permission.)

in cross-sectional area per unit area. These relationships are key to defining the areas of interest to an engineer regarding design, manufacturing, and failure.

Figure 2.5 shows an illustration of a stress–strain diagram for a ductile material.

There are three distinguishable zones on this curve:

0–1 The linear portion of the curve is the design zone. In this zone, the material behaves elastically, and deformation is recoverable.

1–2 In this zone, the material is plastically deforming, and this is generally the zone in which shape-changing manufacturing processes are carried out.

2–3 In this final zone, the material behaves in an unpredictable manner, and this is the zone in which the material begins to come apart and fail.

In a testing laboratory, this graph is generated using a testing machine and a standard sample shape as defined by ASTM E8. The specimen dimensions for the most commonly used bar are shown in Figure 2.6. Other standard bar geometries are given in E8 to accommodate smaller and nonround geometries.

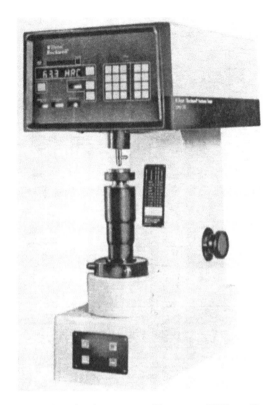

FIGURE 2.3 Digital-readout hardness tester. (Courtesy of Wilson Company.)

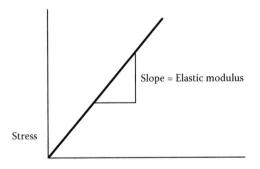

FIGURE 2.4 Stress vs. strain in the elastic portion of the curve.

In a test lab, the data are gathered as load (pounds or kilograms) and elongation (inches or millimeters), but for comparison purposes they are converted to stress and strain and reported as *engineering stress and strain*.

$$\sigma \ (stress) = load/area$$

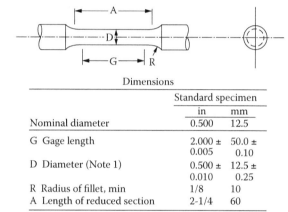

FIGURE 2.5 An illustration of stress–strain diagram for ductile material.

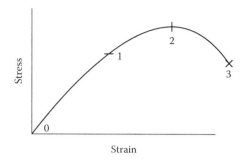

Dimensions

	Standard specimen	
	in	mm
Nominal diameter	0.500	12.5
G Gage length	2.000 ± 0.005	50.0 ± 0.10
D Diameter (Note 1)	0.500 ± 0.010	12.5 ± 0.25
R Radius of fillet, min	1/8	10
A Length of reduced section	2-1/4	60

FIGURE 2.6 Tensile test specimen dimensions per ASTM E8-90 specifications.

where load is in pounds or kilograms; area, measured at the start of testing, is in square inches or square millimeters; and stress is reported in power per square inch (psi) or Megapascal (Mpa). Stress can be calculated at any point on the curve up to the point of necking. The maximum load (highest point on the curve/the original area) is reported as the tensile strength of the material (Point 2 in Figure 2.5).

The definition of Point 1 in Figure 2.5 is quite important, as it is the boarder between the design and manufacturing zones. The point of yielding (onset of plastic deformation) can be defined by the elastic limit or the yield strength. The elastic limit is the precise point at which a material will not return to its original dimension when the load is released. This point is determined by experiment through loading and unloading of the sample until it yields (changes shape). Although this accurately defines the point of transition, this method is not practical in an industrial setting. The qualification of a material and safety factors are set based on the yield strength as determined by the *offset method*.

In the offset method, an arbitrary permanent plastic deformation is used to find a point on the curve that is easily reproducible at any test laboratory. A common offset is 0.2%. Much of the data reported in the literature uses a 0.2% offset.

Example 2.1   Determine the 0.2% offset yield strength for the following metal sample using Figure 2.7. A standard 50-mm gage-length sample is used in Figure 2.6.

Solution   Using the 50-mm gage length of the sample,

$$50 \text{ mm} \times 0.002 = 0.1 \text{ mm offset}$$

The intersection of a line drawn parallel to the elastic portion of the curve at 0.1 mm and the test line defines the *yield point load.*

Yield point load @ 0.2% offset / Area = the 0.2% offset yield strength.

Other offsets can be used in determining the yield point load, but it must be clearly stated in the test results.

The ductility of a sample is measured as percent elongation or percent reduction in area.

$$\% \text{ Elongation} = \Delta \text{ length/original length} \tag{2.1}$$

$$\% \text{ Reduction in area} = \Delta \text{ area/original area} \tag{2.2}$$

These measurements are made after the fracture of the sample in the test zone (gage area) and at the fracture line, respectively.

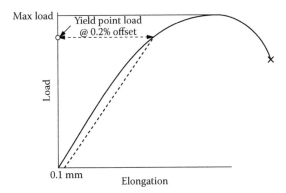

FIGURE 2.7 Tensile curve for a ductile mat'l illustration for the offset method of determining yield point load.

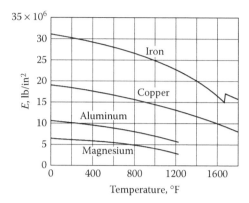

FIGURE 2.8 Elastic modulus vs. temperature for some common metals. (From A. G. Guy, *Elements of Physical Metallurgy,* Addison-Wesley, Reading, MA, 1959. With permission.)

The elastic portion of the graph (0–1) (Figure 2.5) is defined as the stiffness, or modulus of elasticity (E), and is called Hooke's law (Figure 2.4).

$$E = stress/strain$$

E (Young's modulus) does not change significantly for a given family of materials. E is a function of the bonding energy atoms of the material and is only affected by temperature (Figure 2.8). Some common material moduli are listed in Table 2.3.

Any force, no matter how small, will extend the atom spacing if it is a tensile force and compress the spacing for a compressive force. As long as the elastic condition is maintained, the strain is proportional to the stress as expressed by Hooke's law,

$$E = stress/strain \tag{2.3}$$

It will be the same in tension and compression. It is of interest to examine the magnitude of the change in length when subjecting a common engineering material to a range of forces.

Example 2.2 For a cylindrical steel bar 0.505 inches in diameter with a gage of 2 in (ASTM E-8 specification), compute the stress for loads of 60 and 6000 lb.

The cross section of the bar is 0.2 in^2; therefore the stress is

$$Stress = Load/area = 60\ lb/0.2\ in^2 = 300\ psi\ or\ 2\ Mpa$$
$$6000/0.2\ in^2 = 30{,}000\ psi\ or\ 20\ Mpa$$

Using Hooke's law (value from Table 2.3),

$$\varepsilon = \sigma/E = 300/30 \times 10^6 = 10 \times 10^{-6}$$

The value of strain is essentially a dimensionless number of inches/inch or mm/mm. For a 2-in gage, the total change in length becomes

$$\Delta L = 2 \text{ in} \times (10 \times 10^{-6} \text{ in/in}) = 20 \times 10^{-6} \text{ in}$$

This small force will elongate the specimen 2μ in. This is barely measurable in common engineering tests. For a force of 6000 lb–ft, the strain will be 100 times greater and the corresponding strain will be 2×10^{-3} and is measurable with common instrumentation.

This barely measurable force in a common engineering test will elongate the specimen 2μ in. Upon unloading, the sample will return to its original length.

If we carry out a test in which the plastic region is reached in the loading cycle, a permanent change in length is achieved, but upon releasing the load, elastic recovery occurs (Figure 2.9a). From the plastic region the sample returns to an elongated value as illustrated in Figure 2.9b. Upon reloading, the linear portion of the curve is repeated parallel to the original load curve, but the point of yielding (where the slope of the curve changes) is higher as a result of *strain hardening*. Strain hardening is a common method of increasing the strength of a metal and is part of manufacturing processes in which a part's shape is defined by bending, twisting, or stretching. A more extensive discussion of this phenomenon is found later in this book.

TABLE 2.3
Typical Moduli and Poisson's Ratio at 20°C

Material	Average Elastic Modulus,[a] E		Poisson's Ratio
	MPa × 10⁵	psi × 10⁶	
Aluminum alloys	0.71	10.3	0.31
Plain carbon steels	2.0	30	0.33
Copper	1.1	16	0.33
Titanium	1.17	17	0.31
Tungsten	4.0	58	0.27
MgO (magnesia)	2.07	30	0.36
Si_3N_4	3.04	44	0.24
Al_2O_3 (alumina)	3.80	55	0.26
BeO	3.11	45	0.34
Plate glass	0.69	10	0.25
Diamond	10.35	150	—
Nylon	0.025	0.4	—
Polymethyl methacrylate (plexiglass)	0.035	0.5	—

[a] The moduli vary with direction of measurement and with composition, the latter being a small effect and the former being a small effect in most large polycrystalline bodies.

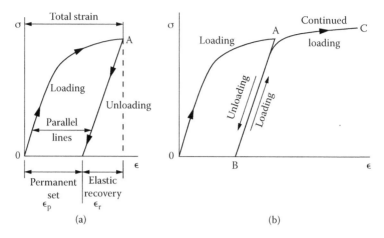

FIGURE 2.9 Strain hardening causes the stress required to maintain plastic flow to increase but does not affect the modulus of activity.

2.3.2 Poisson's Ratio

In the elastic region of a crystalline material a lateral contraction occurs during extension. The ratio of lateral contraction stress to longitudinal extension strain is called the Poisson's ratio, which is expressed by the following:

$$\Delta d/d = -\varepsilon_x = -\varepsilon_y = \mu -\varepsilon_z \tag{2.4}$$

where

d = bar diameter

$\varepsilon_x, \varepsilon_y$ = lateral contraction strains (negative)

ε_z = positive longitudinal strain

μ = Poisson's ratio

Crystalline solids are not isotropic, and neither are most noncrystalline solids. A metal or ceramic body is made of many grains (crystals) of varying orientation such that we are dealing with average values when measuring directionally dependent properties of polycrystalline bodies. For most polycrystalline metals μ is approximately 0.3, and for many glasses μ is 0.25 (i.e., isotropic behavior). It is of little or no practical significance, but it should be pointed out that in the elastic region there will be a slight change in volume when $\mu > 0.5$, and it is always a positive number, which can be expressed by the equation

$$\Delta V/V = \varepsilon_z (1 - 2 \mu) \tag{2.5}$$

The volume change is so small that for practical purposes the modulus can be computed by measuring the change in either length or area. Generally, the strain value is used in moduli determinations since the extension is easier to measure accurately than is the reduction in area.

2.3.3 ENGINEERING VS. TRUE STRESS AND STRAIN

The definition of the elastic portion of the curve is important, as these property values define the *design* region of a material (Line 0–1 in Figure 2.5). In the manufacturing arena, however, the portion of the stress–strain curve between yielding and ultimate tensile stress is the region of focus. Another way of describing this region is the onset of plastic movement (slip) to the onset of plastic instability (necking). In this region the metal can be shaped and the properties increased in a predictable manner. In the engineering stress and strain diagram the curve decreases after the maximum load is achieved because we choose to use the constant area A_o measured at the beginning of the test. The actual force required to maintain extension after the maximum in the engineering curve is reached is decreasing but the true stress is increasing.

Figure 2.10 compares the plots of true stress vs. engineering stress,

$$\sigma_t = \text{True stress} = L \, / \, A_i \tag{2.6}$$

where A_i is the instantaneous cross section and L is the load.

In the plastic region, to the point of necking, the volume stays constant.

$$A_i \, l_i = A_o \, l_o$$

$$\varepsilon = 1 - l_o \, / \, l_o = \Delta l \, / \, l_o \text{ or } \Delta A \, / \, A_o$$

$$\varepsilon_T = \Delta \, l \, / \, l_i$$

Therefore

$$\sigma_t = \sigma_t(\varepsilon + 1) \tag{2.7}$$

True strain is

$$d \, \varepsilon_T = dl \, / \, l \text{ (the incremental change in length)}$$

$$\varepsilon_T = \int_{lo}^{lf} dl \, / \ln = 1_f \, / \, l_o$$

$$\text{True strain} = \varepsilon_t = \ln \, (l/l_o) \tag{2.8}$$

Example 2.3 A 0.505-in-diameter test bar was elongated to a permanent length of 2.3 in. The original gage was 2 in (see Example 2.2).

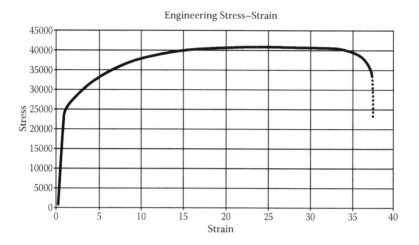

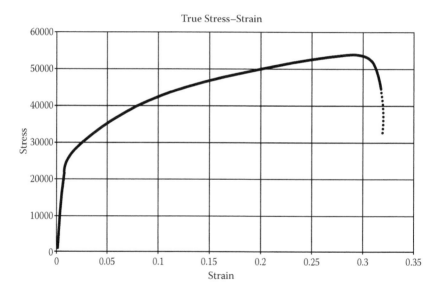

FIGURE 2.10 True stress vs. engineering stress. (Courtesy of W. Riffe, Kettering University.)

(a) What was the total true strain obtained as measured by the change in length?

$$\varepsilon_t = \ln l_f / l_o = \ln 2.3/2.0 = 0.14 \text{ (remember, strain has no dimensions)}$$

(b) What is the engineering strain?

$$\varepsilon = \Delta l / l_o = 0.3/2 = 0.15$$

(c) What is the true reduction in area?

From constancy of volume, $Ai = A_o \, l_o \, / \, l_i = 0.2 \times 2/2.3 = 0.174$ in^2

$$\varepsilon_t = \ln A_o \, / \, A_f = 0.2/0.174 = 0.14$$

(d) What is the engineering reduction in area?

$$\varepsilon = 0.2 - 0.174/0.2 = 0.13$$

The differences in true and engineering strain values become much larger at higher values of strain. Also note that the true strain is the same, whether measured by change in length or area. This will always be true as long as we are dealing with uniform deformation, but not when localized necking occurs.

(e) If the load on this bar required to produce the total strain measured was 1000 lb, what was the final engineering and true stresses in psi and MPa units?

Engineering stress = 1000 lb/0.2 in^2 = 5000 psi (34.5 Mpa)

True stress = 1000 lb/A_f = 1000 lb/0.174 in^2 = 5747.1 psi (39.7 Mpa)

The true stress is higher than the engineering stress in the plastic region. True stress and strain are used in manufacturing process analysis to determine press size and other machine parameters in cold forming processes.

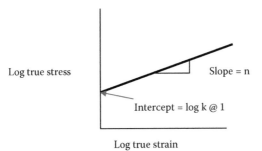

2.3.4 Power Law

If log true stress vs. log true strain is plotted, a relationship describing the plastic portion of the stress–strain curve to necking can be determined.

$$\sigma_t = k \, \varepsilon_t^n \tag{2.9}$$

k = constant
n = strain-hardening exponent

TABLE 2.4
Selected Strain Hardening Components

Material	n	k (ksi)	Crystal Structure
Ti	0.05	175	HCP
Alloyed steel	0.15	93	BCC
Quenched and tempered carbon steel	0.1	228	BCC
Mo	0.13	105	BCC
Cu	0.54	46	FCC
70/30 Brass	0.50	130	FCC
300 series SS	0.52	220	FCC

The importance of the strain-hardening exponent (n) is illustrated in Table 2.4. The general trend for the value of the strain-hardening coefficients relating to material structure is

$$HCP < BCC < FCC$$

If a metal has a high n value, usually it can be strain-hardened to a greater extent before necking.

2.3.5 USEFUL INFORMATION

The strain to fracture (that is, the total observed plastic strain) is called the *ductility* of the material. It can be measured by either the overall change in length or the overall change in area. The measurement is made on the specimen after fracture by meshing together the two broken sections. Therefore, this strain does not include the elastic strain, which was released at the point of fracture. Although some data may be reported in terms of the engineering strain by using either the original length or the original area, it is recommended that true strain to fracture be used.

Which is better — the strain to fracture as measured by change in length or by change in area? When extensive necking occurs, the deformation is highly localized and the strain to fracture as measured by change in length becomes a function of the gauge length. Thus, the strain for a 2-in gauge length will be different from that of a 10-in gauge length. The ductility should represent a material property and not be a function of the specimen dimensions. Therefore, strain to fracture as measured by reduction in area is a better measure of ductility.

The same equations also hold for ceramics, polymers, and composites. There is a big difference in properties of the various materials, especially for ductility and modulus of elasticity. Materials that exhibit little or no plastic strain prior to fracture are said to be brittle. Glasses, ceramics, and gray cast iron are examples of brittle materials.

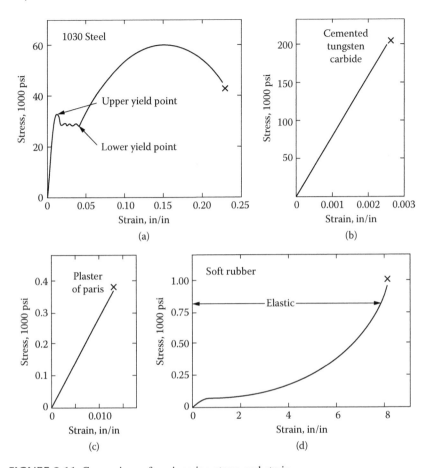

FIGURE 2.11 Comparison of engineering stress and strain.

The room temperature stress–strain curves of some typical materials are compared in Figure 2.11. (See the Suggested Reading for a compilation of stress–strain curves.) Note that in Figure 2.9a two yield points are evident. This is a characteristic of many steels and some alloys. Also note that only the steel shows strain hardening, a process most often present in metals or metallic components of composites. It is seldom seen in ceramics or polymers.

The stress–strain behavior of all materials is strongly temperature dependent. We noted previously that the elastic moduli, and hence the slope of the elastic stress–strain line, decrease with increasing temperature. The yield strength, ultimate tensile strength, and stress to maintain plastic flow also decrease with increasing temperature, while the ductility and toughness increase. Generally, by increasing the temperature sufficiently, one can change a material that is brittle to one that is ductile. The effect of temperature on the shape of the stress–strain curve is depicted in Figure 2.12.

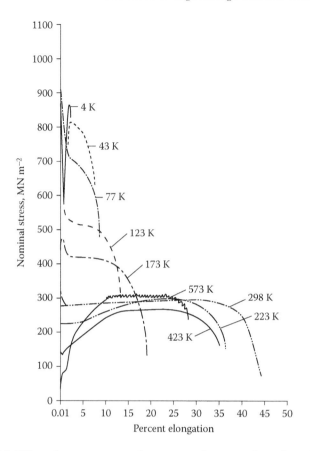

FIGURE 2.12 Effect of temperature on the stress–strain curve of tantalum. (From J. H. Bechtold, *Acta Metall.*, 3, 249, 1955. With permission.)

2.3.6 STRESS–STRAIN RELATIONS IN TORSION

At this point in the discussion, we should briefly examine the elastic stress–strain relations in torsion. Figure 2.13 shows a cylindrical bar being elastically twisted through an angle ϕ by a stress τ.

This is a shear stress, and by convention the symbol τ is used to distinguish it from normal or uniaxial stress, which is usually denoted by the symbol σ. The shear strain $'\delta$ is given by δ/l, which is the tangent of ϕ. The stress is a function of the torque, or twisting moment M, at the point of interest r and the bar diameter D, and can be expressed by

$$\tau = \frac{32M_t}{\pi D^4}\ r\ (2-) \tag{2.10}$$

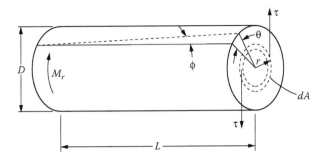

FIGURE 2.13 Torsion in a cylindrical bar.

The maximum stress occurs on the outer fiber, where r = D/2. The shear strain is proportional to the shear stress, with the proportionality constant being the shear modulus G, that is,

$$\tau = G - \delta \tag{2.11}$$

This is simply Hooke's law in torsion. It can be shown that G and E are related by

$$G = E/2(1 + \mu) \tag{2.12}$$

The torsion test can be extended to the plastic range but is seldom used to determine ductility or flow stress.

2.3.7 FRACTURE TOUGHNESS

Although the ductility of a material is an important property, frequently the fracture toughness of a material will also be specified for design purposes. Generally, but not always, a ductile material will also be a tough material. Ductility is measured in a relatively slow strain-rate test. Strain rates on the order of 10^{-3} to 10^{-4} per second are normally used. The ASTM E8 specification states that the strain rate shall not exceed 8×10^{-3} per second.

It is possible for a material to be ductile in a slow strain-rate test but fracture in a brittle manner in a fast-impact test. Toughness is related to the plastic deformation during crack propagation. When a crack propagates, plastic deformation occurs in the vicinity of the crack tip and slows the propagation rate by absorbing some of the energy applied. At fast strain rates and at low temperatures this plasticity is inhibited, thereby favoring brittle or fast fracture, the crack propagating through the body in micro- to milliseconds. Toughness, then, is the ability of a material to resist crack propagation, and it is frequently determined by measuring the energy absorbed during the fracture process while a fast-swinging pendulum is impacting the material. Strain rates in the impact test are on the order of 10^2 to 10^4 per second.

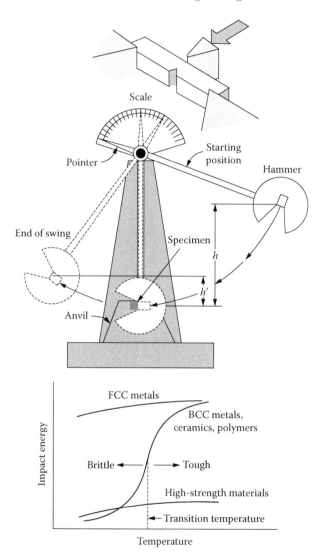

FIGURE 2.14 Schematic drawing of a standard impact tester.

Figure 2.14 shows a schematic of a simple impact-testing machine, often referred to as a Charpy V-notch test. The Charpy specimen shape is shown in the schematic just above that of the impact tester. The specimen is placed on the anvil, and the pendulum hammer is released from a predetermined height, impacting the specimen and causing a fast fracture to occur. The material response can be brittle or ductile or some combination of both, depending on the metal type, processing history, and test temperature.

The temperature dependency of fracture is a serious concern when designing parts subject to low temperature during its service life. This temperature dependency

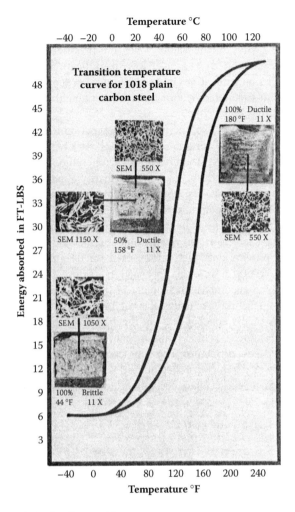

FIGURE 2.15 Brittle–ductile transitive temperature.

of the fracture toughness of steels has resulted in some tragic failures in the past. One of the most striking occurrences of this effect was the fracture of the hulls of Liberty cargo ships in World War II. These ships had been fabricated by welding relatively low-carbon steel plates. About 25% of the ships developed severe cracks. Subsequent investigations showed that these steels became brittle at temperatures on the order of 40°F. Stress concentrations in the original design also contributed to the failures.

Figure 2.15 illustrates a typical impact–temperature curve for a 1018 steel together with optical and scanning electron microscopic pictures of the fracture surfaces. This change in impact energy absorbed during fracture as a function of temperature is known as the ductile–brittle transition, and the temperature at which the largest rate of change occurs is often called the *transition temperature*.

This transition temperature is not a precise value like the melting temperature; it can be defined as the mean value of a temperature range, a fracture appearance, or a percentage of the ductile value. When the material is brittle, the fracture surface appears granular by eye or at low magnifications and shows cleavage markings at higher magnifications in the electron microscope. The ductile fracture appears fibrous at low magnifications and has an abundance of micro voids as seen in the electron microscope.

From a design perspective, the transition temperature varies with strain rate, stress concentration, and certain metallurgical variables associated with the processing of the steel. Fast strain rates, low temperatures, and stress concentrations tend to promote brittle behavior. The Charpy V-notch test represents the worst condition (that is, a very fast strain rate impact on a notched specimen [notches create stress concentrations]). The impact test tells us at what temperature it is safe to use the material in question. The test is a sorting tool to categorize steels in materials selection and qualify materials in a quality control scheme, but it does not provide data that can be used for design purposes. In the late 1950s and early 1960s, a different approach to fracture toughness emerged, the principles of which encompass the subject of fracture mechanics.

A theoretical fracture mechanics approach is beyond the scope of this book. The foundations of the concept are that the size and shape of an existing flaw can act as a stress raiser, which will concentrate the applied force at the tip of the crack to propagate a fracture. The stresses near the tip of a very sharp crack are governed by a parameter called the stress intensity factor K, which is a function of the crack length a and the applied stress . It is expressed by the formula

$$K_1 = y \ \sigma \ \sqrt{\pi} a \qquad (2.13)$$

where

σ = the applied normal stress
a = flaw size
y = a geometric factor describing the flaw

The critical stress intensity that will propagate the crack to failure is called K_{1c} (pronounced "kay-one-see"), and is a true material constant. It is the *fracture toughness* of the material. The fracture toughness has units of MPa $\sqrt{m}$ or ksi $\sqrt{in}$. The conversion factor between the two units is 1.1, so the values in these units for all practical purposes are interchangeable. Attempts have been made to empirically correlate Ki values with Charpy impact tests, but they must be treated with caution. The advantage of the fracture mechanics approach is that one can assume the existence of a certain flaw size, probably just at the limit of that which is detectable by nondestructive testing techniques, and compute an applied stress that would produce fracture. For example, if a crack is detected in a tank wall during a production cycle, it is possible to assess the severity of the problem and determine if immediate action is warranted or if the production cycle can be completed before maintenance is done.

Example 2.4   A 1-m-wide aluminum plate has a fracture toughness of 25 ksi $\sqrt{\text{in}}$ (27.5 MPa$\sqrt{\text{m}}$) and a yield strength of 70 ksi (517.5 MPa) and contains a center crack 0.1 in (2.54 mm) in length. Calculate the fracture stress assuming that Y = I (valid when plate width is very large compared to crack length).

Solution $\sigma = K_{Ic}/\sqrt{\Pi a/2} = 27.5 \text{ Mpa } \sqrt{\text{m}}/\sqrt{2.54\Pi \times 10^{-3}} = 435.6 \text{ Mpa}$

It is common to use $a/2$ if a is a center crack. The crack will propagate with an applied load of about 84% of the yield strength.

2.3.8 Fatigue Life, Fatigue Limit, and Fatigue Strength

Earlier we described the characteristics of the yield strength, a value that is widely used by engineers in their designs as a basis for determining the working loads (yield strength/safety factor) of a material. The design engineer should be aware that in situations where alternating stresses are anticipated, the yield strength values obtained in uniaxial tension tests should not be used. Structural members that have been subjected to an alternating stress for long periods of time have been observed to fail at stresses as low as 25% of their uniaxial tensile yield strength. A rule of thumb for steels is that the failure stress in fatigue is about 50% of the ultimate tensile strength. Of course, such rules cannot be applied in design. Some steels show better fatigue resistance than others. Surface condition and the design geometries also play a major role. Stress concentrators such as scratches, sharp corners, and notches cause failure in fatigue at stresses much lower than that for polished parts and for components where generous radii at reduced sections, threads, and keyways have been provided by the design engineer.

Fatigue has been widely studied in metals because most rotating machine parts and flexing parts such as springs and beams are of metallic construction. The term *fatigue* was coined in the mid-nineteenth century to describe failures in such things as railroad wheel axles. Some of the early reports described fatigue failures as being due to crystallization of the metal, but of course this could not be happening because metals become crystalline bodies during solidification from the molten state. Nor do metals become tired as a result of the stress concentration when subjected to alternating stresses.

Plastic deformation can take place on a microscale at stresses below the yield stress (i.e., in the normal elastic range). The to-and-fro movement under alternating stresses of dislocations, a type of crystalline imperfection that exists in all metals, can cause permanent deformation in localized regions. This deformation can nucleate cracks, but often microcracks, sufficiently small that an electron microscope is needed for their detection, already exist. Alternating stresses, even though in the normal elastic range macroscopically, can, via stress concentration, cause plastic deformation at these crack tips and thereby cause the cracks to slowly propagate. Over a period of time, which may vary from a few months to a decade, the crack will grow until it reaches a critical size sufficient to cause failure of the entire part. As the crack grows, the remaining good material carries

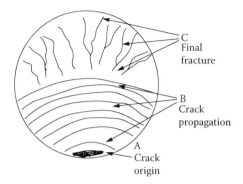

FIGURE 2.16 Schematic description of a fatigue fracture surface of a shaft.

more and more of the load and is thus subjected to a larger stress, even though the overall nominal applied stress remains constant. At a critical stress level, the member fails catastrophically. Because of this long period of time required for failure to occur, the process was called fatigue. One could say that a metal member as a whole becomes weaker with time as the crack grows, but only in the sense that the load-bearing portion is becoming smaller. Since the stress is equal to force divided by area, if the area decreases, the stress must increase.

A typical fracture surface of a shaft that failed by fatigue is shown schematically in Figure 2.16. It is composed of three parts according to the three stages. The first region, denoted by the letter A, is the region where the crack nucleates, or if it already exists, propagates to a size sufficient for detection. This region is a very small part of the total fracture surface, but it may represent a large part of the total life. Region B is the macroscopic propagation stage. It often contains characteristic markings commonly called *beach markings* or *clamshell markings*, for obvious reasons. These markings result from the interruption of the load. The crack then stops propagating, and when the load is resumed at this same value, the crack resumes propagation, probably on a slightly different geometrical level. The crack tip also may become oxidized if the load is removed or reduced below the critical crack propagation level for a period involving hours or days. Such oxidation makes the markings more visible. This propagation stage, which may also be a large portion of the part life, requires that a tensile stress be present. Tensile, not compressive, stresses propagate cracks. The absence of the characteristic markings does not necessarily mean that fatigue has not occurred. If the two fracture surfaces rub together during the propagation stage, they may smear the markings until they cannot be distinguished. Also, if the load is not interrupted during the propagation stage, the markings will not be present. However, their presence proves that fatigue was the cause of the failure.

Finally, the crack attains the length at which the remaining material cannot carry the load, and catastrophic failure occurs in a matter of seconds. This is region C in Figure 2.16. A large final fracture region indicates that the part was subjected to a rather high stress, while a small region indicates a low overall

FIGURE 2.17 Fatigue fracture surface of a shaft.

stress application. Figure 2.17 shows a fracture surface of a shaft that failed in fatigue. Note that the failure initiated in two different points and thus caused two sets of beach markings.

Another fatigue fracture marking that is frequently observed at high magnification is called a *striation* (Figure 2.18). These marks are the result of the plastic deformation that occurs owing to stress concentration at the advancing crack tip. In most cases striations represent one cycle of stress and are seen in-between beach marks.

2.3.8.1 Fatigue Testing

In the laboratory simple alternating tensile and compressive stresses are imposed on a test sample and the time to fracture at a given stress is measured. Fatigue data are usually represented by an S–N curve, a plot of the cycles to failure N vs. the applied stress S. The plot is a stress vs. log N. The relationship between alternating stress data tests and real applications must be linked with some care. In the real applications stresses are not always uniformly alternating around a mean value but are often random. The stress on the axel of a car riding on smooth

FIGURE 2.18 Fatique striations that occurred during crack advancement in an aluminum alloy.

pavement fluctuates uniformly around a mean stress, but if that car is driven on an unpaved road in the wilderness, the stresses on the axel are quite different.

Fatigue is statistical in nature. The severity and distribution of microcracks and microdefects cannot be predicted. Therefore, a number of tests must be run at different stress levels to define the nature of a material. Figure 2.19 shows scatter in the fatigue limit for 10 specimens tested in an identical fashion.

Often data will be represented by several S–N plots, one curve being the mean value and each point on the curve representing the number of cycles for 50% survival. Curves for other percentages, frequently for 5 and 95% survival, are plotted as shown in Figure 2.20.

Fatigue data are not as commonly available as tensile and yield strength data, as they are much more costly to produce and are often, if developed by a company, of a proprietary nature. (See the suggested readings for some good sources.)

Ferrous alloys, BCC structures, have S–N curves that tend to level out at a certain value of stress (i.e., the material will last indefinitely at this stress). This stress would then be a safe design value, applying an appropriate factor of safety. Remember that fatigue data must be treated statistically. If the handbook or other published data do not show the scatter obtained, as shown in Figures 2.19 and 2.20, it is assumed that the curve represents that for 50% survival at each stress value. The stress at which the material would last for an infinite number of cycles is called the *fatigue limit* or *endurance limit*, the former being the preferred term.

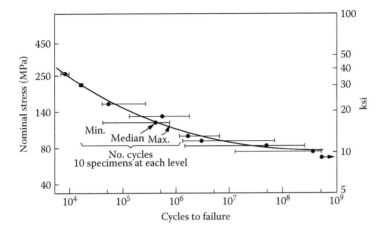

FIGURE 2.19 S–N curve obtained for a 7075-T6 aluminum. (From H. F. Hardrath, E. C. Utley, and D. E. Guthrie, *NASA TN D-210*, 1959. With permission.)

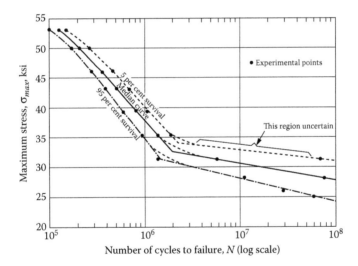

FIGURE 2.20 S–N data for phosphor bronze strip. (From M. N. Torey and G. R. Gohn, *ASTM Proceedings*, 1956. With permission.)

For materials that do not show a well-defined limit (FCC materials such as aluminum and copper show a continuously declining curve), we must choose a specific number of cycles that the material will withstand as the design life and base our design stress at that level.

During the past 40 years we have seen an increase in the use of polymers for machine parts such as gears and bearings. Many of these parts fail by fatigue and often show an S–N curve similar to that for metals. Many polymers show a fatigue limit like those exhibited by ferrous metals, and as with high-cycle, long-life metallic parts, the nucleation of the crack occupies a large fraction of the polymer life. The mechanisms are somewhat different, however. In some cases fatigue failure in polymers occurs via large-scale hysteretic heating that leads to softening of the polymer. Hysteretic heating is due to the stress and strain not always being in phase. In other cases cracks nucleate and propagate in a way similar to metals except that the crack-tip plastic deformation occurs by a different atomic mechanism in polymers.

Fatigue behavior in composites is very complex because of the variation of the properties with direction. In general, the damage done by alternating stresses is similar to that under short-time static loading except that fatigue at a given stress level causes increasing damage with the increase in the number of cycles. Instead of a single crack propagating to failure as in metal fatigue, one speaks of the increase of crack density with the number of cycles. The damage ratio is described as the ratio of crack density at n cycles to the crack density at final failure. In fatigue more cracks occur than during static loading and appear to reach a crack density limit. This limit occurs during the first 20% or so of fatigue life. The rest of the fatigue life is largely occupied by delamination, fiber-matrix debonding, and fiber breakage. Fatigue data for composites are often expressed in S–N curves, just as for metals and polymers, but with the added information reflecting the orientation of the components.

2.3.9 CREEP AND STRESS–RUPTURE

Creep is a very slow plastic deformation of materials that occurs over a long period of time and is most noticeable at elevated temperatures. Like fatigue, it occurs at stresses far below the yield stress for the temperature in question. Creep can occur in all materials discussed in this book, although it was first noted and has been most studied in metals. The lower-melting-point metals such as lead and tin creep at room temperature at stresses well below their room-temperature yield strength. This has become a problem in lead–tin solders in electronic circuits. Stress relaxation by creep of a solder joint can result in an open circuit. (Fatigue is also common in solder joints.) One of the more important cases observed of room- or ambient-temperature creep was that of aluminum cables used in cross-country electrical power distribution. Pure aluminum is needed for high electrical conductivity, but as is the case for most pure metals, it has a relatively low strength compared to the metal in alloy form. Some of the longer spans (for example, in river crossings) sagged under their own weight and thus required frequent tightening. To solve this problem, the high-purity aluminum strands were wound around an aluminum alloy core, and for very long spans, around a steel core. Steel does not creep at room temperature and for most aluminum alloys it is barely detectable. The steel-reinforced cables are described in ASTM specification B232. Their strength-to-weight ratio is about two times that of copper for equivalent direct-current resistance.

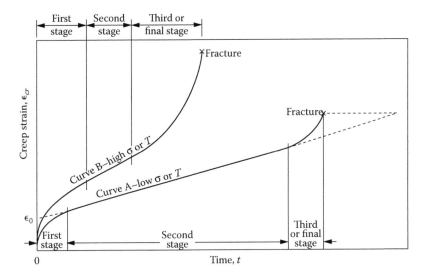

FIGURE 2.21 Schematic creep curves showing the three stages of creep. (From C. W. Richards, *Engineering Materials Science,* Wadsworth, Belmont, CA, 1961. With permission.)

Creep data are usually presented in the form of creep strain (i.e., change in length per unit length) vs. time (Figure 2.21). The curve can be divided into three regions. The first is called primary creep, where the creep rate decreases with time. The second and most important region is composed of the secondary creep, where the creep rate is constant over a long period of time. This region is used for design purposes. When the creep curve is run at the same temperature as the intended application, the life, that is, the time for allowable creep strain to take place, can easily be predicted. Creep data for a given alloy are generally presented at a number of temperatures, and extrapolation to other temperatures is permitted with certain limitations. One rule to remember is that *most creep rates increase exponentially with temperature and linearly with time.* It is much safer to extrapolate downward to lower temperatures than upward to higher temperatures. (See the Suggested Reading section for a compilation of creep data.)

Example 2.5 The initial clearance between the ends of the turbine blades and the housing of a steam turbine is 0.003 in. The blades are 8 in long, and their elastic elongation during operation is calculated to be 0.0008 in.

(a) If it is desired to hold the final clearance to a 0.001-in minimum, what is the maximum percent creep that can be allowed in the blades?

(b) What will be the life of the turbine blade?

Solution From published data it is found that at the intended operating temperature and stress, this alloy will have a primary creep strain of 2×10^{-5} and a secondary creep rate of 6×10^{-6} per hour. Allowable creep strain = 0.003 in − 0.001 in − 0.0008 in = 0.0012 in.

(a) 0.0012 in/8.0 in = 1.5×10^{-4} in = 0.015% max

(b) 0.0012 in − 0.00005 in = 0.00115 in allowable secondary creep strain

(c) 0.00115 in/0.000006 in/in/h = 192 h

In many high-temperature applications, such as the superalloy components inside a jet aircraft engine, we are more interested in the time to fracture than in the creep rate. When data are presented in this fashion, the process is referred to as *creep-rupture*. Whereas the creep rate increases with increasing temperature according to the Arrhenius relationship

$$d\varepsilon/dt = A \exp (-Q/RT) \qquad (2.14)$$

the time to rupture, t, decreases with increasing temperature; that is,

$$t_r = A \exp (Q/RT)$$

In both these expressions, A is a proportionality constant, Q is the activation energy for the process causing creep strains or causing the material to rupture, R is the gas constant, and T is the absolute temperature in Kelvin. Kelvin is used because atoms do not recognize or understand other temperature scales. The activation energy is often the movement of atoms from one position to another in a material. This is a diffusion process in which the atoms are influenced by the applied load.

Polymers, especially thermoplastics, are prone to creep, more so than metals. The chief mechanism of creep in polymers is the uncoiling of polymer chains and the slippage of polymer molecules past one another. These are the same mechanisms that were responsible for plastic flow in the uniaxial tension test. The only difference in creep is that a lower constant load is applied such that plastic deformation proceeds slowly over a long period of time. Polymer creep shows many of the same characteristics as metal creep: an instantaneous elastic response to the load, a fast-creep-rate primary region that decreases with time, and a slow secondary creep process that eventually leads to failure. Data are also expressed as creep-rupture: the lifetime of a polymer at a constant load. A high degree of cross-linking results in a dramatic reduction in molecular mobility and correspondingly the creep rate. The hard elastomers and thermosets do not exhibit extensive creep but rupture after a relatively small amount of creep strain.

2.4 PHYSICAL PROPERTIES

The physical properties of a material generally include electrical and heat conductivity, thermal expansion, magnetic properties, dielectric strength, optical properties, ferroelectricity, and piezoelectricity. In this section, only the determination of electrical and thermal properties of metals, polymers, and ceramics are covered. The other properties will be covered in later sections of this book.

2.4.1 ELECTRICAL CONDUCTIVITY

Electrical conductivity is the reciprocal of electrical resistivity, and these are the properties most interesting to engineers. We use these properties to classify materials. Metals, semiconductors, and insulators are classified today according to where they fall on the electrical conductivity scale of materials. Such a scale for room-temperature values is shown in Figure 2.22.

Materials with the lowest conductivities are at the top of the scale. Note that resistivity is listed on one side of the scale and conductivity on the other. This can easily be accomplished by changing the sign of the exponent since these properties are reciprocals of each other. Also note the wide range of values for all materials, from about 10^{-16} $(\Omega. M)^{-1}$ for diamond conductivity to 10^8 for copper (that is, copper is 10^{24} times more conductive than diamond). Semiconductors, as their name suggests, fall between these two extremes. If we consider super-conductors below their superconducting transition temperature, we see an even wider range of conductivity values. Their conductivities are of the order of 10^{25} $(\Omega. m)^{-1}$, or a factor of about 10^{40} times that of diamond.

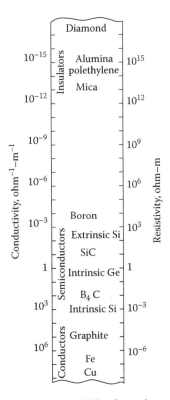

FIGURE 2.22 Range of electrical conductivities for various materials.

Let us examine how the units of conductivity arise. Ohm's law states that the current I passing through a material that is subjected to a potential gradient V is proportional to V and inversely proportional to the material's resistance, R:

$$I = V/R \qquad (2.15)$$

A material's resistance is dependent on its dimensions. If we consider a wire or rectangular specimens, the larger the cross-sectional area, the larger the current flow for a given value of V, and the longer the length of the conductor, the less the current flow. This can be expressed in equation form as

$$R = \rho\, l\, /A \qquad (2.16)$$

where
  R = resistance
  A = area
  I = length
  p = a proportionality constant called resistivity

The resistivity is not a function of specimen dimensions. Since R is expressed in ohms and the dimensions are expressed in meters, the units for p become

$$P = RA/l = \Omega.\ m^2/\ m = \Omega.\ m \qquad (2.17)$$

The conductivity, a, is simply the reciprocal of resistivity. Thus,

$$\sigma = I/\rho\ (\Omega.m)^{-1} \qquad (2.18)$$

The reciprocal of the resistance is called the conductance, G:

$$I\,G = 1/R\ \ \Omega^{-1} = \text{specimens} \qquad (2.19)$$

The conductance can also be expressed as

$$G = \sigma A/l$$

Like resistance, conductance is a function of specimen dimensions.

The conductivity and resistivity of metals are generally measured by the four-point probe technique, where a current of a few amperes is fed through the two outside leads, often to a cylindrical specimen. The voltage drop is picked off via the two inside leads that contact the specimen over some known specimen length between the two current leads. To measure resistivity accurately, the specimen dimensions must be known accurately. Knife-edges for the voltage contact points minimize the error in measurement of specimen length. The resistance is determined from the known I and V values using Equation 2.20, and then the resistivity

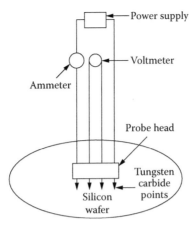

FIGURE 2.23 Four-point probe method of measuring the sheet resistivity of semiconductor wafers.

can be computed via Equation 2.22. The voltage drop is small but can be measured with a microvolt meter or, for high accuracy, a Wheatstone bridge.

The four-point probe method can also be used to measure the sheet resistivity of silicon wafers. The four points — the two current and the two voltage contacts — rest on the surface of the silicon as shown in Figure 2.23.

Since the current path is not directly through the wafer thickness, the value measured is called sheet resistivity rather than bulk or volume resistivity. However, when certain correction factors are applied to account for the wafer thickness and current path, the resistivity value obtained is valid for circuit design purposes. The correction factors are obtained from the solution of a partial differential equation for the current flow under certain boundary conditions. Two assumptions are made: (1) minority carrier injections at the metal–semiconductor current contacts recombine near the electrode, and (2) the boundary region between the current contacts and the silicon is hemispherical and small in diameter compared to the distance between the probes. A satisfactory arrangement used in our laboratories consists of a probe head that contains four tungsten carbide points spaced 0.1 cm apart (probe head and specimen holder were obtained from Signatone Corporation). The outer two carbide points supply the current, usually about 1 mA, while the two inner probes pick up the voltage drop and feed it to a mill volt-range meter. Gallium arsenide resistivity measurements are somewhat more difficult to obtain than those for silicon because of the high ohmic contact values at the probe–semiconductor interface. Indium solder joints minimize these high-resistance contacts. When the appropriate boundary conditions are applied, the solution to the partial differential equation results in the following simple equation for the resistivity:

$$\rho = \pi V \, W / I \ln 2 \qquad (2.20)$$

where W is the wafer thickness.

2.4.2 Thermal Properties Temperature Scales

In the United States, we frequently listed temperature in both Celcius and Fahrenheit. A little history will assist in our understanding and comparison of the scales. Gabriel Daniel Fahrenheit (1686–1736), a German physicist, developed the scale known by his name. He used three major points on his scale: 0°F for the freezing point of a salt solution, 32°F for the freezing point of pure water, and 96°F for the normal temperature of the human body (later found to be 98.6°F). Unfortunately for U.S. residents, Anders Celsius was born after Fahrenheit. If it had been the other way around, the Fahrenheit scale would never have existed and we would not be subjected to the anguish of having to convert from Fahrenheit to Celsius and vice versa. Celsius devised the centigrade scale, which is now officially called the Celsius scale, in 1742. He essentially had two major points on his scale, the freezing and boiling points of pure water. The scale contained 100 divisions between these two points, a much more logical approach. This scale is used just about everywhere but the United States and is used internationally by all scientists.

Another scale that is even more logical than the Celsius scale is that suggested by William Thomson Kelvin in the late nineteenth century, which bears his name. The Kelvin or absolute temperature scale is widely used by scientists and could, and perhaps should, be adopted by engineers. A unit of measurement on the Kelvin scale is the same as on the Celsius scale, but the zero point, theoretically the lowest temperature attainable, is –273°C. Shortly after Kelvin devised his scale, William Rankine introduced the Rankine absolute scale. It is similar to the Kelvin scale except that the unit of measurement is the same as on the Fahrenheit scale. Absolute zero now becomes –459.69°F, and the freezing and boiling points of pure water become 491.69°R and 671.69°R, respectively. If we could do away with the Fahrenheit scale, the Rankine would also cease to exist. These temperature scales are compared in Figure 2.24.

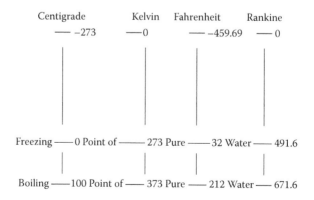

FIGURE 2.24 Comparative temperature scales.

The thermal properties of most interest to engineers are thermal conductivity and thermal expansion. The temperature coefficient of electrical resistivity (i.e., the change of resistivity with temperature) is a very important thermal–electrical characteristic of both metals and semiconductors, but we cover this subject in a later section.

2.4.2.1 Thermal Conductivity

The heat flow through a wall per unit of area is called the thermal flux, J, which is proportional to the thermal gradient, the proportionality constant being the thermal conductivity. In equation form we have

$$J = K \, \Delta T/\Delta X \qquad (2.21)$$

where $\Delta T/\Delta.X$ (i.e., temperature change per unit thickness, X) and K is the thermal conductivity. The thermal conductivity is expressed in U.S. units as BTU-ft/(hr ft^2 °F) and in the metric system as Kcal meters (s m^2 °K). In word form, the thermal conductivity is the quantity of heat flow through unit thickness and unit area per unit time.

2.4.2.2 Thermal Expansion Coefficient

This coefficient represents a change in dimension per unit temperature change. Just as in mechanical strain, thermal expansion and contraction can be expressed by a change in volume, area, or length, with the latter being the most frequently used. The volume coefficient of expansion is three times the linear coefficient. Linear handbook values will usually be given in centimeters/centimeter/degrees Celsius or in inches/inch/degrees Fahrenheit. As the temperature of a material is increased from absolute zero on the Kelvin scale, the atoms begin vibrating around their equilibrium positions, and the amplitude of vibration increases with temperature. This amplitude of vibration is not symmetrical, however. As will be discovered later, a balance of both attractive and repulsive forces between atoms determines the equilibrium position. The repulsive forces increase more drastically as atoms are pushed together than do the attractive forces as atoms are pulled apart. In other words, it is easier to pull atoms apart than to push them together. Thus, the amplitude of vibration is nonsymmetrical, resulting in a net atom positive displacement, the magnitude of which increases with increasing temperature.

There are a few exceptions to this behavior. Uranium, for example, has a negative temperature coefficient of expansion; that is, it contracts in one crystallographic direction. This is a result of the directionality of the bonding forces. In many crystalline materials there also exists a very sharp discontinuous change in dimensions at certain precise temperatures. As the atoms gain energy, at certain well-defined temperatures they prefer to alter their geometrical arrangement from one type of atom architecture to another. This dimensional change (called a phase change) is not related to the normal thermal expansions and contractions and will be better understood after we have examined the various atom arrangements that can exist in crystalline structures. Thermal expansion data are normally reported to three significant figures and can be measured accurately by a number of methods. Dilatometers can be readily purchased for this purpose. Some of the simpler ones mechanically transmit the linear dimensional change from a specimen in a furnace via a zero- or low-coefficient-expansion material to a dial gauge or an electronic sensor. The coefficient varies somewhat with temperature and is frequently reported to be over a fairly narrow temperature (e.g., from 0 to 100°C).

DEFINITIONS

Activation energy: Energy required to initiate a process or chemical reaction.

Brinell hardness number: Hardness number obtained by measuring the diameter of the impression made by a specific-size hard ball under a specific load.

Brittle: Describes a material that shows little or no plastic deformation prior to and during fracture.

Charpy V-notch test: Impact test in which the specimen has a V-shaped notch and in which the energy absorbed during fracture is measured at a specific temperature.

Diamond pyramid hardness (DPH): Hardness value obtained when a pyramid-shaped diamond makes an impression in a material under a specified load and in which the hardness number is determined from the length of the diagonals of the impression.

Ductility: Strain obtained prior to fracture, usually in a uniaxial tension test.

Elastic deformation: Recoverable deformation (change in shape) obtained after the force is removed.

Elastomer: Polymer material that essentially exhibits only elastic deformation when stressed.

Electrical conductance: Reciprocal of electrical resistance. It is a function of specimen dimensions.

Electrical conductivity: Measure of the ease of the passage of electrical current. It is the reciprocal of electrical resistivity and is independent of specimen dimensions.

Electrical resistance: Resistance to current flow. It is the reciprocal of electrical conductance and is a function of specimen dimensions.

Electrical resistivity: Inherent ability of a material to resist current flow. It is the reciprocal of conductivity and is independent of specimen dimensions.

Engineering strain: Change in length or area divided by the original length or area.

Engineering stress: Force (or load) on a body divided by the original cross-sectional area.

Fatigue life: Number of cycles a material will survive at a specified alternating stress.

Fatigue limit: Number of cycles in a fatigue S–N curve where the curve becomes horizontal (i.e., at this stress value the material has an infinite life).

Fatigue strength: Stress level corresponding to a specific fatigue life.

Fracture toughness: Resistance to crack propagation. It is an inherent property of the material.

Hooke's law: Law that states that strain is a linear function of stress in the elastic condition.

Modulus (shear): Proportionality constant between shear stress and shear strain during elastic behavior.

Modulus (Young's): Proportionality constant between normal stress and normal strain in the elastic state. It is determined from the slope of the elastic stress–strain line.

Plastic deformation: Nonrecoverable permanent deformation. It remains when the force is released.

Poisson's ratio: Ratio of the lateral contraction to longitudinal extension when a body is tested in uniaxial tension in the elastic region of behavior.

Proportional limit: Point on the stress–strain diagram where Hooke's law ceases to be obeyed and the graph becomes nonlinear.

Shear strain: Tangent of the angle of twist in torsion; the shear displacement divided by the distance over which it acts.

True strain: Natural logarithm of the ratio of the instantaneous length of a specimen to its original length. In terms of the area, it is the natural logarithm of the original area divided by its instantaneous area.

True stress: Force on a body divided by the instantaneous cross-sectional area on which it acts. Note: Stress at a point is defined by six independent parameters, of which three act normal to the faces of an infinitely small cube; the three shear stresses act parallel to these faces.

Ultimate tensile strength: Highest point in the engineering stress–strain diagram; the engineering stress at which necking begins in a uniaxial tension test.

QUESTIONS AND PROBLEMS

2.1 A 0.505-in-diameter by 2-ft gauge-length standard tensile test bar of 30×106 psi modulus is loaded to an elastic strain of 0.002. What force is required to produce this strain?

2.2 An ASTM E8-90 standard tensile bar of a copper alloy was observed to fracture at a force of 12,000 lb. The true stress at fracture was 80,000 psi. What was the diameter of the bar at the point of fracture?

What were the engineering and true strains to fracture? What was the ductility of this alloy? What was the true fracture stress in MPa?

2.3 An ASTM E8-90 standard tensile bar of steel was loaded to 12,000 lb. The ultimate tensile strength was not attained. After the load was released, it was found that the gauge length had increased to 2.046 in. The modulus of this steel is 30×106 psi and the yield strength was observed to be 30×103 psi.

(a) What was the plastic strain after load was released?

(b) What was the total length of the bar before the load was released?

(c) Graph the stress–strain curve assuming a linear relation between stress and strain in the plastic region.

2.4 List five mechanical properties and five physical properties of materials that are found in most handbooks.

2.5 A specific pure metal has a shear modulus of 28×103 MPa and a Poisson's ratio of 0.3. What is Young's modulus of elasticity?

2.6 The yield strength of steel is defined by ASTM E8-90 as that stress necessary to achieve a 0.002 plastic strain. Why would a different offset be preferred for some materials?

2.7 Verify by solving the following problem that the reduction in area prior to fracture is a better measure of ductility than that obtained by a change in length. Consider a 2-in-gauge-length tensile bar that was elongated such that the center section necked and thereby increased in length from 0.5 in to 1.0 in. The remaining 1.5-in gauge section elongated uniformly by 10%. Now consider a 10-in-gauge-length specimen of the same material in which again the center 0.5-in original-length section necked and thereby increased in length to 1.0 in while the remaining 9.5-in length of gauge section elongated uniformly by 10%. Compute the engineering strain to fracture (the ductility) for both specimens.

2.8 What are the one bad and the one good features of the Rockwell hardness test? What are the advantages of the diamond pyramid scale?

2.9 Approximately how much greater is the strain rate in a Charpy impact test compared to a standard tensile test?

2.10 What is a typical value for the impact energy absorbed in fracture for a specimen of a ceramic material such as alumina ($AI203$)? What is it for a mild steel such as a 1020 steel alloy?

2.11 What does the Charpy impact test really tell us about the design of a steel ship or bridge?

2.12 What is the chief advantage of the fracture mechanics approach to fracture toughness over the Charpy impact test?

2.13 A steel alloy has an ultimate tensile strength of 1500 MPa. What would be a good guess for the alternating stress that it could withstand for 107 cycles?

2.14 What is the difference between a creep test and a creep-rupture test?

2.15 What region of the creep curve is used for design purposes?

SUGGESTED READING

American Society of Metals (ASM) *Metals Handbook*, Vol. 8, Mechanical Testing and Evaluation, ASM International, Metals Park, OH, 2000.

Boyer, H. E., Ed., *Atlas of Creep and Stress-Rupture Curves*, ASM International, Metals Park, OH, 1988.

Boyer, H. E., Ed., *Atlas of Stress–Strain Curves*, ASM International, Metals Park, OH, 1986.

Courtney, T. H., *Mechanical Behavior of Materials*, McGraw-Hill, New York, 1990.

Hertzberg, R. W., *Deformation and Fracture Mechanics of Engineering Materials*, Wiley, New York, 1989.

Keithley Instruments, Inc., *Low Level Measurements*, 1984.

3 Atomic Structure and Atomic Bonds

3.1 STRUCTURE OF THE ATOM

Since atomic structure is covered in introductory chemistry courses, the student should be aware that the atom consists of a nucleus containing neutrons and protons around which electrons orbit. Figure 3.1 shows, for example, the electron configuration of a sodium atom, in which the nucleus consists of 11 protons and neutrons and the outer orbits contain 11 electrons. Orbits are more or less confined to certain radii. We will not be concerned here with the subatomic particles such as quarks and neutrinos. The smaller the diameter of the orbit, the greater the attractive force between the electron and the nucleus and the greater the absolute binding energy. Binding energies, by convention, are considered to be negative, so we are speaking of a large negative binding energy for the innermost orbit. The electrons in the outer orbits are bound less tightly, and the outermost electrons can to some extent be considered to be loosely bound and not necessarily residing in well-defined orbits. The outer electrons are the valence electrons and are involved in the bonding together of the atoms and hence strongly affect all physical, mechanical, and chemical properties of materials. The nature of this bond is what determines whether the substance is a metal (iron, aluminum, and so forth), in which the bond is often between like atoms (Al atoms are bonded to Al atoms in metallic Al), a ceramic (e.g., alumina, silicon nitride), or a polymer material (e.g., polyethylene, polycarbonate), where the bond is between dissimilar atoms (Si to N in silicon nitride or carbon to hydrogen in polyethylene). Before describing these various bond types, we will briefly review the electron configuration and the related periodicity of the elements.

3.1.1 ATOMIC MASS UNIT AND ATOMIC NUMBERS

The mass and the charge on each of the atomic particles of interest are listed in Table 3.1. The *atomic mass* of an element is the sum of the masses of all protons and neutrons gathered in the nucleus and electrons of the orbit. The mass of an electron is almost negligible compared to the masses of protons and neutrons. The mass of an atom is concentrated in the nucleus. Hence, the nucleus of an element controls its density.

The number of electrons and protons must be fixed in a stable atom for each element in order to be electrically neutral, but the number of neutrons can vary somewhat among the atoms of a given element because it does not affect the charge of an atom. Thus, the atoms of some elements have two or more different

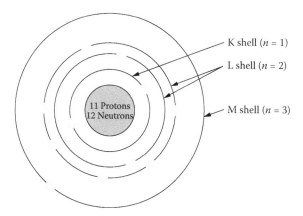

FIGURE 3.1 Electron configuration of a sodium atom.

TABLE 3.1
Masses and Charges of Atomic Particles

Particle Type	Mass in Grams	Charge in Coulombs or Ampere Seconds
Proton	1.673×10^{-24}	$+ 1.602 \times 10^{-19}$
Neutron	1.673×10^{-24}	0
Electron	9.109×10^{-28}	$- 1.602 \times 10^{-19}$

atomic masses. These different atoms are called *isotopes*. For example, iron has been found to contain atoms made of 28, 31, and 32 neutrons.

The *atomic weight* of an element is the weight in grams of 1 mol of atoms. For example, iron has an atomic weight of 55,847 g/mol. One mole contains 6.023×10^{23} atoms, which is *Avogadro's number*. The atomic weight of an element is presented in most periodic charts. The periodic table is given in Figure 3.2.

The *atomic number* of an atom gives the sequence of the elements of the periodic table and represents the number of protons in the nucleus. Hydrogen has the atomic number of 1 and Fe of 26. A neutral atom will have the same number of electrons revolving around the nucleus. If the atom has more or fewer electrons than in the neutral state, it is said to be *ionized*, and we now call the resulting atom an *ion*. If it contains more electrons than the neutral atom, it is a negative ion, called an anion (because in an electric field this negative ion would move to the anode, which is the positive pole). Hence, if fewer electrons orbit the nucleus, it becomes a positive ion or cation (because it would move to the negative pole, the cathode). This is important in corrosion, where aggressive media dissolve metal atoms in an electrolyte, where the metal atoms are becoming cations.

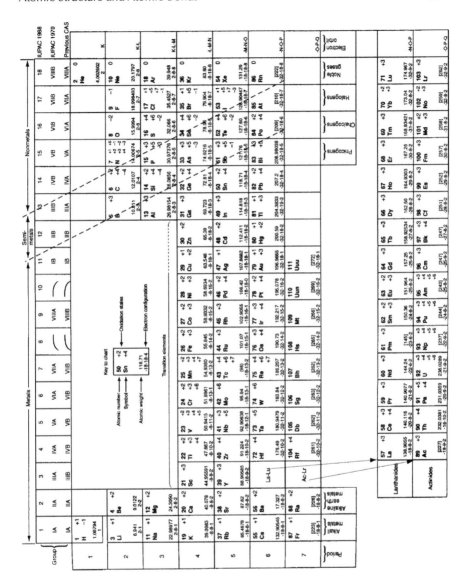

FIGURE 3.2 Periodic table of elements. (From H. Baker, *Metals Handbook, Desk Ed.*, 2nd ed. *Structure and Properties of Metals*, ASM International, Materials Park, OH, 1998, p. 87. With permission.)

Example 3.1 Calculate the weight in grams of a sodium atom:

(a) From the information listed in Figure 3.2

(b) By the weights of each nucleus particle

(c) Note the difference.

Solution
(a) The gram-molecular weight of sodium is found from Figure 3.2 to be 22.9898 g/mol.

The weight of one sodium atom is

22.9898 g/mol/6.02 × 10^{23} atoms/mol = 3.819 × 10^{-23} g

This is the weight of one sodium atom.

(b) Na contains 11 protons, 11 neutrons, and 11 electrons. Summarizing the weight of those 33 atomic particles, the weight will be 3.6838 10^{-23} g

This is obviously different from the former calculation.

(c) The difference in the two values is a result of the various isotopes that are present in natural sodium.

Example 3.2 Compare the density of aluminum and tungsten and derive the number of protons from the atomic number.

Aluminum has the atomic number of 13 and hence 13 neutrons and protons. The density is 2.7 g/cm^3. The atomic number of tungsten is 74 (74 protons and neutrons). Tungsten has a density of 19.32 g/cm^3. The large amount of atomic particles lead to a much higher density of W. Even though we neglected the isotopes of W and Al, this is a clear indication that the atomic particles of the nucleus stand for the density of materials.

3.1.2 ELECTRON CONFIGURATION AND PERIODIC TABLE

Important material properties that are controlled by the configuration of electrons around the nucleus are:

- Chemical behavior
- Bonding behavior
- Magnetic properties

The simplest atom to use to describe the electron configuration is that of hydrogen, since it has only one electron revolving around a proton. Niels Bohr described the hydrogen atom in this way in 1913. The theory of the hydrogen atom may be adopted to other atoms. Not all phenomena can be explained with this model, but it helps in understanding the basics of chemical bonding.

Bohr developed his theory for the hydrogen atom, in which the electron was suspected to revolve around the proton in a spherical orbit. These orbits are exactly defined; we refer to these as energy levels. They surround the nucleus and in this

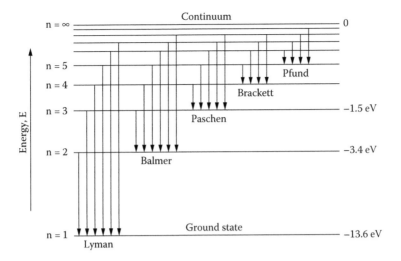

FIGURE 3.3 Energy-level diagram determined from the spectra of radiation emitted from excited hydrogen.

book they will be called K, L, M, N ... or a number $n = 1, 2, 3, 4$ ... with increasing distance.

All atoms in the fundamental or ground states have electrons occupying the most stable energy-level configuration, usually that of the lowest energy. There exist an infinite number of allowable energy levels, but in hydrogen only one level will be occupied since there is only one electron available. If hydrogen gas is subjected to a spark discharge, the electrons can be excited from the innermost stable orbit of $n = 1$ to higher orbits (e.g., where $n = 2, 3, 4$ ...). Therefore, we need energy to move the electron from the ground level to higher orbits; we have to overcome the attractive electrostatic forces between the nuclei and the electron. Those attractive electrostatic forces lead to what we call Coulombic attraction.

The electron energy-level diagram for a hydrogen atom is given in Figure 3.3, that is, the energies to excite the electron from the stable level to the outer orbits $n = 1, 2, 3$ ... . If the electron springs back to an inner orbit, it emits light quants, practically used in neon lights, where electric currents excite the electrons. The frequency of the spectral lines that arise from a transition from level 2 to 1 is given by

$$v = \Delta E/h = (E_{n1} - E_{n2})/h \qquad (3.1)$$

where

v = frequency of spectral line

ΔE = energy difference (for example, for an electron excited to level 2 from the Lyman series E = -3.4 eV $- (-13.6$ eV$) = 10.2$ eV

H = Planck's constant, 6.626×10^{-34} Joule-second (Js).

An electron volt, eV, is the energy given to an electron by accelerating it through 1 volt of electric potential difference. For an engineer this becomes more obviously an energy term by converting electron volts into Joules or Newton meter (Nm): $1\ eV = 1602 \times 10^{-19}\ J$ ($1\ J = 1\ Nm$).

The energy required to remove the electron completely from the atom is the ionization energy, and for the hydrogen atom it is 13.6 eV.

The excited incandescent gas emits a spectrum of radiation composed of lines at definite wavelengths. These characteristic spectra were named after those who discovered them and are noted in Figure 3.3 along with the computed energy levels for various values of n. Practically, this can be used for analyzing a material's composition by spectral analysis. Exciting a material by thermal energy excites the electrons to the outer orbits, and by spring-back, each element emits a definite wavelength. Measuring the wavelength and the intensity of the light quants is the basis for the spectral analysis technique.

The Bohr model is the best method available for specifying the energy levels of the electron in a complex atom. For an engineer, the understanding of this model is more or less sufficient to understand the bonding behavior of various materials, but note that this is a simplified model. Famous physicists and chemists (for example, Sommerfeld, Schrödinger, Pauli, and so forth) applied quantum mechanic calculations for deriving atomic models. Therefore, we now tend to think more in terms of an electron cloud or charge density as depicted in Figure 3.4. Here the orbit of the electron for $n = 1$ for the hydrogen atom is shown as a precise circle, whereas the shaded region represents the electron shaded cloud as per quantum mechanical calculations.

3.1.2.1 Periodic Table

The periodic table of Figure 3.2 lists all the elements from hydrogen, atomic number 1, to lawrencium, atomic number 103. Elements that have comparable chemical behavior are listed vertically in groups. For example, main groups 1 and 2 are alkali and earth alkali metals, and main group 8 is inert gases. The horizontal rows are the periods; the first period contains only 2 elements, next 2

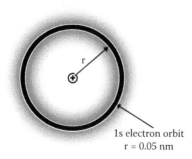

1s electron orbit
r = 0.05 nm

FIGURE 3.4 Electron charge cloud surrounding the 1s orbit for the hydrogen atom.

rows contain 8 elements, and the following rows have 18 and even 32 elements. The periodic table contains important data such as atomic mass, melting point, boiling point, electronegativity of the elements, and sometimes lattice structures. Today the periodic table can be easily downloaded from the Internet.

The periodic table is well ordered by increasing atomic number, that is, by increasing the number of protons. Consequently, the number of electrons will increase. Hydrogen has one proton and one electron, and He has two of each. The electrons occupy the K-orbit ($n = 1$), which contains a subshell $1s$. This subshell can contain a maximum of two electrons. The next element, Li, therefore must fill a third electron in a second shell, L ($n = 2$), which refers again to a discrete energy level. The third electron occupies the $2s$ subshell. The L shell has a total capacity of $2n^2$ electrons. Consequently, Be fills up the $2s$ level and has four electrons. Boron, the following element, has five electrons and must fill the fifth electron in the $2p$ subshell. Neon then has 10 electrons, where the L-mainshell is completely occupied by electrons:

$$Ne \ 1s^2 \ 2s^2 \ 2p^6$$

The $2p$ subshell is completely occupied. This rule of filling main and subshells works for the first 18 elements, and the electron structure of all elements is given in Table 3.2.

Electrons occupy discrete energy levels within an atom. Each electron possesses a specific energy, with no more than two electrons having the same energy. The energy levels to which a specific electron belongs is determined by four quantum numbers. The first three quantum numbers are:

- Principle quantum number
- Azimuthal quantum number
- Magnetic quantum number

They determine the number of possible energy levels. The fourth quantum number is the spin quantum number described below.

The principal quantum number n is described by integers that can take on all integral values from 1 to ∞. It represents the main shell or orbit, where there is a high probability of finding electrons. The azimuthal or I quantum number specifies the subenergy levels that exist within the main shell and can be viewed as more closely defining the position of the electron. However, only the allowable states for which electrons are available will be filled. In the absence of any excitation, the filled levels will always be those nearest the nucleus.

The I number can have values of $I = 0, 1, 2, \ldots, n$. The letters $s, p, d,$ and f are used to define these sublevels in the shorthand notation. The magnetic, or m_l, quantum number has permissible values of "-1" to "$+1$" including zero. When $I = 1$, the permissible values for m_l are "-1," "0," and "$+1$," and this pattern continues. Thus, in general, there are $2I + 1$ allowable values for m_l. In terms

TABLE 3.2
Electron Structure of the Elements

		K(n = 1)	L(n = 2)		M(n = 3)			N(n = 4)				O(n = 5)				P(n = 6)			Q(n = 7)
Shell																			
Subshell		s	s	p	s	p	d	s	p	d	f	s	p	d	f	s	p	d	s
Capacity		2	2	6	2	6	10	2	6	10	14	2	6	10	14	2	6	10	2
1	H	1																	
2	He	2																	
3	Li	2	1																
4	Be	2	2																
5	B	2	2	1															
6	C	2	2	2															
7	N	2	2	3															
8	O	2	2	4															
9	F	2	2	5															
10	Ne	2	2	6															
11	Na				1														
12	Mg				2														
13	Al				2	1													
14	Si		Filled		2	2													
15	P		(2; 2–6)		2	3													
16	S				2	4													
17	Cl				2	5													
18	A				2	6													
19	K							1											
20	Ca							2											
21	Se						1	2											
22	Ti						2	2											
23	V		Filled				3	2				Transition elements							
24	Cr		(2; 2–6; 2–6)				5	1				(Manganides)							
25	Mn						5	2											
26	Fe						6	2											
27	Co						7	2											
28	Ni						8	2											
29	Cu							1											
30	Zn							2											
31	Ga							2	1										
32	Ge		Filled					2	2										
33	As		(2; 2–6; 2–6–10)					2	3										
34	Se							2	4										
35	Br							2	5										
36	Kr							2	6										

TABLE 3.2 (Continued)
Electron Structure of the Elements

Shell		K(n = 1)	L(n = 2)		M(n = 3)			N(n = 4)				O(n = 5)				P(n = 6)			Q(n = 7)
Subshell		s	s	p	s	p	d	s	p	d	f	s	p	d	f	s	p	d	s
Capacity		2	2	6	2	6	10	2	6	10	14	2	6	10	14	2	6	10	2
37	Rb											1							
38	Sr											2							
39	T		Filled							1		2							
40	Zr		(2; 2–6; 2–6–10; 2–6)							2		2							
41	Nb									4		1		Transition elements					
42	Mo									5		1		(Technetides)					
43	Tc									6		1							
44	Ru									7		1							
45	Rh									8		1							
46	Pd									10									
47	Ag											1							
48	Cd											2							
49	In		Filled									2	1						
50	Sn		(2; 2–6; 2–6–10; 2–6–10)						*Empty*			2	2						
51	Sb											2	3						
52	Te											2	4						
53	I											2	5						
54	Xe											2	6						
55	Cs											2	6			1			
56	Ba											2	6			2			
57	La											2	6	1		2			
58	Ce										2	2	6			2			
59	Pr										3	2	6			2			
60	Nd										4	2	6			2			
61	Pm										5	2	6			2			
62	Sm		Filled								6	2	6			2			
63	Eu		(same as above)								7	2	6			2			
64	Gd										7	2	6	1		2			
65	Tb										9	2	6			2			
66	Dy										10	2	6			2			
67	Ho										11	2	6			2			
68	Er										12	2	6			2			
69	Tm										13	2	6			2			
70	Yb										14	2	6			2			
71	Lu										14	2	6	1		2			

TABLE 3.2 (Continued)
Electron Structure of the Elements

Shell	K(n = 1)	L(n = 2)		M(n = 3)			N(n = 4)				O(n = 5)				P(n = 6)			Q(n = 7)	
Subshell	s	s	p	s	p	d	s	p	d	f	s	p	d	f	s	p	d	s	
Capacity	2	2	6	2	6	10	2	6	10	14	2	6	10	14	2	6	10	2	
72 Hf													2		2				
73 Ta													3		2				
74 W				Filled									4		2				Transition
75 Re				(2; 2–6; 2–6–10; 2–6–10–14; 2–6)									5		2				elements
76 Os													6		2				(Rhenides)
77 Ir													7		2				
78 Pt													8		2				
79 Au															1				
80 Hg															2				
81 Tl															2	1			
82 Pb				Filled											2	2			
83 Bi				(2; 2–6; 2–6–10; 2–6–10–14; 2–6–10)										Empty	2	3			
84 Po															2	4			
85 At															2	5			
86 Rn															2	6			
87 Fr				Filled											2	6		1	
88 Ra				(same as above)											2	6		2	
89 Ac															2	6	1	2	
90 Th															2	6	2	2	
91 Pa											2				2	6	1	2	
92 U											3				2	6	1	2	
93 Np				Filled							5				2	6		2	
94 Pu				(same as above)							6				2	6		2	
95 Am											7				2	6		2	
96 Cm											8				2	6	1	2	
97 Bk											9				2	6		2	
98 Cf											10				2	6		2	

of the *s*, *p*, *d*, and *f* notations, it means there exist one *s* orbital, three *p* orbitals, and seven *f* orbitals for each allowed *s*, *p*, *d*, and *f* subenergy level.

Finally, the spin quantum number, the m_s quantum number, is pictured as defining the spin of the electron and by some will be used as a clockwise or counterclockwise spin, and by others as a parallel or antiparallel spin. If two electrons occupy a specific allowable level, they must be of different spin.

A very important principle determines how the allowable energy levels for each atom in the periodic table are filled with electrons. This is the *exclusion principle* first described by Wolfgang Pauli, and it states that "in a single atom no two electrons can have the same set of four quantum numbers." This principle is more often stated in the form "only two electrons can occupy a state specified by the quantum numbers n, l, and m_l, and they must be of opposite spin." The exclusion principle plus the four quantum numbers and their restrictions will be used to describe how the energy states are filled with electrons as we proceed from hydrogen through the periodic table as it was described above.

The overall filling pattern is called the *electron configuration* of the elements and in shorthand notation can be written as follows:

$$1s^2\ 2s^22p^6\ 3s^23p^63d^{10}\ 4s^24p^64d^{10}4f^{14}\ 5s^25p^65d^{10}5f^{14}\ 6s^26p^66d^{10}6f^{14}$$

In writing the electron configuration of any element, the principal quantum number n is written first, followed by the letter or letters for any l subshells, with these letters being written with a superscript number indicating the number of electrons in each subshell. Table 3.2 gives the order in which energy levels in an atom are filled with electrons, and Table 3.3 gives the individual electron configuration of each element.

3.1.2.2 Subshells

We can see that the inert elements Ne, Ar, Kr, Xe, and Rn have filled outer subshells as well as filled inner main shells.

Some rather obvious inconsistencies in the way the shells are filled must be addressed. These discrepancies are first noticed in the elements of atomic numbers 19 through 28 and then again in the 37-through-45 group, and in many of the rare earth elements and others of high atomic number. These exceptions can be

---

TABLE 3.3
Order in Which Stable Energy Levels in an Atom Are Filled with Electrons

Main Shell	Subshell				Total Capacity
	s	p	d	f	
(K) $n = 1$	2				2
(L) $n = 2$	2	6			8
(M) $n = 3$	2	6	10		18
(N) $n = 4$	2	6	10	14	32

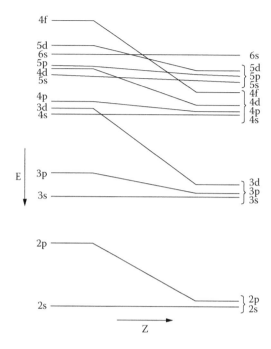

FIGURE 3.5 Dependence of energies of electron orbitals on the nuclear charge Z. (After W. J. Moore, *Physical Chemistry*, Prentice-Hall, Englewood Cliffs, NJ, 1955, p. 291. With permission.)

presented, and perhaps better explained, by referring to Figure 3.5, where the energy levels are plotted as a function of the atomic number Z. At a certain atomic number, in the first case that for potassium, the $3d$ energy level is lower than that of the $4s$ orbital. This difference is predicted from a quantum-mechanical picture and will not be covered here in detail. The cause of the lowering of the $3d$ level can be attributed to the shielding effect of the inner filled shells. The $4s$ orbital can be viewed as penetrating in toward the nucleus, and thus the $4s$ electrons are more tightly bound to it because of the high positive charge on the nucleus. Elements that portray this characteristic are known as *transition elements*. One of the major consequences of this filling sequence is that there now exists an imbalance of spins for the electrons in the unfilled d shell, which has a profound effect on the magnetic behavior of some of these elements.

Example 3.3

(a) Aluminum has an atomic number of 13. Write its electron configuration without referring to Table 3.2. What is its valency?

(b) Iron is a transitional element with an atomic number of 26. Its electron configuration is $1s^2\ 2s^22p^6\ 3s^23p^63d^6\ 4s^2$. Cobalt is also a transition element and has an atomic number of 27. Write its configuration without using Table 3.2.

Solution

(a) $1s2 \, s^22p^6 \, 3s^23p^1$. Aluminum generally behaves as a valency of 3 (i.e., both the s and p electrons become involved in bonding, even though the s level is filled). Given the atomic number, the student should be able to write the electron configuration of any element other than the transition elements.

(b) $1s^2 \, 2s^22p^6 \, 3s^23p^63d^7 \, 4s^2$. In the transition elements, usually the partially filled subshell, in this case the d shell, will fill up with the addition of one electron with each increase in atomic number.

A final look at the periodic table is now in order. The elements that are in the same vertical row have similar outer-shell configurations and hence similar properties. The horizontal rows all begin with a very active, or as some prefer, reactive, metal on the left and end with an inert element on the right. The transition elements are concentrated in the center. The active elements in the vertical rows on the left tend to form stable compounds with those elements on the right, excluding the inert elements. The terms *electropositive* and *electronegative* are used to describe the elements that tend to form these compounds, with the electropositive elements on the left being metallic and giving up electrons in forming these compounds, while the electronegative elements on the right receive the electrons. The most electronegative elements are in groups 6A and 7A of the periodic table. *Electronegativity* can be defined as the tendency of the atom to attract electrons. All of this information is pertinent to our discussion of atomic bonding in the following section. As can be seen from Figure 3.6, electronegativity increases in one row from left to right and decreases from row 1 to 2 to 3 and so on.

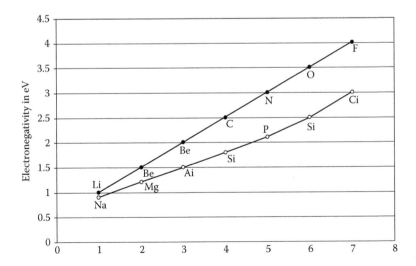

FIGURE 3.6 Electronegativities of 2nd and 3rd row elements of the periodic table.

3.2 BOND TYPES

The most stable electron configuration is the configuration of the inert gases with filled outer *sp*-levels with eight electrons. This is the electron configuration that dissimilar atoms try to achieve if they are brought together, and they do it because they want to. Hence, the number of electrons in the outer s-p-levels that can be removed or filled up in order to get a filled *s*- and *p*-subshell determines the *valency* of the element. In a technical and less colloquial way of expressing this concept, they bond together because they can reduce the total energy of the system. Chemical bonds are classified as strong and primary bonds (covalent bonds, ionic bonds, and metallic bonds) and secondary, or *weak, bonds* (van der Waals or polarization bonds [hydrogen bonds]).

In the following sections, each bonding type will be discussed individually. However, in materials there are often mixed bond structures (such as a combined covalent and metallic bond) and not, for instance, 100% covalent bonding. Table 3.4 shows bonding types of various materials.

High bonding energies and strong bonds correspond with high Young's moduli. The melting point of materials increases with increasing bond strength.

3.2.1 COVALENT BONDS

Purely covalent bonds appear in diamond, silicon, and germanium. These are the dominant bond in silicates, carbide, and nitride ceramics, and they contribute to the bonding of some metals such as W, Mo, and Ta. They are bonds in polymers, too, linking the carbon atoms to each other along the polymer chain.

TABLE 3.4
Materials with Strong Bonding Energies and Given Amount of Bonding Nature

Substance	Bond
MgO	70% ionic
NaCI	59% ionic
Al	Metallic
Cu	Metallic
W	Metallic
Ti	Metallic
Si	100% covalent
C (diamond)	100% covalent
SiC	88% covalent

Covalent bonds arise when two atoms share electrons in an attempt to fulfill their needs in the sense of completely filling the *s, p, d,* and *f* suborbitals. The most stable states are those when the *s*-orbitals contain 2 electrons; the *p* orbitals, 6 electrons; the *d* orbitals, 10 electrons; and the *f* orbitals 14 electrons. The simplest case of the covalent bond is that of two hydrogen atoms coming together to form a combined *s* orbital that contains two shared electrons, one being contributed by each hydrogen atom. This is the classic example in chemistry texts, where the electron-dot notation is used to show the bond as follows (the two dots represent the shared electrons):

H:H

Perhaps a better way to visualize this bond is by illustrating the overlap of the two 1*s* orbitals, as depicted in Figure 3.7. Yet another way to describe this bond is that the electrons resonate between the two atoms and essentially trick the atoms into thinking that each hydrogen atom always has two electrons in its 1*s* shell. Whatever the model, experiments show that the reduction in energy is about 4.5 eV when the hydrogen molecule is formed. Remember, an electron volt, eV, is the energy given to an electron by accelerating it through 1 volt of electric potential difference, and 1 eV = 1602×10^{-19} J or Nm. This also means that 4.5 eV must be supplied to separate the two atoms (i.e., to dissociate the molecule). In higher- atomic-number common gases, such as nitrogen, oxygen, and fluorine, the atoms attain a stable shell of eight electrons by sharing outer-shell electrons. In nitrogen this is accomplished by the sharing of three 2*p* electrons of one atom with three 2*p* electrons from another nitrogen atom. Similarly, two oxygen atoms share two 2*p* electrons, and fluorine atoms share one 2*p* electron from each atom, as shown by dot notation in Figure 3.8.

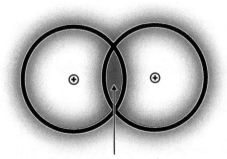

Region of high electron charge

FIGURE 3.7 Hydrogen molecule bond.

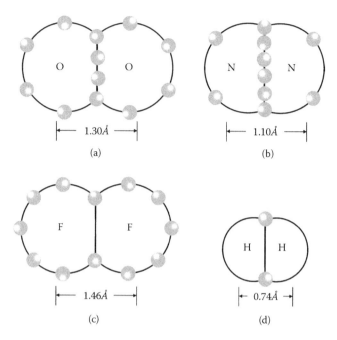

FIGURE 3.8 Covalent bonds of diatomic molecules. (From D. W. Richardson, *Modern Ceramic Engineering*, 2nd ed., Marcel Dekker, New York, 1992. With permission.)

3.2.1.1 Hybridization Bonding Process

The covalent bond is fairly common in organic molecules and occurs in many cases by the "hybridization" of the s and p orbitals. The s shell is more or less spherical, whereas the p shell is shaped more like a dumbbell. When they hybridize, a different shape of charge density occurs.

Let us examine pure carbon first. Carbon exists in two modifications: diamond and graphite. The electrons involved in the hybridization are the $2s^2 2p^2$ electrons, which are changed to $2s^1 2p^3$ states by excitation of a $2s$ electron to the $2p$ level. In pure carbon we now have four equivalent sp^3 hybrid orbitals, as depicted in Figure 3.9. Now each of these four orbitals shares an electron with identical orbitals from four other carbon atoms to achieve an eight-electron outer shell. Carbon in this form is crystalline (diamond structure), as shown in Figure 3.10. Even though energy is required for an s electron to be excited to the p level, there is a net decrease in energy when the carbon atoms are bonded together. This is an extremely strong bond and it is why diamond is so hard, which is reflected in its high Young's modulus, the highest of all materials. In this type of bond the electrons resonate between the atoms and have a very high probability of being in the dark regions in Figure 3.10. Covalent bonds of this type are very directional and result in a directionality of properties. The strength of diamond, for example, is dependent on the crystallographic direction in which it is measured. All

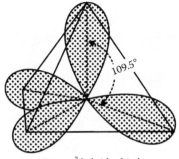

Four sp^3 hybrid orbitals

FIGURE 3.9 Four sp^3 hybrid orbitals for the carbon atom.

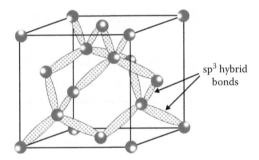

sp^3 hybrid bonds

FIGURE 3.10 Diamond crystal structure formed from four sp^3 orbitals that combine with four sp^3 orbitals from four other carbon atoms.

electrons are related to bonds, and we have no free electrons to move in an electric field. Therefore pure covalent structures are electric isolators.

When carbon atoms combine to form graphite, a different type of hybridization occurs. Graphite has a hexagonal layered structure (Figure 3.11). Here three electrons hybridize in bonds at 120° to each other, leaving one electron only weakly bonded to the others. This electron accounts for the high electrical and thermal conductivity of graphite compared to diamond. The layers are bonded together by weak secondary bonds, which allow them to slide past each other easily, thereby producing the lubricating properties of graphite.

3.2.2 IONIC BONDS

Ionic bonds are formed by a complete transfer of electrons from one atom to another, in contrast to the sharing found in the covalent bond. Another distinct difference between the two is the lack of directionality in the ionic bond that was so prevalent in the covalent bond. Sodium chloride shows ionic bonding, as do other alkali halides (e.g., LiF), oxide ceramics (e.g., alumina, magnesia), and

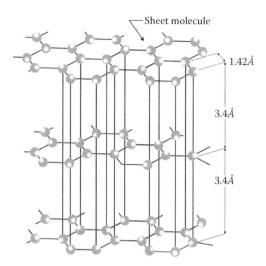

FIGURE 3.11 Sheet structure of graphite. (From C. W. Richards, *Engineering Materials Science,* Brooks/Cole, Monterey, CA, 1961, p. 45. With permission.)

cement (e.g., hydrated carbonates and oxides). Let us examine the bond in NaCl. Sodium of atomic number 11 has an electron configuration of

$$\text{Na } 1s^2 \, 2s^2 \, 2p^6 \, 3s^1$$

while chlorine has 17 electrons in its configuration.

$$\text{Cl } 1s^2 \, 2s^2 \, 2p^6 \, 3s^2 \, 3p^5$$

Chlorine completes its third shell of eight electrons by receiving the $3s^1$ electron from the sodium when the two atoms are brought together. For the moment let us assume that this bond is one of 100% ionic character. Then there would be a total charge transfer resulting in a sodium ion with +1 charge (+1.602 $\times 10^{-19}$ C) and a chlorine ion with an equal quantity of negative charge as depicted in Figure 3.12. The Coulombic attractive force between the two ions is proportional to the product of the charges divided by the equilibrium interatomic distance, r_o, squared; that is,

$$F_{\text{attractive}} = a \, (Z_1 \, Z_2 \, e^2/r_o^2) \tag{3.2}$$

where a is the proportionality constant. The repulsive force, owing primarily to the inner electron shells of the two ions, is of the form

$$F_{\text{repulsive}} = - \, (nb/r_0^{n+1}) \tag{3.3}$$

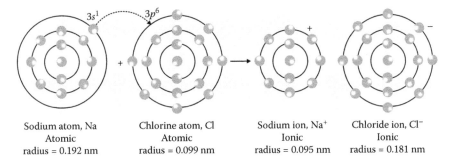

Sodium atom, Na   Chlorine atom, Cl   Sodium ion, Na⁺   Chloride ion, Cl⁻
Atomic   Atomic   Ionic   Ionic
radius = 0.192 nm   radius = 0.099 nm   radius = 0.095 nm   radius = 0.181 nm

FIGURE 3.12 Formation of a sodium chloride ion pair by transfer of negative charge from a $3s1$ sodium orbital to a $3p$ chlorine orbital. (From W. F. Smith, *Principles of Materials Science and Engineering*, McGraw-Hill, New York, 1986, p. 36. With permission.)

where n and b are constants. At the equilibrium interatomic distance these forces will be equal in magnitude and opposite in sign, resulting in a net zero force.

Ionic-bonded materials such as oxide ceramics are hard, and usually they have low electrical conductivity. The isolating properties of ionic-bonded oxide ceramics can be explained by there being no free electrons that can move in an electric field. All electrons are used for the bonding process.

3.2.3 METALLIC BONDS

In the metallic bond, which is the dominant bonding mechanism in solid and liquid metals and alloys, generally we are talking about the bonding together of like metallic atoms. It is obvious that we cannot explain the electrical conductivity of metals with the mechanism of covalent or ionic bonding. Therefore, a model for a different bonding mechanism must be applied. In metallic materials each atom gives up its valence electrons to form a cloud of negative charge (sometimes called electron gas or sea of electrons) in which the positive ions that remain behind are immersed, somewhat as depicted in Figure 3.13. After the bonding the valence electrons do not belong to a specific ion anymore. The forces that hold the ions together are somewhat akin to the ionic bond in that there exists an electrostatic attraction between the negative charge cloud and the positive ions, with the repulsive forces being primarily those produced by the inner electron shells.

The valence electrons are free to roam throughout the body of the material. They are constantly in rapid motion at temperatures above 0 K and in essence resonate among all atoms. These *free electrons* account for the high electrical and thermal conductivity of metals and to some extent their plasticity. Alloys of two or more metals in solid solution also have a metallic bond, the bond strength of the unlike atom pairs being somewhat different, but not drastically so, than that of the like atoms.

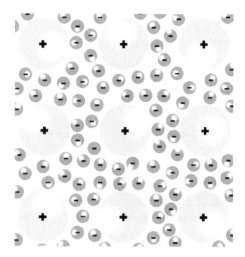

FIGURE 3.13 Metallic bond formed by the attraction of the positive nuclei to the valence electron space charge.

Unlike metallic alloys, intermetallic compounds or intermetallic phases such as the nickel and titanium aluminides (Ni_3Al, $NiAl$, Ti_3Al, and $TiAl$) are stoichiometric compounds between two metals and are currently being seriously studied for high-temperature applications. They can be considered to be of a mixed bonding nature with some degree of ionicity and covalency mixed with the metallic character. The greater the difference in the electronegativity of the elements, the larger the ionic character. The titanium aluminides would be expected to have more ionic character than the nickel aluminides, which have only about 2.2% ionic character. These compounds are brittle, somewhat like the ionic and covalent compounds, but their electrical and thermal conductivities are more like those of metals.

3.2.4 WEAK POLARIZED SECONDARY BONDS

Weak secondary bonds are by far weaker than the primary bonds, but they are still very important. H_2O molecules in water are attracted by secondary bonds: hydrogen bonds. The weak bonding leads to a melting point for water of 0°C and a boiling point of 100°C. Without those secondary bonds water would boil at −80°C. Polymers are also bonded by this mechanism. Polyethylene (PE) forms long molecule chains of a $-C_2H_2-$ monomer. Whereas the bonds in the chain are covalent, the attraction forces between the chains are weak secondary bonds. Owing to those weak bonds (and the chain structure), the material shows viscous behavior at comparatively low temperatures and can be easily manufactured by extruding and injection molding. Furthermore, gas atoms and gas molecules such as nitrogen are held together by van der Waals bonds. The weak bonding forces

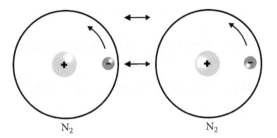

FIGURE 3.14 Electron probability distribution in nitrogen. Two molecules form a fluctuating dipole that is responsible for the attractive forces in N_2 gas.

can be easily derived by the low melting point of N_2 at $-198°C$ at atmospheric pressure. Let us now explain the bonding mechanisms.

In 1873, J. D. van der Waals proposed a new thermodynamic equation of state of a real gas that took in account the weak attractive forces between gas molecules, for example, between N_2 molecules. These weak forces are a result of both permanent and fluctuating dipoles. On approaching each other, even the inert gases experience this attraction. A dipole moment exists when there is a nonsymmetrical charge distribution; this means the electron probability distribution is not homogenous, as shown in Figure 3.14 for nitrogen. The electron probability is accidentally higher on one side of the N_2 molecule, inducing a dipole in the neighbor molecule. The molecules are held together by the dipole charge distribution. The dipole in N_2 is not permanent, but it is a fluctuating dipole that is formed accidentally and creates attraction sufficient to cause liquefaction of gases at low temperatures.

Other types of weak physical forces are stronger than the van der Waals forces but much weaker than the strong ionic, covalent, and metallic bonds. These bonds arise owing to a *permanent* nonsymmetrical charge distribution in the molecule (i.e., they have permanent dipoles). Water is the most common example. In this molecule the hydrogen atoms are bonded to the oxygen atom at an angle that places the positive hydrogen atoms on one side of the molecule (Figure 3.15).

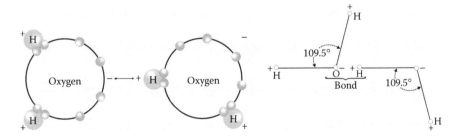

FIGURE 3.15 Hydrogen bond between two water molecules. (From C. W. Richards, *Engineering Materials Science,* Brooks/Cole, Monterey, CA, 1961, p. 26. With permission.)

This causes the positive side of the molecule to be attracted to the negative side of a neighboring molecule.

As water vapors are cooled, the attractive forces are sufficient to cause the vapor to condense to a liquid at 100°C (373 K). This bond is called the *hydrogen* bond. (Be aware that vaporization separates the hydrogen molecules. Cracking the covalent bond of the hydrogen molecule requires much higher energy.) The hydrogen bond is a prevalent bond between polymer molecules and biological molecules such as DNA.

Roughly, we can compare primary bonds to hydrogen and van der Waals bonds by a factor of 100:10:1. Bond strengths are given in kilojoule/mole (kilocalorie/mole). Primary bonds have values of about 50 to 1000 kJ/mol, hydrogen bonds of about 30 kJ/mol, and van der Waals bonds of about 2 to 8 kJ/mol.

3.3 INTERATOMIC DISTANCES

The equilibrium spacing between atoms is determined by a balance between attractive and repulsive forces no matter what type of bond is involved. The bonding force between atoms may be represented approximately by the general equation

$$F = F_{attr} - F_{rep} = A/r^M - B/r^N \ (N > M) \tag{3.4}$$

where r is the center-to-center spacing between atoms, and A, B, M, and N are constants. The first term represents the attractive force and the second term, the repulsive force. When these two are equal, $F(r) = 0$, the atoms are said to be at their equilibrium spacing, r_0. Any change in these two variables will alter the spacing accordingly.

Let us examine how these forces vary with distance by referring to the graph in Figure 3.16. The shapes of the attractive and repulsive curves depend on the values of the constants. We will assume that one atom is fixed at the position $r = 0$ and that another atom is made to approach it from some large distance, infinity if you wish. As the atoms approach, they are drawn together by the attractive force, which outweighs the repulsive force. If they are made to approach farther than the equilibrium spacing, the repulsive force predominates and tries to push them back toward their equilibrium position. Note that the repulsive force curve has a much steeper slope than that of the attractive force curve. The inner filled electron shells of each atom provide a very strong repulsive force between the atoms as they approach each other. It is more difficult to push atoms together than to pull them apart. Hence, the second part of Equation 3.4 must increase more rapidly for diminishing values of r than does the first term, and N is necessarily greater than M.

The shape of the attractive force curve is the same as that for the attraction between any unlike charges. According to Coulomb's law, this force is inversely proportional to the square of the distance between the charges. Thus, M has the value

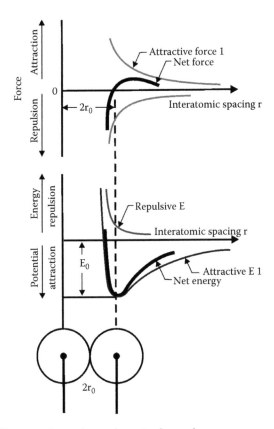

FIGURE 3.16 Forces and energies vs. interatomic spacing.

of 2 for the strong interatomic bonds, but M may be as high as 7 for the weak intermolecular bonds. The value for N is more difficult to establish and is a function of the type of bond for both strong and weak bonds. Experimentally, it has been found that N varies from about 7 to 12 for strong bonds, with metallic bonds generally falling in the lower part of this range and the ionic and covalent bonds in the upper part. For example, if the attractive force is proportional to r^{-2} the repulsive force is proportional to r^{-10}. Thus, the attractive forces predominate at large atom separations, while the repulsive forces take over at small interatomic spacings.

The equilibrium interatomic distances for most materials is on the order of 0.4 nm (10^{-10} m = 0.1 nm).

In the net force curve of Figure 3.16, the slope dF/dr at $r = 2r_0$ is related to the modulus of elasticity. The modulus is determined from a straight-line relationship (compare the stress–strain curve). In most elastic deformations the atoms move less than 10% of their equilibrium spacing. For this reason we can say that the elastic moduli in tension and compression are essentially the same.

Higher attractive forces lead to higher binding energies. The equilibrium binding energy E_0 is given at $r = 2r_0$ and has a negative value. If the binding energy is higher (more negative), the slope of the net force curve, $F = dU/dr$, is steeper, which corresponds with a higher Young's modulus.

3.3.1 BOND ENERGY VS. INTERATOMIC DISTANCE

In most cases it is better to deal with energies than with forces. The potential energy between the two atoms, which is the energy of interest here, is found by integrating the force equation as follows:

$$U(r) = \int F(r)\, dr \qquad (3.5)$$

Without derivation this leads to

$$U(r) = -\,a/r^m + b/r^n \quad (n > m) \qquad (3.6)$$

where a and b are new constants related to A and B, $m = M$, and $n = N - 1$. The equilibrium spacing is that of lowest energy, which also occurs when $F(r) = 0$, as can be illustrated by taking the derivative of the energy equation; thus,

$$U(r) = \int F(r)\, dr$$

$$dU/dr = F(r)$$

Both the force and the derivation of the energy curve, dU/dr, are zero at the same value of r (i.e., for $r = 2r_0$) (see Figure 3.16).

Let us finally consider how the energy curve looks if the bonding energy increases. Higher attractive forces lead to higher binding energies. The equilibrium binding energy E_0 is given at $r = 2r_0$ and has a negative value. If the binding energy is higher (more negative), the slope of the net force curve, $F = dU/dr$, is steeper, which corresponds to a higher Young's modulus. The $F(r)$ curve is given by the derivation of the $U(r)$ curve and the Young's modulus by its second derivation

$$E \sim dF/dr = d^2U/dr^2 \qquad (3.7)$$

This is an atomistic interpretation of the Young's modulus. The easy expression is, the higher the binding energy between the atoms of a material the higher is the Young's modulus.

DEFINITIONS

Anion: Negatively charged nonmetallic ion that results from an atom gaining an electron.

Atomic mass unit (AMU): Measure of atomic mass; 1/12 of the mass of a carbon 12 atom.

Atomic number (Z): Number of protons in the nucleus of an atom.

Atomic weight: Mass in grams of a mole of atoms of an element, including all naturally occurring isotopes.

Avogadro's number: Number of atoms or molecules in a mole of a substance. It has the numerical value of 6.023×10^{23}.

Azimuthal quantum number: Quantum number l, which is related to the ellipticity of an orbiting electron.

Cation: Positively charged metallic ion that results from a loss of an electron from an atom.

Coulombic force: Force between charged particles that varies inversely with the square of the distance between the particles.

Electron volt: Unit of energy that is equivalent to that acquired by an electron when it moves through an electric potential of 1 V.

Electronegative: For an atom, the tendency to accept electrons when forming a compound.

Electronegativity: Measure of the tendency of an atom to attract electrons.

Electropositive: For an atom, the tendency to lose electrons when forming a compound.

Hybrid bond: Bond formed when the s and p electrons combine and act as if they were in equivalent energy levels.

Intermetalic compound: Stoichiometric compound formed between two or more metallic atoms.

Ion: Atom that has either an excess or a deficiency of electrons with respect to the number of protons in the nucleus.

Mole: Quantity of a substance corresponding to 6.023×10^{23} atoms in the case of an element, or molecules in the case of compounds.

Pauli exclusion principle: Principle that only two electrons can occupy a given energy level and they must be of opposite spin.

Photoelectric effect: Emission of electrons from a metallic surface when the surface is exposed to radiation in the visible, ultraviolet, or near-visible range of frequencies.

Planck's constant (h): Universal constant that has a value of 6.63×10^{-34} Js.

Quantum numbers: Set of four numbers that defines allowable electron energy states. Three of these numbers describe the size, shape, and orientation of the electrons' probability density; the fourth defines the electron spin.

Transition elements: Elements whose s energy level is less than the 3d in one series of elements. In the higher-atomic-number elements the s levels are less than the f levels.

Valence electrons: Electrons in the outermost shells that become involved in the bonding process.

van der Waals forces: Weak forces between atoms or molecules that arise from dipoles, permanent or induced.

QUESTIONS AND PROBLEMS

3.1　Write the electron configuration of copper (atomic number 29) without referring to Table 3.2.

3.2　List some intermetallic compounds that are formed from the transition metals of atomic numbers 22 through 28. Many inorganic compounds are listed in the *Handbook of Chemistry and Physics* (CRC Press, Boca Raton, FL). Explain their valencies in terms of their electron configurations (i.e., how many $3s$ and $3d$ electrons are involved).

3.3　Compute the weight of an iron atom based on the average weight of all isotopes.

3.4　What types of bonding would you expect in (a) Si, (b) Al_2O_3, (c) W, and (d) SiC?

3.5　Which of the following metals would you expect to have the smallest bonding energy in the pure metallic state of each: Pb, Al, Cu? Explain how you arrived at your selection.

3.6　List two properties that are indicative of bond strength.

3.7　How might the energy required to separate two atoms to an infinite spacing be determined from the force-vs.-atom separation distance (Figure 3.15)?

3.8　How many bonds are formed in the semiconducting element germanium? How many electrons are involved in the bonding process? What type of bond exists?

3.9　How many atoms are there in 1 g of nickel? In 1 lb?

3.10　What are the allowed values of each of the four quantum numbers?

3.11　What is the percent by atomic weight of Ti in the Ti_3Al intermetallic compound?

3.12　What is valency of Ti in the Ti_3Al compound?

3.13　Why do metals have a high electrical conductivity?

SUGGESTED READING

Miller, F. M., *Chemistry: Structure and Dynamics,* McGraw-Hill, New York, 1984.

Smith, W. F., *Principles of Materials Science and Engineering,* 2nd ed., McGraw-Hill, New York, 1990.

4 Structure of Crystalline Solids

4.1 INTRODUCTION

From the previous chapters we have gained an understanding as to the structure and bonding of atoms. We now need to consider larger structures that consist of an aggregate of many atoms. For engineering purposes materials can be classified as crystalline (atoms or molecules containing long-range order) or amorphous (atoms or molecules without long-range order). The formula for common salt is NaCl, which tells us the ratio of atom species present in the material and as a solid the repeating atomic unit, which makes up a crystal of salt. If the salt is dissolved in water it dissociates to its ionic species Na^+ and Cl^- and the container defines the shape of the liquid.

Metals, ceramics, polymers, and composites are most commonly used in the solid state and consist of many individual atoms bonded together to form a useful piece of material. The physical state of the material and the mechanical, electrical, and chemical properties are dependent on the atomic species in the material and how they were processed.

Understanding the basics of the arrangements of the atoms and molecules in solids can help us determine the boundary conditions under which they can be used and modified.

4.2 BASIC TERMS AND DEFINITIONS

A crystal is a solid with long-range order. That is, the atoms are arranged periodically in three-dimensional space. To illustrate the structures of crystals, atoms are considered to be spheres with a definite radius. The electrons around an atomic nucleus are spread out in a "cloud," although there is a most probable distance of each electron from the nucleus. In spite of this atomic structure, the "hard sphere" model for crystal structure is an accurate way to describe the structure. The nuclei of the atoms have fixed positions that can define the crystal structure.

Having defined a crystal as a solid in where the atoms (molecules or ions) are arranged periodically in three-dimensional space, the next question to be answered is how are they arranged. A seventeenth century scientist named Bravais discovered that there were 14 unique ways to arrange solid spheres with three-dimensional periodicity. These are called the Bravais Lattices. The smallest complete lattice is called the unit cell. Thus, by knowing and understanding the unit cell, one can begin to predict the properties of a crystalline solid.

In this chapter, we will focus on cubic crystal structures. Other structures exist, it is important that you know they exist, but the details are beyond what is needed for a general engineer. A cube can be completely defined by one measurement: the length of its edge. This is referred to as the lattice parameter, and is given the symbol a. Knowing a, one can completely analyze the unit cell of a cubic structure.

4.3 CUBIC CRYSTAL STRUCTURES (METALS)

Metallic bonding, where electrons are delocalized, or shared among many atoms, is the most complicated. However, metals have the simplest crystal structures. There are three types of cubic crystal structures seen in metals:

- Simple cubic, in which the atoms touch along the edge of the cube.
- Body-centered cubic, in which the atoms touch along the body diagonal of the cube.
- Face-centered cubic, in which the atoms touch along the face diagonal of the cube.

4.3.1 THE SIMPLE CUBIC LATTICE

Only one metal, Po, above room temperature is known to have the simple cubic structure, however, the structure is worth discussing, as it can be used to illustrate the principles of crystallography. The simple cubic unit cell is shown in Figure 4.1. The atoms touch along the cube's edge. In this case the lattice parameter, or edge length, is therefore twice the atomic radius: $a = 2r$, therefore for Po,

$$a = 2r = 2(0.140/nm) = 0.280/nm$$

In the simple cubic lattice there is only one atom per unit cell. Each atom is represented by a sphere in a corner, and only one-eighth of this sphere is a part of this cube. Another way of looking at this is to say that each atom is in eight other unit cells. Therefore, the number of atoms in the unit cell is: $8(1/8) = 1$.

In this structure each atom touches six others. The number of atoms an individual atom touches is referred to as the coordination number. The coordination number in a simple cubic unit cell is 6.

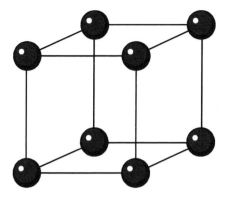

FIGURE 4.1 Simple cubic unit cell.

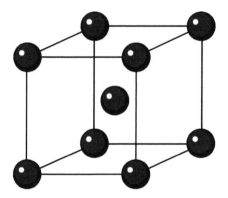

FIGURE 4.2 The body-centered cubic (BCC) unit cell.

4.3.2 The Body-Centered Cubic (BCC) Lattice

A common cubic structure is BCC, as shown in Figure 4.2. Here, eight atom centers are on the corners of the unit cell cube, and one is at the center of the cube. The coordination number is thus 8. There are two atoms in two BCC cells, one from the corners and one in the center of the cell. The atoms touch along the diagonal of the cube, so that the relation between the atomic radius and lattice parameter is:

$$a = \frac{4r}{\sqrt{3}}$$

The lattice parameter of α-Fe is thus

$$a = \frac{4r}{\sqrt{3}}$$

$$= \frac{4(0.124nm)}{\sqrt{3}}$$

$$= 0.286nm$$

Metals that have the bcc structure include chromium (Cr), vanadium (V), niobium (Nb), tungsten (W), molybdenum (Mo), the alkali metals (Group I), and iron below 910°C (α-Fe).

4.3.3 The Face-Centered Cubic (FCC) Lattice

The most common metallic crystal structure is FCC, shown in Figure 4.3. In the face-centered structure there are atom centers at each of the eight corners of the cube and at the centers of each of the six faces of the cube. The eight atoms at the cube corners, one-eighth of the unit cell, and each of the six face-centered atoms are half in the unit cell, making a total of four atoms in the unit cell. The coordination number in FCC is 12.

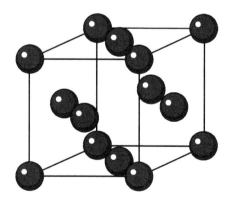

FIGURE 4.3 Face-centered cubic (FCC) unit cell.

The atoms in the face-centered cubic unit cell touch along the face diagonal in this structure, and therefore,

$$a = \frac{4r}{\sqrt{2}}$$

The lattice parameter for Cu is therefore,

$$a = \frac{4r}{\sqrt{2}} = \frac{4(0.128nm)}{\sqrt{2}} = 0.362\,nm$$

Some metals that have an FCC structure are aluminum (Al), copper (Cu), gold (Au), lead (Pb), nickel (Nu), platinum (Pt), and iron above 910°C (γ-Fe).

4.4 IONIC SOLIDS

Structures of ionic compounds such as MgO are more complicated than those of crystals containing only one kind of atom. As stated earlier, a single molecule of MgO(s) contains millions or billions of Mg ions and O ions. In ionic solids the ions arrange themselves in such a way that charge neutrality is maintained. Frequently, this involves a long-range three-dimensional pattern, and thus many ionic solids are crystalline. We will describe some of the most common compound structures: CsCl (cesium chloride), NaCl (sodium chloride), ZnS (zinc blende), calcium fluorite (CaF_2), and barium titanate ($BaTiO_3$).

4.4.1 COORDINATION NUMBER

The coordination numbers of ions in ionically bonded structures are related to the ionic radii of the ions. A positively charged ion will try to surround itself with as many negatively charged ions as possible. However, the negatively charged ions repeal one another. Thus, there are two competing factors. The coordination number will depend upon the ratio of the two radii. The limiting case will be where the anions just touch one another. The "ideal" radius ratio for a particular coordination number can be calculated from a structure of closest packing for that radius ratio. These ideal ratios are given in Table 4.1.

Consider sixfold coordination; that is, each cation (+ charge) is surrounded by six anions (– charge). This means the cation must be large enough to just separate six negatively charged anions. Sketching this case, as in Figure 4.4, one can isolate a triangle as shown. Note that this is an isosceles right triangle. The length of the two equal sides must equal the sum of the radius of the cation and the radius of the anion. The length of the other side is twice the radius of the anion.

TABLE 4.1
Minimum Radius Ratios
for Given Coordination
Numbers in Ionic Crystals

CN	Minimum r/R
3	0.155
4	0.225
6	0.414
8	0.732
12	1.000

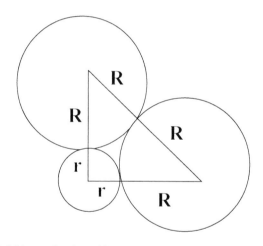

FIGURE 4.4 Sixfold coordination of ions.

Using the Pythagorean Theorem,

$$2(r + R)^2 = (2R)^2$$

$$\sqrt{2}(r + R) = 2R$$

$$(r + R) = \sqrt{2}R$$

$$r = (\sqrt{2} - 1)R$$

$$\frac{r}{R} = (\sqrt{2} - 1) \text{ or } 0.414$$

In order for the cation to have a coordination number of 6, the cation must be large enough to separate six anions or

$$\frac{r}{R} \geq 0.414$$

Thus, the ratio at which all particles touch is 0.414. If r/R is larger than 0.414, the ions can arrange themselves with sixfold coordination and the anions will not touch. Table 4.1 shows the minimum radius ratio for a given coordination number in ionic solids.

4.4.2 CESIUM CHLORIDE STRUCTURE (CsCl)

The coordination number of CsCl can be found from the ratio of the ionic radii,

$$\frac{r}{R} = \frac{r(Cs^+)}{r(Cr)} = \frac{0.165nm}{0.181nm} = 0.912$$

Therefore, the coordination number is 8, and each ion is surrounded by eight oppositely charged ions. The CsCl unit cell is shown in Figure 4.5. It is inappropriate to call this bcc, because the central particle is not the same as those on the corners. In fact, this is a type of simple cubic structure. Like simple cubic there is only one CsCl molecule per unit cell. In this structure there is only one ion pair, for example chloride (Cl⁻) is at the cube corners and the other (Cs⁺) is at the center of the cube. Thus, the structure of the individual ions is simple cubic.

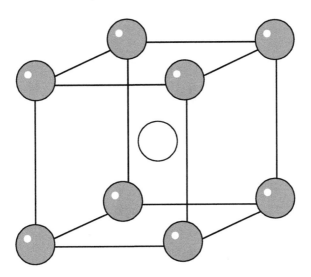

FIGURE 4.5 Cesium chloride structure (CsCl).

The ions touch along the body diagonal, $\ell = a\sqrt{3}$, one of each ion, so the lattice parameter is,

$$a = \frac{2}{\sqrt{3}}(r + R)$$

4.4.3 SODIUM CHLORIDE STRUCTURE (NaCl)

In NaCl the ratio of ionic radii is,

$$\frac{r^+}{R^-} = \frac{r(Na^+)}{r(Cl)} = \frac{0.098nm}{0.181nm} = 0.541nm$$

and thus the coordination number is 6.

This sodium chloride structure is cubic and is shown in Figure 4.6. The unit cell can have either sodium or chloride ions at the corners of the cube; in Figure 4.6 the chloride ions are on the corners. In addition, chloride ions are at the face centers, so that the structure of the chloride ions by themselves is face-centered cubic. Sodium ions are between the corner chlorides; the sodium ions are on a face-centered cubic lattice. Therefore, the Bravais lattice for this structure is considered to be fcc, as the arrangement of both Na and Cl ions are face-centered cubic. The ions touch along the edge of the cube, and therefore

$$a = 2(r + R)$$

and the x-ray pattern resembles that of an fcc metal. Like a lattice, there are four NaCl molecules per unit cell.

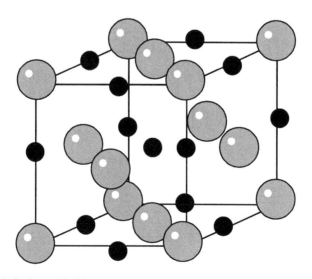

FIGURE 4.6 Sodium chloride structure (NaCl).

4.4.4 Zinc Blende Structure (ZnS)

In a ZnS the ratio of the ionic radii is

$$\frac{r}{R} = \frac{r(Zn^{+2})}{R(S^{-2})} = \frac{0.073\ nm}{0.174\ nm} = 0.41$$

Therefore, CN = 4 and the ions have a tetrahedral arrangement. It is possible to arrange the tetrahedra into a cube as shown in Figure 4.7. Note that each ion pair touches along the body diagonal but, that this distance accounts for only 25% of the length. Thus,

$$\frac{a\sqrt{3}}{4} = (r + R)$$

$$a = \frac{4(r + R)}{\sqrt{3}}$$

where r represents the radius of the cation, and R represents the radius of the anion.

4.4.5 Calcium Fluorite Structure (CaF$_2$)

The CaF$_2$ structure is shown in Figure 4.8. It resembles the zinc blende structure. The difference between the calcium fluorite structure and the zinc blende structure is the presence of additional ions. There are eight anions in the unit cell and four cations. Note that the stoichiometry of the unit cell is precisely the same as the stoichiometry of the chemical formula. There are four cations, (Ca^{+2}), and eight anions (F$^-$). This must be the case, as the unit cell represents the compound.

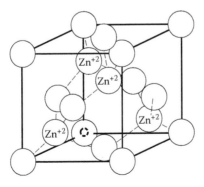

FIGURE 4.7 Zinc blende (ZnS) structure. Each Zn ion is bonded to four S ions. The ions are shown to be the same size for clarity.

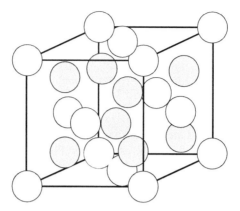

FIGURE 4.8 Calcium fluorite (CaF_2) structure. The F^- ions are colored gray. For clarity all ions are shown to be the same size.

The geometric relationship between ion size and lattice parameter is the same. The ions still touch along the body diagonal, and this distance is one-quarter the length of the body diagonal. Thus,

$$\frac{a\sqrt{3}}{4} = (r + R)$$

$$a = \frac{4(r + R)}{\sqrt{3}}$$

4.4.6. Perovskite (BaTiO₃) Structure

The perovskite structure is a cubic structure based on three ions. The lattice is similar to both the fcc and NaCl lattices. O^{2-} ions occupy the face-centered positions, Ba^{2+} ions occupy the corner positions, and a Ti^{4+} ion occupies the body-centered position. This means there are three oxygen anions, one barium cation, and one titanium cation. Thus, the stoichiometry of the system is preserved. The center of this structure is shown in Figure 4.9. Note that the anions and the cations must touch, the question is, which set of ions touch? By assuming the O^{2-} ions touch the Ba^{2+} ions, it can be shown that the Ti^{4+} ion is slightly too large to fit in the center of the cell. This means that the cell is slightly stretched. If one applies a force to the cell, its dimensions will change in one direction. Assume, for example, that one pressed on the top and bottom surface of the cell. The z-dimension would then decrease and the Ti^{4+} ion would be pushed off-center. This will induce a charge imbalance, or electric dipole in the cell as shown in Figure 4.7. Thus, by imposing a force on the cell, one causes an electric field to be induced. This is how a microphone works. One's voice (or the displacements in the air associated with sound waves) causes a force to be induced on the microphone, and this generates an electric field that can be analyzed, manipulated, and reproduced. Conversely, an electric field will cause the Ti^{4+} ion to be displaced

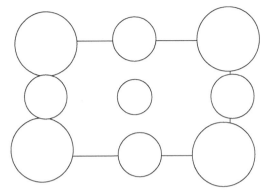

FIGURE 4.9 Center of the perovskite structure.

from the center of the cell. This will cause a dimensional change in the unit cell. The positive ion still attracts the negative ions. This is how a speaker works. An electric field causes the crystal to change dimension, which causes vibrations in the membrane of a speaker. These materials are called piezoelectrics.

4.5 STRUCTURE OF COVALENT NETWORK SOLIDS

4.5.1 DIAMOND CUBIC

Covalent network solid is a carbon (diamond), silicon, germanium, and SiO_2 are covalent network solids. That is, the atoms are covalently bonded to one another in such a way that the molecular pattern repeats itself. This is most commonly seen where sp^3 hybridization occurs. Therefore, the molecular geometry is tetrahedral. As with the Zinc blende (ZnS) structure. The tetrahedral structure can be arranged in a cubic structure except that all atoms are identical. The diamond cubic structure is shown in Figure 4.10. Note there is only one interstitial atom

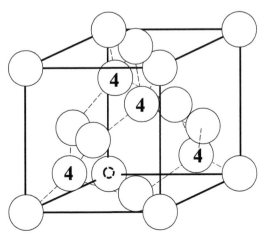

FIGURE 4.10 Diamond cubic unit cell.

per body diagonal. In the figure each interstitial atom is indicated by a 4, and each atom is bonded to one corner atom and three face-centered atoms.

There are some differences between the fcc unit cell and the diamond cubic unit cell. The fcc unit cell has four atoms, the diamond cubic unit cell has eight. The four atoms from the fcc lattice and the four additional atoms internally, giving eight total. In addition, each atom is bonded to four other atoms. The atoms touch along the body diagonal. This means the relationship between the lattice parameter and the radius of the atom will be different than for an fcc structure. As with ZnS, the atoms touch along the body diagonal and this distance is only one-quarter of the length. However, since the atoms, and thus their radii are identical the relationship between the lattice parameter and the atomic radius is,

$$\frac{a\sqrt{3}}{4} = 2r$$

4.5.2 SILICA

Many rocks and minerals are silicate compounds. Many silicates are important engineering materials. The simplest silicate is silicon dioxide, SiO_2. The Lewis structure is shown in Figure 4.11. Each oxygen ion is bonded to two silicon ions, and each silicon ion to four oxygen ions. The crystal structure is shown in Figure 4.12.

4.6 CRYSTAL FEATURES

Within a given unit cell are lattice positions, lattice directions, and lattice planes. Because this text is focused on the general engineering student, not the materials engineering student, our focus will be limited to the most important directions and planes in cubic systems.

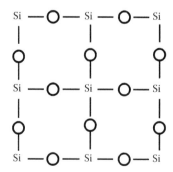

FIGURE 4.11 The Lewis structure of SiO_2.

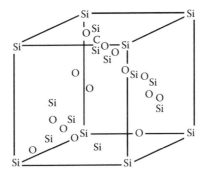

FIGURE 4.12 The structure of β-cristobalite, the cubic form of silica.

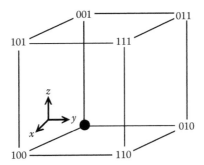

FIGURE 4.13 Lattice positions in a cubic unit cell.

4.6.1 LATTICE POSITIONS

Lattice positions are taken relative to an origin, usually fixed at the lower, left, rear corner of the unit cell; and measured in terms of the edge length of the cube. Thus, the position one unit in the "x-direction" from the origin is labeled 100 (note that this is not pronounced "one-hundred," but "one-zero-zero") and corresponds to the lower, left, front corner. The position which is one unit in the "x-direction" and one unit in the "y-direction" is labeled 110. The most important lattice positions for cubic crystal structures are shown in Figure 4.13.

4.6.2 LATTICE DIRECTIONS

In describing crystal structures, is it useful to have a method of designating directions in the unit cell or crystal lattice. A direction in the crystal lattice is a vector from the origin of the coordinate system to a point in the lattice. A more formal way of defining a direction is

A direction [uvw] is the vector that extends from the origin to the point uvw.

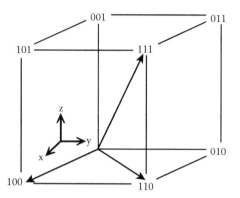

FIGURE 4.14 Important directions in cubic crystal systems.

The mechanical and electrical properties of a crystal often depend on the crystal's orientation. As the distance between atoms varies from direction to direction,

- The [100] represents the edge, the vector from the origin to 100.
- The [110] represents the face diagonal, the vector from the origin to 110.
- The [111] represents the body diagonal, the vector from the origin to 111.

These directions are shown in Figure 4.14. This is most useful in determining which direction will be strongest in a unit cell. For example, the face diagonal will be the strongest direction in FCC, as it is the direction in which the atoms touch.

Comparison of Directions in fcc Crystals

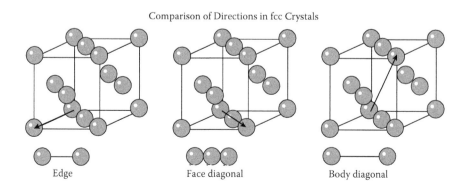

4.6.3 LATTICE PLANES

The plane (hkl) is the plane that intersects the points

$$1/h \ 00, \ 0/1k \ 0, \ \text{and} \ 00 \ 1/l.$$

The simplest plane to construct is the (111) plane, or for our purposes better referred to as the triangle cut. It intersects the points: 1/1 00, 0 1/1 0, and 00 1/1, or 100, 010, and 001. This is shown in Figure 4.15.

There are two other planes of primary interest,

- the (110) or diagonal cut, and
- the (100), or face.

The construction of these planes is shown in Figure 4.16 and Figure 4.17, respectively. Note for the (110), or the diagonal cut, the points of intersection are: 1/1 00, 0 1/1 0, and 00 1/∞. This means the plane intersects the points: 100 and 010, but never intersects the z-axis. This means the plane is parallel to the

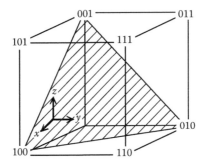

FIGURE 4.15 Triangle cut in a cubic crystal structure.

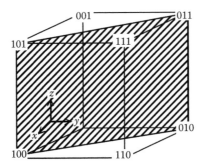

FIGURE 4.16 Diagonal cut in a cubic crystal structure.

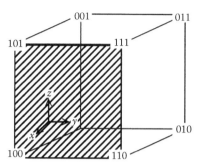

FIGURE 4.17 Face plane in a cubic crystal structure.

z-axis. For the (100) or face, the points of intersection are: 1/1 00, 0 1/∞ 0, and 00 1/∞. This means the plane intersects the points: 100, but never intersects either the *y* or *z* axes. Thus, it is parallel to the *y* and *z* axes.

4.6.4 Bulk Properties

Density (also referred to as bulk or volume density), ρ, is defined as mass per unit volume. The density of a unit cell is therefore the mass of the unit cell divided by the volume of the unit cell. These quantities can be calculated for any unit cell from a knowledge of how many atoms are in the cell. To find the number of atoms in the unit cell, one must visualize how much of each atom is in the cell.

For example, we can calculate the density of aluminum knowing only its atomic radius, 0.143 nm, its crystal structure, fcc, and its atomic weight of 27.0 grams/mole. The unit cell contains four atoms, therefore the mass of the unit cell is the mass of four aluminum atoms. The mass of the unit cell can be determined from,

$$M = \frac{4MW}{N_A} = \frac{4(26.95 \text{ g/mole})}{6.02 \times 10^{23} \text{ atoms/mole}}$$

Since aluminum is fcc,

$$a = \frac{4r}{\sqrt{2}} = \frac{4(0.143 \text{ nm})}{\sqrt{2}} = 0.404 \text{ nm}$$

and

$$\rho = \frac{M}{V} = \frac{\frac{4MW}{N_A}}{a^3} = \frac{\frac{426.95 \text{ g/mole}}{6.02 \times 10^{23} \text{ atoms/mole}}}{(0.404 \times 10^{-7} \text{ cm})^3} = 2.72 \text{ g/cm}^3$$

The packing factor of a structure is defined as the ratio of the volume of the atoms in the unit cell to the volume of the unit cell itself. In the face-centered cubic structure, there are four atoms, so the volume occupied by the atoms is simply four times the volume of an atom, $4 \times 4\pi R^3/3$ and the volume of the unit cell is a^3. However, there is a relationship between a and r is $a = \dfrac{4r}{\sqrt{2}}$. The packing factor is:

$$\frac{16\pi R^3/3}{16\sqrt{2}R^3} = \frac{\pi}{3\sqrt{2}} = 0.7406$$

For the body-centered unit cell, the packing factor is 0.68. Simple cubic 0.52 and diamond 0.34. For ionic solids this will depend on each system as the r/R will vary from compound to compound.

4.7 NONCUBIC STRUCTURES

The most important thing to remember about noncubic crystal structures is that they exist and metals or ceramics with these structures are brittle. The most important noncubic structures for the general engineer to be aware of are the hexagonal-close-packed (hcp) and body-centered-tetragonal (bct). Metals with the hcp structure include the rare-earths, magnesium, and zinc. These metals are brittle. While a thorough understanding of these structures is necessary for materials engineers and metallurgists, they are beyond the scope of an introductory course for general engineers.

4.8 SUMMARY

The arrangement of atoms in solids is the basis for all materials property management. In this chapter, we have discussed cubic crystal systems, which are the basis of many engineering materials.

DEFINITIONS

Amorphous: State of matter that is without long-range order.
Anisotropic: State of a material in which the properties are dependent on the direction in the body of the material in which they are measured.
Atomic packing factor: Density of atom packing in crystalline structures.
Bravais lattice: Arrangement of points in space that is used to describe a crystal structure. There are 14 such arrangements.
Crystalline: State of matter in which there is a high degree of long-range order with respect to atomic or molecular positions.

Lattice: Construction of points that extends through a single-phase region of a crystalline body and is used to describe atom positions.

Lattice parameter: Distance a unit cell must move in a lattice to find itself in exactly the same atom arrangement in which it existed prior to the move.

Phase: Region of a substance that is structurally homogeneous and in which the properties are uniform throughout. Phases will be separated by phase boundaries.

Unit cell: Smallest repeat unit of a crystalline lattice structure.

QUESTIONS AND PROBLEMS

4.1   Aluminum has a FCC structure. The distance of closest approach of the atoms (distance between nuclei in the direction of closest packing) is 2.862 Å (28.62 nm). What is the lattice parameter of aluminum?

4.2   Would you expect the properties of a small-grain-size polycrystalline metal to be isotropic or anisotropic? Why?

4.3   What is the diameter of the largest atom that will just fit in the largest interstitial site for BCC iron? The lattice parameter for BCC iron at room temperature is 2.8664 Å.

4.4   Calculate the planar density for the {111} and the {100} planes of BCC iron.

4.5   Copper has a FCC structure with a lattice parameter of 3.615 Å and an atomic weight of 63.54 g/mol. Calculate the density of copper.

4.6   Four spheres of diameter (d) are arranged in a pyramid touching each other. Show that the diameter of the largest sphere that will fit into the voids between the spheres is 0.155 d.

4.7   Iron changes from BCC to FCC at 912°C. What is the volume change in the lattice?

SUGGESTED READING

ASM Metals Handbook, Desk Edition, ASM International, Metals Park, OH, 1985

Callister, W. D. Jr., *Materials Science and Engineering*, Wiley, New York, 1990.

Cullity, B. D., *Elements of X-Ray Diffraction*, 2nd ed., Addison-Wesley, Reading, MA, 1978.

Guy, A. G., *Elements of Physical Metallurgy*, Addison-Wesley, Reading, MA, 1959.

5 Diffusion and Plastic Deformation

5.1 INTRODUCTION

It is important to understand some of the basic management techniques for the control of metallic properties. How atoms move and how to retard or prevent this movement is key to getting a material to perform as planned. The mechanisms of diffusion and plastic deformation in crystalline materials are dependent on the presence of imperfections in the lattice. In a material, if all the atoms were in place and all the bonds were 100% active, the material would be very strong and nearly impossible to form into a marketable product. We know, however, that deformation processing is a common occurrence in the manufacturing world. Imagine, if you will, zipping your jacket; then when it comes time to take it off, you grab the lapels and pull perpendicular to the axis of the zipper to open the jacket. We all know the cloth will most likely tear before the zipper is separated (assuming it is a good metallic zipper), and it will take a high level of effort to accomplish this task. However, if we take the zipper pull and unhook each of the zipper teeth one at time, in the normal manner, the zipper is opened with little effort and the garment stays intact. In a perfect lattice, movement and deformation will require large forces, but with defects in the structure this force is reduced to a manageable level.

The imperfections in the crystalline lattice are referred to as *defects*. This term does not infer anything deleterious. It is simply a way of talking about the major classifications of crystalline imperfections, which are called point, line, and planar defects.

5.2 POINT DEFECTS

A point defect consists of either a vacant lattice site or a foreign atom. The foreign atom can occupy a site normally occupied by an atom of the main body of the material of interest or it can be shoved in-between or among the *host* atoms interstitially. The terms *host*, *parent*, and *solvent* have essentially the same meaning and are interchangeable. They are the major constituent. The term *solvent* as used in the solid state may be a little confusing, so let us elaborate on that concept a bit. The term *solid solvent* arises simply because in solids we can have *solid solutions*, somewhat comparable in concept to single-phase miscible-liquid solutions or single-phase liquid solutions that contain dissolved solid solutes, such as sugar dissolved in coffee. Liquid solutions, being a single phase, are uniform in composition throughout this phase, but we can dissolve only so much solid sugar in liquid coffee. The amount of sugar that can just be dissolved and have no solid residue at the bottom of the cup is called the *solubility limit* of sugar for that temperature of the solvent coffee. If we exceed this quantity of sugar, we will have a two-phase system consisting of a liquid solution containing dissolved sugar (no longer a solid) and a second phase, which is the undissolved solid sugar.

Similarly, we can have a solid solution of a solvent containing foreign atoms called the *solute*. The foreign atoms can occupy a host atom site or reside in an interstitial site between or among the host atoms, but there is no liquid in this solution. When the added foreign atom content exceeds the solubility limit, a solid precipitate, usually a compound or second phase forms. We now have a system consisting of a solid solution plus a solid precipitate, a two-phase system very similar in concept to the liquid solution containing a solid precipitate. The solid solution, being a single phase, is uniform in composition and crystal structure throughout this phase.

5.2.1 VACANCIES

An empty lattice site is called a vacancy. Why do they exist? Would not a perfect crystal, with no vacant lattice sites, be at a lower-energy state and thus more than one with unoccupied sites? Why do the vacancies, which could be considered as accidents of growth of the solid phase from the liquid during solidification, not just ooze out (diffuse) to the surface and escape the crystal, which really should not want this defect anyway? The answer is related to the *entropy* of the system. Entropy is a measure of randomness. Gases have a high entropy because there is no order, and perfect crystals have low entropy because everything is organized. Systems given the energy to allow atomic movement will seek to adjust themselves to the lowest energy state. In chemical thermodynamics, the lowest energy state is the one with the lowest free energy as defined below:

$$F = E - TS \qquad (5.1)$$

where
    F = free energy
    E = internal potential energy
    T = temperature in Kelvin
    S = entropy

5.2.2 SUBSTITUTION SOLUTE ATOMS

A substitution solute atom is a replacement or substitute of the host atom by a foreign atom. An example of this is illustrated in Figure 5.1. In this process the replacing atom must meet some basic requirements to be successful:

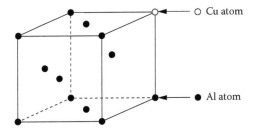

FIGURE 5.1 Substitutional solid solution of aluminum in copper.

1. It has to fit. A size difference of ±15% is most conducive for extensive mixing of atoms.
2. The crystal structures have to be compatible.
3. The valence electronics must be satisfied.
4. No reaction to form a compound will occur.

Probably the two most critical of this set of rules are 1 and 4. Numerous examples can be sited to illustrate this concept. Copper and nickel are completely soluble in each other and can be mixed in any proportion to create an alloy, their atomic size difference falls within the ±15% size difference, they have similar crystal structures, and no reaction takes place. By contrast, lead is a very large atom, and only a few of them will fit into the copper lattice before Pb is rejected and is seen in the microstructure as a second phase. Aluminum will only accept a small amount of copper before it reacts to form a $CuAl_2$ phase. This compound forms when the limit of copper solubility in Al is exceeded. Oxygen, however, will react with aluminum to form Al_2O_3 rather than dissolving in the Al to form any alloy. The ability to replace "fit" is also dependent on temperature as shown in Figure 5.2. The result of adding substitution atoms to a structure is internal disruption and strain fields that retard the atoms' ability to move. We refer to this behavior as *solid solution strengthening*. The amount of strain and the effect of each alloying element on strength not only depend on the amount added, but also on the combination of elements added.

Solute atoms and vacancies have a large effect on the strength of metals, but this effect is even more profound on the electrical properties of semiconductors.

5.2.3 INTERSTITIAL SOLUTES

Atoms that are very small, such as hydrogen, carbon, nitrogen, boron, and sometimes oxygen, can fit in-between the host atoms. They occupy the interstitial sites. In most cases they are forced into these positions. As a result of this force fit, considerable strain results in the host lattice. Again, solubility increases with temperature, and when the limit is reached at whatever temperature, a second phase or compound is formed.

A good practical example of this is interstitial carbon atoms in the body-centered cubic (BCC) iron lattice. Figure 5.3 shows the solubility limit curve as

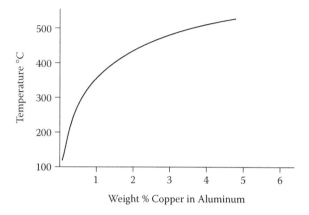

FIGURE 5.2 The solid solubility limit for copper in aluminum increases with increasing temperature.

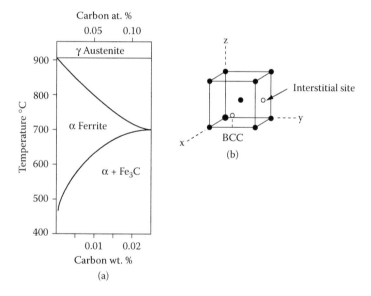

FIGURE 5.3 (a) Solid solubility limit as a function of temperature for carbon in BCC iron; (b) the BCC crystal structure showing the interstitial sites (interstices).

a function of temperature along with the interstitial solid solution sites. In BCC iron the tetrahedral sites are the larger of the two and can accommodate an atom of 0.36 Å radius (i.e., 0.29R, where R is the radius of the host atom). The carbon atom is about 0.7 Å in radius and introduces considerable strain into the surrounding lattice. The carbon atom seems to prefer the smaller octahedral site because of the directionality of the elastic moduli in the BCC lattice. In either case the strain is large enough to limit the maximum solid solubility of carbon in BCC (alpha) iron to about 0.02 wt%. The compound Fe_3C forms when the solid solubility is exceeded. Again, this compound, which has an orthorhombic structure, is a distinct second phase and not a part of the parent iron lattice. We will find later that the strength of iron–carbon alloys, called steels, can be changed by a factor of 3 or more just by changing the size, shape, and distribution of these carbide particles, without changing the amount of carbon. This strength increase is accomplished through heat-treating processes, which we will explore in detail later.

The defects discussed above are the basic point defects that are found in metals, but identical ones exist in ceramics and semiconductors. Solute atoms and vacancies have an even more profound effect on the electrical conductivity of semiconductors than they do on either the mechanical or physical properties of metals. Silicon and gallium arsenide, and all semiconductors for that matter, have been intentionally injected, by one means or another, with minute quantities (<0.01 atomic %) of foreign solute atoms, often called *dopants*. We say that the host material has been doped with foreign atoms. Most of these foreign atom dopants reside as substitutional solutes in sites normally occupied by the host atom. Some electronic ceramics, such as the oxide superconductors, ferri- and ferroelectrics, and dielectric materials, are also affected by minute quantities of solute atoms, sometimes added intentionally to achieve the desired properties.

5.3 DIFFUSION

At $0°K$ all atomic vibration stops, and one can theoretically predict the exact position of the atoms in a lattice. At any temperature above that, the atoms vibrate in a space determined by the atomic bonding and the temperature. As the atoms vibrate at a temperature, there is a natural tendency to reorganize and to minimize the randomness of the structure, lowering the free energy. This movement of the atoms in a structure is called *diffusion,* and the rate of movement is exponentially dependent on temperature. The atoms move through vacancy and atom interchange, coordinated movement of several atoms (ring mechanism) or atoms' switching places. The most prominent mechanism is the atom vacancy exchange because it requires the least energy to initiate. The eventual end point is reached with an equilibrium distribution of solvent and solute atoms that is totally random.

5.3.1 Substitutional Solute Diffusion

The driving force for the diffusion process is the concentration gradient (dc/dx). To a first approximation, the diffusion process is strictly statistical in nature (i.e., assuming that there is no chemical attraction between the unlike atoms). Cu–Ni substitutional alloys approach this condition and, furthermore, there is no solubility limit (because both atoms are of nearly the same size and the same crystal structure). We can put as many nickel atoms into copper as we wish, and vice versa, without forming a compound or second phase of any type. At nickel atom concentrations below 50% of the total, the nickel is called the solute, and when the nickel becomes the majority atom in number, copper is the solute. Both pure nickel and pure copper have face-centered cubic (FCC) structures, so the alloy consists of some sites occupied by nickel and the remaining sites by copper atoms. Incidentally, when the nickel atoms are in the majority, the alloys are called Monels. They are widely used in marine environments because of their good corrosion resistance to seawater. When copper is the majority element, they are called cupro-nickels.

Let us now consider a bar of copper welded along a straight-line interface to a bar of nickel. This arrangement is called a diffusion couple and is used to study atom movement, as shown in Figure 5.4. These exists a certain probability that after a period of time any given atom of nickel will reside on the copper side of the interface in a copper lattice position, and similarly, that a copper atom will cross into the nickel section of the diffusion couple. The atom makes its journey across the interface and into the other metal via the atom–vacancy interchange mechanism. The vacancies are constantly moving around, and when they appear adjacent to a solute atom, an interchange occurs. The higher the temperature, the higher the amplitude of atom vibration, the larger the concentration of vacancies, and the greater the probability of the copper atom being on the nickel side of the interface. This probability and the corresponding diffusion rate increase exponentially with temperature. The atoms can migrate by switching places with other atoms or by a coordinated movement of several atoms (ring mechanism), but the atom–vacancy interchange requires the least energy and accordingly is the most prominent substitutional solute diffusion mechanism. A certain activation energy is required for the atoms to move from one stable position to another.

Returning to the statistical nature of diffusion, since the atoms are moving randomly in all directions, and since there is a 100% concentration of copper atoms just one interatomic spacing to the left of the interface in Figure 5.4, statistically speaking, there must be a net flow of copper atoms to the right and

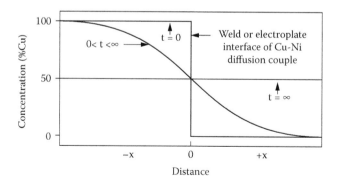

FIGURE 5.4 Copper–nickel diffusion couple showing the copper concentration at time = 0, time = some finite value, and time = ∞.

a net flow of nickel atoms to the left. This is expressed mathematically by *Fick's first law* of diffusion:

$$J = -D \, dc/dx \tag{5.2}$$

where

J = flux of atoms crossing any particular plane per unit time
D = proportionality constant called the diffusivity
dc/dx = concentration gradient at this particular plane

We can solve for the units of D as follows:

$$J(\text{atoms/cm}^2/\text{s}) = D \, (\text{atoms/cm}^3/\text{cm})$$

Therefore, the units of D centimeters^2/second (cm^2/sec).
There is also a *Fick's second law* of diffusion, written as

$$\frac{dC_x}{dx} = \frac{d}{dx}\left[\frac{DdC_x}{dx}\right] \tag{5.3}$$

The second law takes into account the fact that D changes with concentration — thus the subscript x on C. This concentration effect is based on diffusion being not strictly a statistical process; as discussed above, each atom type behaves differently in the presence of unlike atoms (consider this an atomic traffic issue).

Note that in Figure 5.4, the concentration as a function of distance x is shown for $t = 0$, $t =$ some finite time, the curve shape being a function of temperature and time, and for $t =$ infinity. Theoretically, since these elements are mutually soluble in all concentrations, the concentration throughout the bar will eventually be a uniform 50% Ni–50% Cu alloy, although at room temperature the time to establish this uniform concentration could be millions of years. At temperatures near the melting point of copper, it may require only a few hours to attain uniformity. The diffusivity D varies exponentially with temperature according to the Arrhenius equation,

$$D = D_0 e^{-Q/RT} \tag{5.4}$$

where

D_0 = a constant of the same units as D, but independent of temperature
Q = activation energy for the process, usually expressed in calories/mole
R = gas constant, approximately 2 cal/mol/K
T = absolute temperature in Kelvins

Diffusivity measurements are performed by materials specialists and involve measurement of the concentration of the diffusing species as a function of time t and distance x at a constant temperature T and then computing the values of D from Fick's law. For systems in which radioactive isotopes are available, the diffusivity can be measured quite accurately. The self-diffusivity (e.g., Cu in Cu) cannot be measured without radioactive isotopes. Taking the natural logarithms of Equation 5.4 gives

$$\ln D = \ln D_0 e^{-Q/RT} \tag{5.5}$$

This is the equation of a straight line. Thus, by plotting $\ln D$ vs. $1/T$, as shown in Figure 5.5, the intercept on the y-axis becomes $\ln D$ and the slope equal to $-Q/R$. Sometimes diffusion data will be presented in this fashion, as shown in Figure 5.6, but more often in handbooks the constant D_0 and Q are given and D must be computed for the temperature of interest. Some selected values for these constants for some common metal systems are listed in Table 5.1.

Example 5.1   The values of the constants for copper atoms diffusing into nickel at a concentration of 45.4% Cu–54.6% Ni are given as $D_0 = 2.3$ cm²/s and $Q = 60.3$ kcal/mol. Compute the diffusivity D for copper in nickel at this concentration (values vary with concentration) at temperatures of (a) 0 and (b) 1000°C.

Solution
(a) 0°C = 273 K

$$D = 2.3 \ \text{cm}^2/\text{s}^{-1} \exp\left[-\frac{60,300}{2(273)}\right]$$

$$= \frac{2.3}{\exp(110.4)}$$

$$= \frac{2.3}{9.2} \times 10^{-47} \ \text{cm}^2/\text{s}$$

$$= 2.5 \times 10^{-48} \text{cm}^2/\text{s}$$

(b) 1000°C = 1273 K

$$D = 2.3 \ \text{cm}^2/\text{s} \ \exp\left[-\frac{60,300}{2(1273)}\right]$$

$$= \frac{2.3}{\exp(23.7)}$$

$$= 1.96 \times 10^{-10} \ \text{cm}^2/\text{s}$$

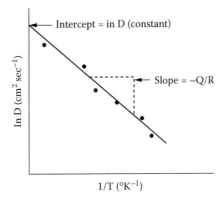

FIGURE 5.5 Schematic of the $\ln D$ vs. $1/T$ relationship, showing how the activation energy Q and constant D_0 are obtained.

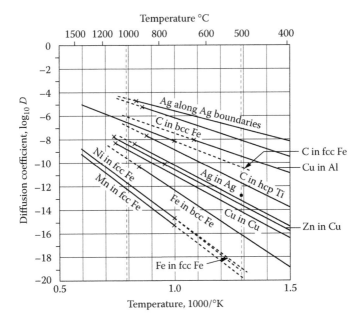

FIGURE 5.6 Diffusion coefficients vs. temperature. (From L. H. Van Vlack, *Elements of Materials Science and Engineering*, 3rd ed., Addison-Wesley, Reading, MA, 1975, p. 127. With permission.)

When one wants to determine the concentration of a solute at a given distance x and for time t, knowing the diffusivity and initial concentrations, solutions to the partial differential Equation 5.3 must be found. Frank's *The Mathematics of Diffusion* (see Suggested Reading) lists solutions for most common geometries of diffusion couples (e.g., gases diffusing through cylinder walls, surface

TABLE 5.1
Diffusion Data for Selected Metals

Solvent	Solute	D_0 (cm²/s)	Q (kcal/mol)
Cu	Cu (self-diffusion)	0.78	49.3
Cu	Zn	0.73	47.5
Al	Cu	0.65	32.3
Al	Mg	0.06	27.4
Cu	Ni	3.8	56.8
Ni	Cu	0.57	61.7
Fe(a) (350–850°C)	C	6.2×10^{-3}	19.2
Fe(g) (900–1600°C)	C	0.1×10^{-6}	32.4

Source: After E. A. Brandes, *Smithell's Metals Reference Book*, 6th ed., Butterworth, Stoneham, MA, 1983. (With permission.)

carburization of steel, and many other situations encountered frequently in industry). For example, the diffusion of carbon into iron for case hardening is given by the equation

$$C_s - \frac{C_x}{C_s} - C_0 = \operatorname{erf}\left(\frac{x}{2\sqrt{Dt}}\right) \tag{5.6}$$

where

C_s = surface concentration of carbon
C_x = concentration of carbon at distance x
C_0 = initial concentration of carbon in iron
D = diffusivity of carbon
t = time (sec)

Error function (erf) values can be found in most standard math tables. The same equation would apply, for example, to chromium-coated steel. In this case the penetration of chromium into steel at room temperature would be negligible, but if an errant worker happened to heat this component to several hundred degrees Celsius, the chromium atoms could diffuse into the steel to the extent that undesirable brittle phases would be formed.

5.3.2 INTERSTITIAL SOLUTE DIFFUSION

Interstitial atoms diffuse through the parent material by jumping around from one interstitial site to another. Since these atoms are relatively small, the diffusivities are much larger than those for substitutional solute, often of the order of

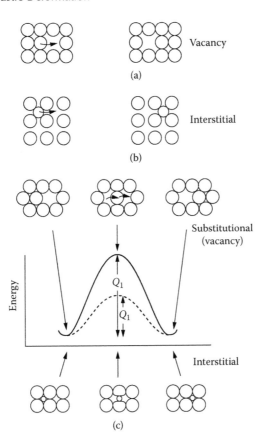

FIGURE 5.7 Diffusion mechanisms: (a) vacancy-atom interchange; (b) interstitial diffusion; (c) corresponding activation energies. (Adapted from D. R. Askeland, *The Science and Engineering of Materials*, 2nd ed., PWS-Kent, Boston, 1989, pp. 121, 123. With permission.)

105 times greater (Table 5.1 and Figure 5.6). Similarly, the activation energies for interstitial diffusion are much smaller than those for substitutional diffusion. The substitutional and interstitial mechanisms, along with their required activation energies, are shown schematically in Figure 5.7. All of the same diffusions apply for interstitial atoms as were presented above for the substitutional solute diffusion situations. Self-diffusion in pure metals occurs by the same vacancy–atom interchange. This would occur, for example, when annealing a pure metal, and in some cases significant diffusion can occur at room temperature.

Example 5.2 A 1010 steel (0.10 wt % C) is being gas-carburized at 927°C. Calculate the time it takes for the carbon content to achieve a 0.40 wt % C content at a depth of 1/16 in (1.5 mm). Assume that the surface carbon concentration is 0.90 wt %.

From the *Smithell's* handbook we find that the diffusivity of carbon in BCC iron (which would be about the same for a 1010 steel) is 1.3×10^{-7} cm²/s at 927°C.

Solution

$$C_s - \frac{C_x}{C_s} - C_0 = \text{erf}\left(\frac{x}{2\sqrt{Dt}}\right)$$ (5.7)

$$C_s = 0.9\% \quad C_0 = 0.1\% \quad C_x = 0.4\%$$

Substitution of these values gives

$$\frac{0.9 - 0.4}{0.9 - 0.1} = \text{erf}\left(\frac{0.015}{2\sqrt{1.3 \times 10^{-7} \text{cm}^2/\text{s}}}\right)$$

$$0.625 = \text{erf}\left(\frac{210}{\sqrt{t}}\right)$$

Let $z = (210/\sqrt{t})$. Thus, $0.625 = \text{erf } z$. We need a number for z whose error function (erf) = 0.625. From handbook tables this number is found to be $0.623 = z$. Therefore, $0.623 = 210/\sqrt{t}$.

$$\sqrt{t} = \frac{210}{0.623} = 337$$

$$t = 337^2 = 113,569 \text{ s} = 31.5 \text{ h}$$

Error Function Table

Z	erf (z)	Z	erf (Z)
0	0	0.6	0.604
0.05	0.056	0.7	0.678
0.1	0.112	0.8	0.742
0.15	0.168	0.9	0.797
0.2	0.223	1	0.843
0.25	0.276	1.2	0.91
0.3	0.329	1.4	0.952
0.4	0.428	1.6	0.976
0.5	0.52	1.8	0.989

The approximation formula for computing time when D and x are known or computing x when t and D are known is

$$x = \sqrt{Dt} \qquad \text{or} \qquad t = \frac{x^2}{D} \qquad (5.8)$$

What does x mean in this case? It is often said that it represents the depth to which appreciable diffusion occurs. We can be a little more precise than this. The x in Equation 5.8 represents the distance at which the concentration becomes half of the difference between the initial and final values.

Example 5.3 Assume that we would like to have a significant carbon content (say, 0.4 wt % C) at a depth of 1.5 mm (0.15 cm). Approximate the diffusion time required at 927°C. (D can be obtained from Example 5.2.)

Solution

$$t = (0.15)^2 \text{ cm}^2/1.3 \times 10^{-7} \text{ cm/s} = 1.7 \times 10^5 \text{ s} = 48.1 \text{ h}$$

Example 5.4 Let us now apply this approximation to Example 5.1 and estimate the time required for significant penetration of copper into nickel at (a) 0 and (b) 1000°C to a depth of 0.01 cm (0.004 in).

Solution
(a) At 0°C (273 K),

$$t = \frac{x^2}{D} = \frac{10^{-4}}{2.5 \times 10^{-48}} = 10^{40} \text{ h}$$

which confirms our earlier suspicion. Even Methuselah would not see significant penetration of copper into nickel at 0°C in his life span.

(b) At 1000°C (1273 K),

$$t = \frac{x^2}{D} = \frac{10^{-4}}{1.96 \times 10^{-10}} = 5.1 \times 10^6 \text{ s} = 142 \text{ h}$$

5.4 LINE DEFECTS: DISLOCATION

Dislocations are the only line defects that exist in crystalline solids. In the strict geometrical sense they are cylindrical defects of about five atom spacings in diameter. They thread their way through the crystal in all sorts of directions, not usually as straight lines. They are one of the more important crystalline imperfections, even though we have known about their existence for only a little more than 70 years. They were predicted in a series of separate papers by E. Orowan, M. Polyani, and G. Taylor in 1934 to explain the mechanism of plastic flow in metallic crystals, but we now know that their presence in semiconductors can be very deleterious to the performance of such devices. Extreme care is taken in the growth of single semiconductor crystals in order to minimize the presence of dislocations. In typical as-grown metallic single crystals, their concentration is on the order of 10^6 per square centimeter. This means that if a cross-sectional thin slice of 1 cm^2 was cut from a metal crystal, 1 million dislocations would intersect this surface. This has been verified experimentally by using special etching techniques that cause a pit to form at the point where each dislocation line emerges from the surface (Figure 5.8). These pits can be seen through an optical microscope at magnifications on the order of 100 to 1000 times.

Since dislocations are usually not straight-line defects, a better way of expressing their density is in terms of the centimeters of length of dislocation lines per cubic centimeter of volume of the material. This definition gives the same units as the etch pit technique (i.e., number/square centimeters). These dislocation lines in the volume of the material can be viewed in a thin section of a crystal by transmission electron microscopy (Figure 5.9), but it is not practical to measure their length per unit volume in this way. The dark lines that we see are a result of the strain fields surrounding the dislocation line. The dislocations form during solidification and recrystallization. They can also be generated in significant quantities during plastic deformation. An annealed metal that contains a dislocation density of 10^6/cm^2 will contain about 10^{12}/cm^2 after the metal has been severely deformed (e.g., by 80% reduction, called *edge* and *screw* types).

5.4.1 EDGE DISLOCATION

The structure of the edge dislocation is depicted in Figure 5.10. It can be viewed as if a half-plane of atoms has been removed from the lattice and then the neighboring atoms collapsed to fill the planar void. One might think that this would be a planar defect, but not so. Look at the atom positions. At points only a few atom spacings away from the missing half-plane of atoms, all of the other atoms are in near-perfect registry, all around the dislocation core. If one mentally projects a distance of about 10 atom distances, the perfect registry can be visualized. The disregistry results from the collapse of the planes.

The configuration shown here is by convention a positive edge dislocation, but for every positive edge dislocation there exists a negative edge dislocation where the missing half-plane of atoms has been removed from the part above the plane denoted by the letters SP in Figure 5.11. We chose SP since we will discover later that this is the slip plane on which plastic deformation occurs. The dislocation lines here are perpendicular to the plane of the paper but lie in the slip plane. Not all dislocations lie in slip planes, but at this point we are concerned only with those that do since these are the ones that permit our understanding of plastic flow in crystalline bodies.

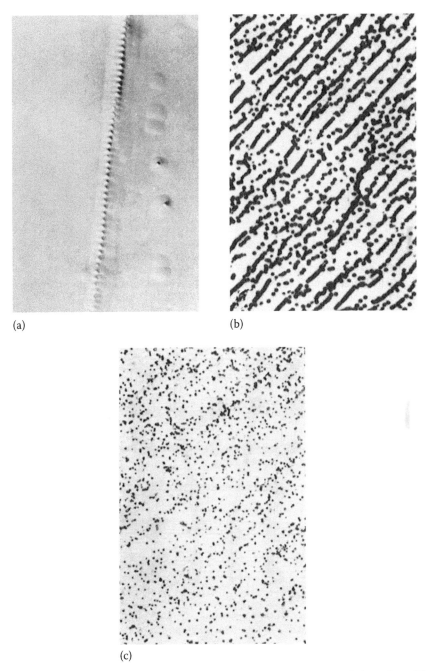

(a)   (b)

(c)

FIGURE 5.8 Dislocation etch pits in germanium (a) in a small angle grain boundary; (b) in subboundaries produced by polygonization; and (c) aligned on traces of slip planes in germanium that has been deformed at a low temperature. (From *Metals Handbook*, Vol. 8, 8th ed., ASM International, Metals Park, OH, 1973, p. 148. With permission.)

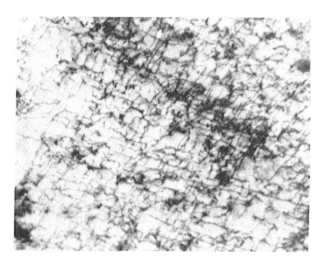

FIGURE 5.9 Uniform dislocation structure in iron foil deformed 14% at −19°C as viewed in a transmission electron microscope (×40,000). (From *Metals Handbook*, Vol. 8, 8th ed., ASM International, Metals Park, OH, 1973, p. 219. With permission.)

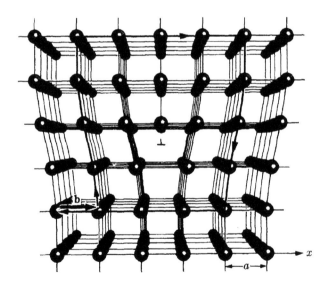

FIGURE 5.10 Structure of an edge dislocation. (From A. G. Guy, *Elements of Physical Metallurgy*, Addison-Wesley, Reading, MA, 1959, p. 110. With permission.)

5.4.2 SCREW DISLOCATIONS

In the screw dislocation there is no missing half-plane of atoms. It can be viewed as if a narrow ribbon of material in two parallel planes had been twisted with respect to each other (Figure 5.12). All along this ribbon the atoms are out of

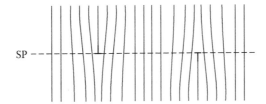

FIGURE 5.11 Negative and positive edge dislocations on the same slip plane. The dislocation lines are in a direction perpendicular to the plane of the page.

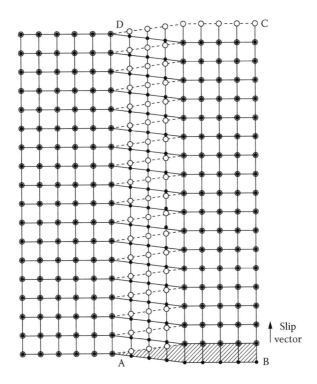

FIGURE 5.12 Arrangement of atoms around a screw dislocation. The plane of the figure is parallel to the slip plane. (From W. T. Read, *Dislocations in Crystals*, McGraw-Hill, New York, 1953, p. 17. With permission.)

registry, again resulting in a cylinder of dislocated atoms about five atoms in diameter. This cylinder threads its way through the crystal, and just like the edge type, the screw dislocation is considered to be a line defect.

5.4.3 DIFFUSION ALONG DISLOCATIONS

As stated previously, diffusion takes place by atom–vacancy interchanges in the cases of substitutional solute and self-diffusion and by intersite hopping for interstitial solute migration. One could expect diffusivities at 400 to 500°C along dislocations to be comparable to that at around 700°C via the common lattice atom–vacancy interchange mechanism. Of course, the volume area of the total dislocations present is a much smaller volume than the lattice proper, so the total diffusant of solute is quite small. However, in semiconductors, where one would like to have a uniform front of diffusing atoms, the dislocations cause long spikes and nonuniform distribution of dopant atoms. Such a nonuniform distribution of dopant atoms interferes with the operation of semiconductor devices — hence the need to have a very low dislocation count in semiconductor chips.

At temperatures below those where normal diffusion takes place, there also exists a tendency for dislocations to attract solute atoms. A solute that is larger than the solvent creates a region of compressive strain. Beneath the slip plane of a positive edge dislocation (Figure 5.10) a tensile strain exists. Thus, the larger solute atom is attracted to this region of tensile strain to reduce the overall total strain. This binding of solute atoms to the dislocations restricts their motion. The dislocation is said to be locked in place and a stress higher than the solute exists, but once the dislocation is released, it can move at a lower stress. We then have a yield point, very common for carbon atoms in iron, where there exists an upper yield stress followed by a more normal lower yield stress to move the dislocations after they have become unlocked.

5.5 PLASTIC DEFORMATION OF SINGLE CRYSTALS

Many years before dislocations were conceived, it was thought that the plastic deformation of crystalline materials occurred by the sliding of parallel planes of atoms over one another, much as a card or a thin packet of cards could be extended beyond the other cards of a deck of playing cards, somewhat like the schematic of Figure 5.13. The experimental evidence suggested such a mechanism because carefully controlled experiments on stressed crystals did show surface markings, as if a thin section of the crystal protruded from the remaining portion (Figure 5.14). These markings are even more evident where slip has been caused by a

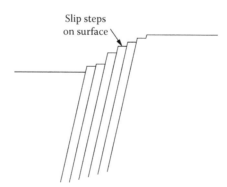

Slip steps
on surface

FIGURE 5.13 Schematic of slip planes protruding from the crystal after slip has occurred.

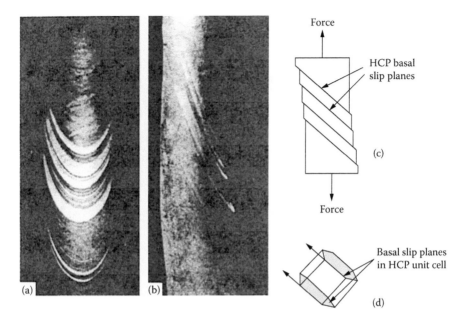

FIGURE 5.14 Slip bands on the surface of a plastically deformed zinc crystal. (Courtesy of Earl Parker, University of California, Berkeley.)

hardness impression on a polished metal surface (the dark region of Figure 5.15 is the impression). These markings are called *slip lines*.

5.5.1 METAL CRYSTALS

From the markings of the type shown in Figures 5.14 and 5.15, and in conjunction with x-ray diffraction crystal orientation determinations, it was found that in metal crystals the slip planes consist of the most closely packed planes and that the slip direction is that of the highest linear atom density. The combination of a slip plane and a slip direction within that plane comprises a *slip system*. The primary slip systems, the most densely packed planes and directions, are the ones requiring the lowest applied shear stress for slip to occur. Some slip systems and the shear stresses required to cause slip are listed in Table 5.2. This stress that causes slip is called the *critical resolved shear stress* (CRSS). It is primarily a function of the atom bond strength and the atom spacing in the crystal.

Although BCC and FCC have the same number of systems, FCC is a closed packed structure and requires lower shear forces to move the atoms than BCC. Hence, FCC structures are considered more ductile (Table 5.3).

Any force applied to a body can be resolved into a force normal to a plane and into shear components along a plane. The method of resolving a longitudinal force on a cylindrical body is depicted in Figure 5.16. The resolved shear stress, τ_r, along a slip plane (shaded plane in Figure 5.16) and in a slip direction resulting from a stress s along the bar is given by

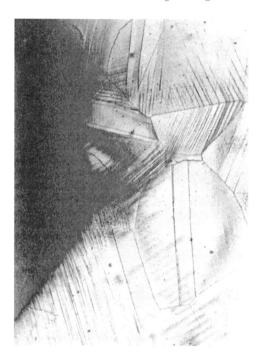

FIGURE 5.15 Slip lines produced by a hardness indentation on a polished and etched surface of an Inconel 600 specimen (×80). Slip lines in annealing twins are at an angle to each other that demonstrates the mirror image of atoms in an annealing twin. Note the intersecting slip lines near the dark edge of the hardness indentation.

TABLE 5.2
Critical Resolved Shear Stress for Selected Metals[a]

Metal	Crystal Structure	Purity (%)	CRSS MPa	(psi)
Al	FCC	99.994	0.7	(100)
Cu	FCC	99.98	0.95	(138)
Ni	FCC	99.98	5.0	(725)
Mg	HCP	99.99	0.7	(100)
Zn	HCP	99.999	0.25	(36)
Fe	BCC	99.96	20.0	(2900)
MO	BCC	—	49.0	(7100)

[a] The values are averaged from a variety of sources. The BCC values are much higher than those for the FCC and HCP close-packed structures.

TABLE 5.3
Slip Systems in Crystal Structures

Crystal Structure	Number of Slip Systems	Element
BCC	12	Fe, W, Mo
FCC	12	Cu, Ni, Au
HCP	3	Zn, Mg

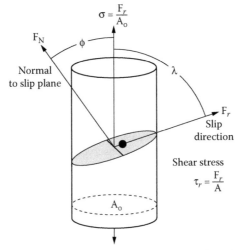

FIGURE 5.16 Computation of the resolved shear stress on plane A (shaded) caused by an axial load F.

$$T_r = \frac{F_r}{A} \tag{5.9}$$

where F_r is the resolved force along the slip plane resulting from the applied force F along the bar axis, and A is the cross-sectional area of the slip plane. In the figure the angle ϕ is that between the normal to the slip plane and the direction of F, and the angle λ is that between the slip direction and the direction of the applied force F. Now

$$F_r = F \cos \lambda$$

$$A = \frac{A_0}{\cos \phi}$$

$$s = \text{normal stress} = \frac{F}{A_0}$$

Therefore,

$$T_r = \frac{F \cos \phi}{A_0 / \cos \phi} = s \cos \phi \cos \lambda \qquad (5.10)$$

The resolved shear stresses can be obtained along any plane from any direction and magnitude of force F. The material does not even have to be crystalline. The reader may have encountered such an operation in a statics course. Here we choose the slip plane and direction because we are interested in the resolved shear stress of the slip system, since when this stress attains the critical value (CRSS), slip occurs. The concept of CRSS is an important concept, but in real polycrystalline metals the interaction of alloys' gain boundaries and other phases makes this concept interesting but overshadowed by many other factors. For engineering materials, **average values** are used to describe the behavior.

5.5.2 DISLOCATION MECHANISM OF SLIP

Plastic deformation of metallic crystals takes place by the movement of dislocations. Substantial plastic deformation is observed at temperatures as low as $0.2T_m$, where T_m is the melting point in Kelvin. The stress required for slip generally increases with decreasing temperature (Figure 5.17), because the modulus (that is, the bond strength) increases with decreasing temperature. When a dislocation moves, it causes plastic deformation to occur by breaking bonds along a single row of atoms simultaneously instead of breaking all bonds simultaneously across a large plane of atoms. Furthermore, the atoms in a row along a dislocation are not directly above or below another row of atoms, but in-between two rows on the plane below it, such that the stress to break these bonds, to a first approximation, is zero.

This will become apparent by observing the movement of an edge dislocation in a crystal by a sequence of steps as pictured in Figure 5.18. Just a small applied stress will move the dislocation because the atoms in the row marked A want to line up with those in row B, since this is the stable lattice configuration. Thus, the atoms in row B attract the atoms in row A, but because of the atom arrangement in a dislocation, the row of atoms go on to the position past row B as depicted in the second step of the sequence of Figure 5.18. The atoms only move one interatomic distance from a point of low energy over an energy hump to another position of low energy. In these low-energy positions, called saddle points, the atoms are attracted equally by the rows in front and back of row A, so the dislocation continues to move from saddle point to saddle point under a very low stress, much like a wave moving through the water. The wave moves to the shore, but the molecules of water move only part of a wavelength. The stress to move a dislocation τ_p (the Periels–Nabarro stress) is given by

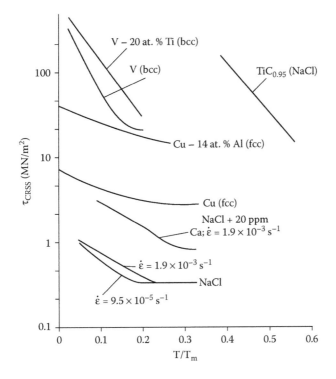

FIGURE 5.17 Variation of the CRSS value with temperature for a variety of materials. Note that the temperature is expressed in terms of the fraction of the melting point in Kelvin. In this way the variation of the CRSS with the types of bonds can be compared. Note that the BCC metals have higher CRSS values and a greater temperature dependency. (From T. H. Courtney, *Mechanical Behavior of Materials*, McGraw-Hill, New York, 1990, p. 141. With permission.)

$$\tau_p = \frac{2G}{1 = \mu} \exp\left[\frac{2\mu a}{b(1-\mu)}\right] \tag{5.11}$$

where

G = shear modulus
μ = Poissons ratio
a = distance between atom planes
b = distance between atoms in the slip direction

Since the modulus increases with decreasing temperature, the stress to move dislocations increases with decreasing temperature, although other factors are also involved, but to a lesser degree. The atoms, after the dislocation has moved through the lattice, are again in perfect registry, except that the last row extends beyond the bottom row one interatomic distance. This is the mechanism of plastic

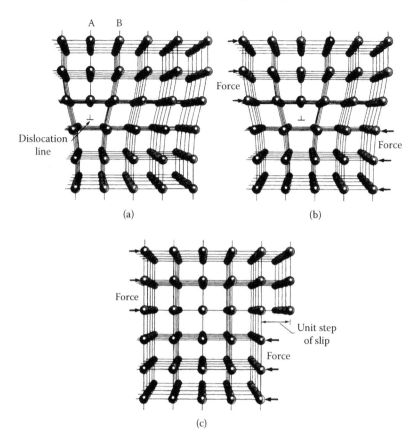

FIGURE 5.18 Motion of an edge dislocation and the resulting unit slip. (From W. T. Read, *Dislocations in Crystals*, McGraw-Hill, New York, 1953, p. 109.)

deformation. You might wonder how a metallic crystal could be twisted, bent, elongated, compressed, cupped, and subjected to whatever other deformation is required to achieve a certain desired shape simply via a slip-plane effect, but there are trillions of dislocations moving in many directions because there are many slip systems. Each of these dislocations produces an offset of one inter-atomic distance in its direction of movement so that we have many offsets in many directions. In plastic deformation, we do not really squash the crystals into an amorphous state. The crystals are squashed into whatever shape we desire, but the material internally remains as an orderly array of atoms. Plastic defor-mation does not destroy crystallinity. Screw dislocations also move out of the crystal and produce an offset as shown in Figure 5.19. They both produce an offset in the same direction, but the dislocation lines of the screw and edge move perpendicular to each other. Remember that in a large crystal the dislocation line is in the form of a loop, consisting of edge, screw, and mixed components. Under

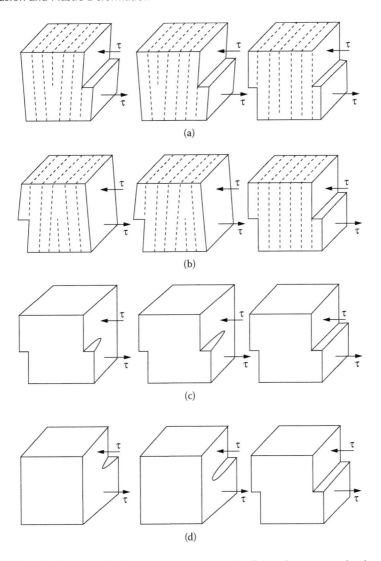

FIGURE 5.19 Ways that the four basic orientations of a dislocation move under the same applied stress to produce unit single slip. (From R. E. Reed-Hill and R. Abbaschian, *Physical Metallurgy Principles*, 3rd ed., PWS-Kent, Boston, 1992, p. 154.)

an applied stress the loop expands and moves completely out of the crystal, producing an offset as depicted in Figure 5.20.

The student may not be able to visualize the above without a crystalline model simulations or an exceptional amount of concentration. For the nonmaterials engineer, it is not important to understand the mechanism, but, instead, just have a little faith that it does happen in this fashion. This mechanism has been verified experimentally, and for all but the doubting Thomases this should suffice. The

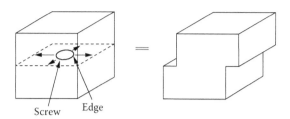

Screw   Edge

FIGURE 5.20 Slip produced by the expansion of a dislocation loop.

more curious students can check the references listed at the end of this chapter for more detailed presentations.

5.5.3 STRESS–STRAIN CURVES OF METAL CRYSTALS: STRAIN HARDENING

The stress–strain behavior of single crystals has little practical value, but it has been examined extensively in research labs in an effort to better understand the more complex behavior of polycrystalline bodies. In portraying the stress–strain curves for single crystals (Figure 5.21), it is more informative to use the resolved shear stress along the slip plane than the normal applied stress. We will also use the amount of slip plane displacement (shear glide) in computing the shear strain, but just as the figure for polycrystalline metals showed strain-hardening effects, so do single-crystal bodies. In all cases the shear stress increases with strain, as

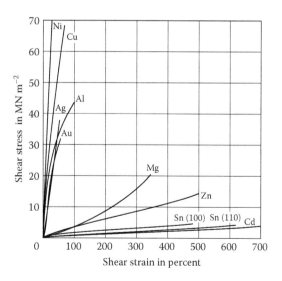

FIGURE 5.21 Shear stress vs. shear strain for different metal crystals. (After E. Schmid and W. Boas, *Kristallplastizitat*, Springer-Verlag, Berlin, 1935.)

one would expect if strain hardening occurs during the slip process. The hexagonal metal crystals (Mg, Zn, Cd) in Figure 5.21 show very little strain hardening, because the crystals have been oriented so that slip occurs on only one set of parallel planes (that is, one slip system). Remember, hexagonal crystals have only three slip systems, so it is easy to orient the crystal so that the resolved shear stress on only one slip system exceeds the CRSS.

For the FCC crystals there are 12 primary systems possible for slip, so perhaps 4 or maybe even 6 or more systems will be oriented for slip at the stress applied. These slip planes intersect, as can be viewed by the slip traces in Figure 5.15. Without going into detail, suffice to say that intersecting slips are one of the prime sources of strain hardening. It requires a larger stress to move dislocations when they must pass through other dislocations. We will return to this subject later when polycrystalline strain hardening is discussed.

5.6 PLASTIC DEFORMATION

The plastic deformation of a metal is key to formation of a part and is a way to manage the mechanical properties. Most metals are polycrystalline (many randomly oriented grains) or polyphase polycrystalline (more than one randomly oriented phase). When subjected to an external mechanical force, atoms in the metal move. This movement is called slip or plastic deformation. It is the way the metal responds to a force above the elastic limit. This response is dictated by the basic crystal structure. FCC metals have more slip planes than HCP, the alloys added and the grain size. When the level of force exceeds the shear stress on the slip plane, lines of dislocation move and in the process create more dislocations. Descriptions of these mechanisms are found in the literature as pining mechanisms and Frank–Read mills. References are given at the end of the chapter for further study.

As dislocations are formed, they can move in many directions and often interfere with each other. This interaction blocks further movement at that stress level, and subsequent slip occurs at a higher load and in different directions as the metal attempts to accommodate the load without fracture. This is called secondary slip. As a result of this interaction and entanglement, more energy is required to get the metal to move. Experimentally, it has been found that the shear stress to maintain plastic flow increases as the square root of the dislocation density. From a practical point of view, the metal is simply appraised as becoming a stronger material, the term *work hardening* or *strain hardening* is used to describe the process.

The engineering application of single-grain materials is limited both by application and cost. The most notable single-grain application is a single-crystal turbine blade used in jet aircraft. If we now look at a common polycrystalline metal, a new factor is added: grain boundaries. Grain boundaries add complications to the atom slip and dislocation generation. Grain boundaries are barriers to the flow of forces in the metal. Even in metals with many available slip planes, forces must be readjusted in crossing a grain boundary. Commonly, dislocation density increases more rapidly with increasing strain in the vicinity of the grain boundry. This effect is illustrated in Figure 5.22. Grain boundaries create barriers

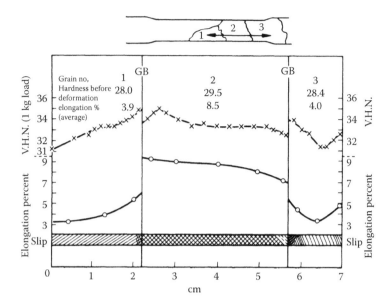

FIGURE 5.22 Inhomogeneity of plastic deformation near the grain boundaries in aluminum as revealed by hardness measurements. A higher dislocation density is formed near the boundaries during deformation. (From R. W. K. Honeycombe, *The Plastic Deformation of Metals*, 2nd ed., ASM International, Metals Park, OH, 1984, p. 224. With permission.)

for the smooth transfer of force from one grain to another in a mechanical working process. In a polycrystalline metal the number of grains in the material has a marked effect on the strength and ductility of a material. The more changes in direction of the force because of grain boundaries, the more energy is used in the process of deformation and the stronger the metal is considered. This can be seen in Figure 5.23.

This is a brief explanation of a very complex three-dimensional process resulting in a physical shape change of the metal piece. The significance of the shape change process is key to manufacturing a part and the control of properties. In its simplest form the reshaping of metal takes place to generate a shape such as a sheet or square by rolling or extrusion. More complex manufacturing processes are used to form automobile parts such as body panels and axels. Properly controlled, these cold-forming processes can yield the proper geometry and the desired physical properties in the same operation.

5.6.1 Property Management by Cold Working

The control of properties by strain hardening is considered in four stages:

- Cold work
- Recovery

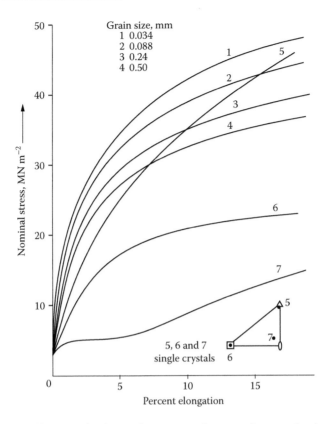

FIGURE 5.23 Effect of grain size on the stress–strain curves for pure aluminum. (From R. W. K. Honeycombe, *The Plastic Deformation of Metals*, 2nd ed., ASM International, Metals Park, OH, 1984, p. 238.)

- Recrystallization
- Grain growth

5.6.2 Cold Work

Cold working = Shape Change

Cold working is typically done at room temperature and can be demonstrated in the laboratory by passing a brass strip through a rolling mill to reduce the thickness and increase the hardness or by bending a wire back and forth until it fractures. In doing so, the metal's hardness is increased and the ductility decreased until the metal becomes brittle and fractures (Figure 5.24). The hardness vs. cold-work curve is shown in Figure 5.25. In this case the cold work is measured by the reduction in cross-sectional area.

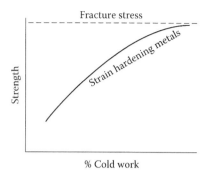

FIGURE 5.24 Strength increases during strain hardening.

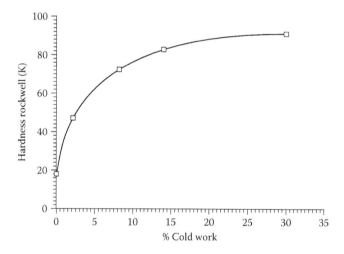

FIGURE 5.25 Hardness as a function of the percent cold work for rolled 70% Cu–30% Zn brass.

$$\% \text{ cold work} = (\text{area initial} - \text{area final }/\text{area initial}) \times 100 \qquad (5.12)$$

Cold working is an addition of energy to the metal in the from of strain.

The property change (shape of the curve) illustrated in Figure 5.26 is typical of a working process. The relative position of the curve is dependent on the family of material being worked (i.e., copper vs. steel) and the alloy content. For example, commercially pure copper vs. brass (70% Cu–30% Zn) is illustrated in Figure 5.27.

An example of where cold working is used to improve the strength in commercial application is a common nail. In the manufacture of a nail, a steel wire is cold upset (head) and sheared (point) (Figure 5.27). This process is carried out at a rate of several hundred operations per minute. The cold heading flattens one

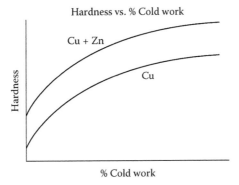

FIGURE 5.26 Schematic of increase in hardness for pure and alloyed materials.

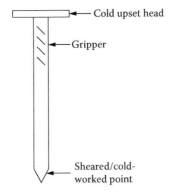

FIGURE 5.27 Schematic of the effect of cold work in forming a nail.

end of the wire to create a strong, hard, flat striking area, and the shearing process used to determine the length cuts the wire and is a cold-working process that provides a hard driving point. The properties of the center section of the nail are unchanged and allow the nail to bend when misstruck.

Cold-forming processes are a key component in the manufacturing world because they can be used to make products with final shapes that have accurate dimensions and excellent surface finishes, often in a single step. Properties can be controlled in specific areas to enhance structural rigidly and comply with assembly and service requirements. The designer must keep in mind that as strength and hardness are increased by cold work, the ductility is reduced.

Other examples of cold-working processes are drawing of axels for automobiles, shot blasting of leaf springs to create residual compressive stress on the tensile side of the spring, and the stamping of gears for lawn mowers. Cold working and the respective property changes are evident in many products that are used daily.

5.6.3 RECOVERY

In this first stage of the annealing process sufficient energy is added to the metal to allow movement of highly strained areas to produce a less stressed arrangement. The internal disruption of the atomic structure is reduced. There is no visible change in the microstructure and only a slight decrease in the hardness and strength. The metal becomes more resistant to corrosion, and some improvement in the electrical conductivity is seen. Because of the readjustment of the stresses, this process is sometimes referred to as a stress-relief anneal.

5.6.4 RECRYSTALLIZATION

As the name implies, in this stage of the management process new grains (crystals) are nucleated and grow in a strain-free (to reduce dislocation) environment. The purpose of this process is to recover the properties of the material as seen before cold working. During recrystallization, strength and hardness decrease and ductility increases (Figure 5.28). This is a thermally activated process, and typically the recrystallization temperature is between 0.35 and 0.5 T_m. In addition, the amount of strain energy stored in the sample during cold working will influence the precise recrystallization temperature (Figure 5.29).

Energy of recrystallization = energy from thermal source
+ energy from strain (cold work)

The nucleation of the strain-free grains requires an incubation period in which the atoms diffuse to the site of the new crystal grain. As a standard measure, the recrystallization temperature is given as the temperature for complete recrystallization in 1 hour. This is a diffusion process, and the diffusion rate and the

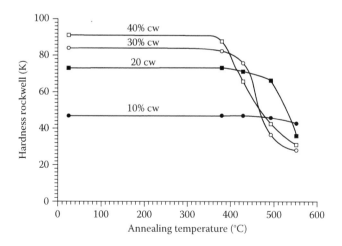

FIGURE 5.28 Reduction of hardness of cold-worked brass during annealing.

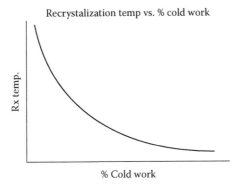

FIGURE 5.29 Recrystallization temperature as a function of cold work.

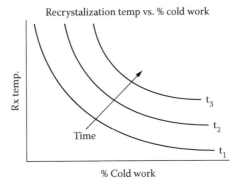

FIGURE 5.30 Recrystallization temperature as a function of different holding times (t) where $t_1 > t_2 > t_3$.

recrystallization rate are exponential functions of temperature. Raising the temperature will decrease the time (t) required for the recovery of properties (Figure 5.30), operating at a temperature lower than T_{RX1hr} but significantly above room temperature can cause recrystallization in a longer period of time. This becomes important in choosing the service temperature for a part. If the operating temperature is too high, the part may lose properties and fail in a period shorter than the design life. Forging, or hot working, occurs at temperatures well above the T_{RX}. At this level the energy supplied by the outside source is so great that no energy is stored from the working process, and recrystallization is continuous. Figure 5.31 shows the cycle. In Figure 5.31a the deformed grains appear to have directionality. In Figure 5.31b recrystallization has occurred. The microstructure consists of very small uniformly sized equiaxed grains and has the smallest grain size of the annealing process. The properties of the metal are considerably reduced compared to the previous micrograph but may be stronger than the original material if its grain size were larger. When all strain energy has been released, recrystallization is complete (Figure 5.31c).

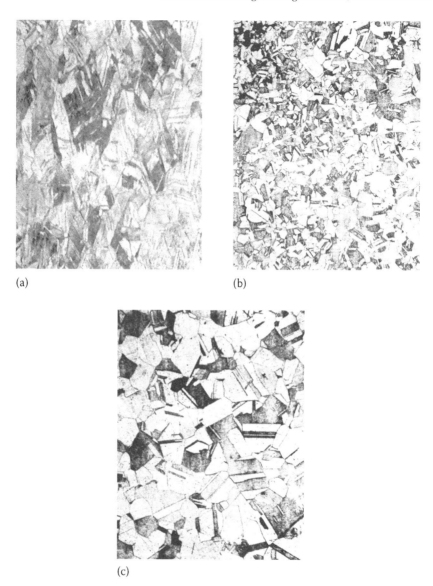

(a)

(b)

(c)

FIGURE 5.31 Cold-worked and recrystallized structures of brass: (a) cold rolled 32% R.A.; (b) small grain size just at the end of the recrystallization process; (c) large grains after the grain growth stage (×200). (Courtesy of John Harruf, Cal Poly.)

5.6.5 GRAIN GROWTH

As the metal remains at an elevated temperature, the grain size will increase. The driving force for this diffusion-based process is the reduction of grain boundary energy through the reduction of the surface area-to-volume ratio. The hardness does not change much initially in this stage, but as the grains become larger, the yield strength and hardness theoretically decreases. In this stage the large grains "swallow" the small grains. With sufficiently long annealing times, one can obtain a structure with a few large grains via grain growth.

The cold-working process followed by annealing is a major management tool for the control of shape and properties. Cold working as illustrated in Figure 5.26 increases the strength and hardness of the metal; an annealing process restores the properties. In many cases metal from a primary refiner is given a large hot reduction to break up and refine the as-cast structure and its poorer mechanical properties. The final shape and properties are attained by cold working because of the more precise dimensional characteristics and the better surface quality of a cold-formed product. Typical cold fusion products are aluminum chalkboard rails, window channels, and small gears.

5.6.6 GRAINS AND PROPERTIES

The cold-work and recrystallization, hot-working, and grain growth processes are effective ways of managing the properties of a material through grain size manipulation. Larger grains tend to be more ductile, while smaller grains tend to be stronger. The measure of grain size is an important metric in quality control. Commonly, grain size is measured by the American Society for Testing and Materials (ASTM) method:

$$N = 2^{n-1} \tag{5.13}$$

where
N = the number of grains/square inch at $100\times$ magnification
n = the grain size number

The measurement of grain size is carried out by a variety of methods. In the simplest format, an image of the microstructures is projected onto a viewing screen and compared to images of grains of known size. A statistical method is to impose a grid, line, or circle on the subject image and count the grain boundry intercepts with the line. Using this information and the line length, a statistical average grain size can be calculated. A third method is to use a computer image analysis system and appropriate software to determine the grain size.

Grain size can be related to properties, rates of chemical reactions, and physical appearance. A relationship exists between grain size and the yield strength of the material. The Hall–Petch relationship is written as

$$\sigma_y = \sigma_o + k_y \, d^{-1/2}$$

where
σ_y = yield strength
σ_o = a constant for the material
k_y = grain boundary parameter
d = average grain diameter

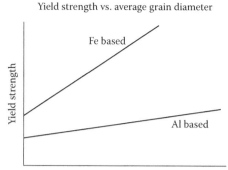

FIGURE 5.32 Yield strength *V* vs. average grain diameter.

This relationship allows an estimation of the change in yield strength as a function of grain size change (Figure 5.32).

DEFINITIONS

Activation energy: Energy necessary to overcome a resistance to a chemical reaction, such as the diffusion of an atom from one position of stability to another stable position.

Annealing: Generic term to denote the heating of a piece to cause a certain process to occur (e.g., recrystallization).

Brittle fracture: Cleavage type of crystalline fracture where a crack propagates with little macroscopic or microscopic plastic deformation. In environmental and amorphous material fractures, which are also brittle, a slightly different definition will be used.

Cold deformation (working): Processing conducted below the recrystallization temperature of metals.

Concentration gradient: Slope of the concentration–distance curve. The driving force for the diffusion process.

Critical resolved shear stress: Stress necessary to cause planar slip (dislocation motion).

Diffusion: Migration of atoms through a lattice.

Diffusion coefficient: Proportionality constant between the diffusion flux and the concentration gradient in Fick's first law.

Dislocation: Linear crystalline defect consisting of a disregistry of atoms. They consist of edge, screw, and mixed components. Dislocation motion causes plastic deformation.

Doping: Intentional addition of certain foreign elements to semiconductors to increase their electrical conductivity.

Driving force: Impetus or power behind a reaction.

Ductile fracture: That occurring in conjunction with substantial plastic deformation. Microvoid coalescence is observed in SEM electron fractographs.

Entropy: Measure of randomness.

Fracture toughness: Resistance of a material to crack propagation.

Free energy: Minimum energy state at equilibrium. The energy available to power a relation to attain equilibrium.

Glassy structure: Amorphous structure usually applied to describe glass ceramics, most polymers, and very rapidly cooled metals.

Grain: Planar defect that separates grains.

Grain growth: Increase in grain size that occurs during annealing at temperatures substantially above the recrystallization temperature or for longer times at the recrystallization temperature.

Grain size number: ASTM method for expressing grain size.

Host atom: Major element in a polyatomic structure. The parent atom. In a solid solution, the solvent atom.

Hot working: Processing conducted above the recrystallization temperature where strain hardening is absent.

Interstice: Interstitial position in a lattice.

Interstitial solid solution: Solid solution where small atoms occupy interstitial positions between or among the solvent or host atoms.

Planar defects: Disregistry of atoms occurring between grains (crystals) or phases.

Point defects: Lattice vacancies or foreign atoms in the host lattice.

Precipitate: In the solid state it is a phase, usually a compound, that forms when the solubility limit has been exceeded. In liquid solutions it is some type of solid phase.

Primary slip system: Slip system with the lowest value of the CRSS required for slip to occur (*see* Slip system).

Recovery: First stage of the annealing of a cold-deformed metal where realignment and dislocation disentanglements occur accompanied by a slight hardness decrease and an improvement in corrosion resistance.

Recrystallization: Second stage of the annealing process of a cold-deformed metal where new grains are formed as the internal strain is removed. A substantial decrease in hardness occurs.

Resolved shear stress: Stress parallel to any plane resulting from any arbitrary force applied to a body.

Secondary slip system: Slip system that has a higher CRSS than the primary slip system.

Slip lines: Lines that appear on a crystal surface resulting from the slipping of one or more planes over other parallel planes. What we see are clusters of slipped planes.

Slip system: Combination of a slip plane and a slip direction that lies within the slip plane.

Solubility limit: Maximum amount of solute that can be dissolved in the solvent at any given temperature.

Solvent: Major constituent of a system.

Substitutional solid solution: Solution in which the solute atom occupies a host atom site.

Vacancy: Vacant lattice site that is normally occupied by a host atom.

QUESTIONS AND PROBLEMS

5.1 Why do imperfections of atomic size have little or no effect on the mechanical properties of amorphous materials?

5.2 Distinguish among the following: host atom, parent atom, solvent atom.

5.3 Vacant lattice sites exist after solidification of a crystalline body. Are these vacancies accidents of growth of the solid crystal? Explain.

5.4 Explain how crystalline grain boundaries are formed during solidification of a molten metal. Sketches will be helpful.

5.5 Using the data from Table 5.1, compute the diffusivity for copper in aluminum at 300°C.

5.6 What is the solubility limit for copper in aluminum at 300°C (Figure 5.2)? What compound is formed when this solubility limit is exceeded?

5.7 What is the crystal structure of Fe_3C? What is the solubility limit of carbon in iron at 727°C?

5.8 Will carbon diffuse faster in nickel than it does in copper? Explain.

5.9 Estimate the distance that carbon will significantly diffuse into iron after 100 h at 900°C. At 700°C.

5.10 Find the diffusivities of carbon in BCC iron at 200, 500, and 700°C. Plot D vs. $1/Q$ and determine the activation energy and the constant D_0.

5.11 Explain the difference between edge and the screw dislocation. Does this difference have any significant effect on plastic flow in metals? Explain.

5.12 What is a typical density of an annealed metal? What would you expect would be the maximum approximate density that one could obtain in a metal? How could this density be achieved?

5.13 By what factor would the flow stress increase for a dislocation density increase of four times?

5.14 A cylindrical single-crystal bar is being pulled in a direction coinciding with the bar axis. The normal to the slip plane of this bar is 15° from the direction of the bar axis. The slip direction in this plane is 35° with respect to the bar axis. If the critical resolved shear stress of this material is 500 psi, what force must be applied to cause plastic slip to occur?

5.15 List the typical values for the CRSS values of FCC, BCC, and HCP crystals. Generally, the CRSS values of BCC crystals are higher than those of the others. There are two reasons for this. Name at least one reason.

5.16 List two reasons why ceramic crystals are less ductile than metallic crystals.

5.17 How does the grain boundary affect the strength of metals?
(a) At temperatures of $0.2T_m$
(b) At temperatures of $0.7T_m$

5.18 Diffusion is (faster, slower) along grain boundaries than through the lattice proper.

5.19 Discuss and explain the mechanism of the increase in dislocation density via plastic deformation.

5.20 If 100 grains per square inch were observed at 100× magnification, what would be the ASTM grain size number?

5.21 If the grain size number decreases by a factor of 4, what would be the factor of increase in the yield strength?

5.22 Sketch the curve of the decrease in hardness as a cold-worked material is annealed at increasingly higher temperatures and describe what is occurring at the atomic level in the various stages of this curve.

SUGGESTED READING

American Society for Metals, *Diffusion*, ASM International, Metals Park, OH, 1972.

Frank, J., *The Mathematics of Diffusion*, Clarendon Press, Oxford, 1975.

Honeycombe, R. W. K., *The Plastic Deformation of Metals*, 2nd ed., ASM International, Metals Park, OH, 1984.

Liebowitz, H., Ed., *Fracture*, Academic Press, New York, 1972; see the paper, "Microstructure Aspects of Fracture in Ceramics," by R. J. Stokes, p. 157.

Reed-Hill, R. E., *Physical Metallurgy Principles*, 2nd ed., D. Van Nostrand, Princeton, NJ, 1973; see Chapter 4, "Dislocations and Slip Phenomena."

Shewmon, P., *Diffusion in Solids*, 2nd ed., TMS Publications, 1989.

6 Property Management of Metallic Materials

6.1 INTRODUCTION

The management of a material's property is dependent on understanding how that material will react to changes in the environment and composition. What state the material is in at any given time corresponds to what one would expect the material to be capable of. In the study of metals and ceramics, phase diagrams are used to depict the conditions of the material with respect to temperature and composition. These diagrams give us a picture of temperature ranges for forging and casting operations, solubility limits for increasing properties by alloying, and understanding the effect of contaminating elements. They also help interpret the microstructure for quality control or failure analysis.

6.2 PHASE EQUILIBRIA

From our discussions in earlier chapters, a *phase* and the many possible atom arrangements within phases should now be familiar concepts. We have also discussed the coexistence of phases in both equilibrium and nonequilibrium systems. These examples included the solubility limit curves of sugar and coffee and of the solid solution–compound systems. The coexistence of phases in equilibrium occurs when the solubility limit is exceeded (Figures 5.2 and 5.3). The solubility limit curves are used to construct *phase diagrams,* or, as they are frequently called, *equilibrium diagrams. Equilibrium* means there is no macroscopic change with time in a phase or system of phases. In the water–water vapor enclosed system, for example, molecules continue to leave the liquid and enter the vapor phase and vice versa. At equilibrium the forward and reverse reactions occur at the same rate, as indicated by the equivalent length of the arrows in the following:

$$H_2O_{liquid} \leftrightarrows H_2O_{vapor} \qquad (6.1)$$

A phase diagram is simply a map. At any point on this map, unless the point lies exactly on a line (a rare occurrence), the phase diagram indicates what phases exist for a particular set of conditions. A familiar example would be the phase diagram for water (Figure 6.1). As long as the water does not dissociate into H and O, this diagram is valid for temperature and pressure conditions.

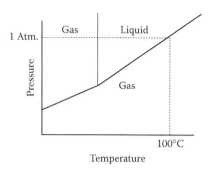

FIGURE 6.1 Unary diagram of water.

This unary (one component) diagram for water should be familiar. At 1 atmosphere of pressure and 100°C, a transition from liquid to gas takes place (boiling). We also know that if the pressure is decreased, as in high-altitude cooking, water will boil at a lower temperature. In a radiator, where the system pressure is increased with a radiator cap, the boiling temperature is increased. Similarly, if the temperature is lowered to 0°C, another transition takes place and water freezes. The diagram similar to this is used commercially in making coffee by atomizing the coffee into a cold chamber to make "brown snow" and reducing the pressure to sublime off the water and create a product that we recognize as freeze-dried coffee. Although this system is dynamic, the concept is based on the use of a similar diagram.

In most metallic and ceramic systems the influence of pressure is not as important as the effects of composition. Consequently, if pressure is maintained at 1 atmosphere (standard thermodynamic conditions) and a compositional variable is added, the standard binary (two-component system) is seen. Pressure is important in a few situations such as the formation of diamond (Figure 6.2), the space shuttle alloy experiment, and the melting of specialty alloys.

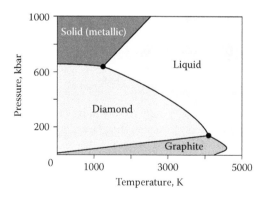

FIGURE 6.2 Pressure–temperature phase diagram for carbon. (From C. G. Suits, *Am. Sci.*, 52, 395, 1964. With permission.)

Many of the commonly used alloy systems are easily represented by a binary phase diagram (those consisting of two elements), even though there are additions of small amounts of other elements for purposes other than altering the phase quantities. More complex systems containing three or more elements are represented by ternary diagrams or other forms of phase diagrams.

6.3 SOME COMMON PHASE DIAGRAMS AND ALLOYS

6.3.1 COMPLETE SOLID SOLUBILITY SYSTEMS

6.3.1.1 Cu–Ni System

Not many common alloy systems have complete solid solubility at all temperatures and compositions in the solid state. In such systems the elements must be of similar size and possess in the pure form the same crystal structure. The Cu–Ni system is the classic example, and this phase diagram is depicted in Figure 6.3. There is only one solid solution phase in the solid state, and it is of the substitutional type. You can call it a solid solution of nickel in copper (usually done up to 50% nickel) or a solid solution of copper in nickel when nickel is the major element. Both copper and nickel have face-centered cubic (FCC) structures, so an atom of one element is replaced by one of the other elements. Note that 100%

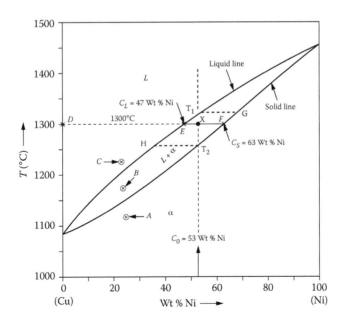

FIGURE 6.3 Cu–Ni phase diagram. (Adapted from K. M. Ralls, T. H. Courtney, and J. Wulff, *Introduction to Materials Science and Engineering*, Wiley, New York, 1976, p. 317. With permission.)

copper represents the 0% nickel composition point on the left side of the diagram, and 100% nickel–0% copper is placed on the right side according to our alphabetical way of arranging elements at the terminus of a phase diagram. In this diagram the composition is plotted in weight percent nickel. In some cases it may be of interest to know the composition in atomic percent (many handbooks show both). A 50% Cu–50% Ni weight percent alloy means that if the total quantity was 100 g, the alloy would contain 50 g of Cu and 50 g of Ni. A 50 at % Cu–50 at % Ni alloy contains equal numbers of Cu and Ni atoms. A mole of Cu atoms weighs 63.54 g compared to 58.71 g/mol of nickel atoms. Thus, the atomic percentages and weight percentages are similar. A 50–50 wt % alloy converts to an approximate 52% Cu–48% Ni on an atom percent basis.

For comparison let us choose a point on our Cu–Ni map at 1300°C and 53 wt % Ni, which we will label with the letter X. The diagram tells us that at this point we are in a two-phase region consisting of a liquid solution and a solid solution. The horizontal 1300°C tie line that intersects the liquid–phase boundary, called the *liquidus*, at point E tells us that the liquid phase has a 53 Cu–47 Ni wt % composition. Similarly, the intersection at the *solidus* phase boundary at point F yields a 37 Cu–63 Ni wt % solid solution phase. Only a liquid solution exists above the liquidus, and only a solid solution is present below the solidus line. In both the liquid and solid solution phases the composition is uniform throughout.

Example 6.1   A 100-g lot consisting of 47 g of Cu and 53 g of nickel was melted and cooled to 1300°C. How many grams of copper are there in (a) the liquid phase and (b) the solid phase?

Solution
(a) First we must determine the quantity of each phase. This is done by what is often called the inverse lever rule. Let the total length of the tie line represent the total quantity of material. Then the ratio of the segment of line from our point of interest A to the solidus intersection point F to the total tie-line length gives us the fraction of the liquid phase, that is,

$$\frac{F - X}{F - E} = \text{liquid fraction of total}$$

Similarly, the ratio

$$\frac{X - E}{F - E} = \text{solid fraction of total}$$

We could measure these line lengths very accurately, but we can accomplish the same result by reading off the compositions at these three points from the axis and let them represent our line segment lengths; thus,

$$\frac{F-X}{F-E} = \frac{63-53}{63-47} = \frac{10}{16} = 0.625 \text{ liquid} = 62.5 \text{ g liquid}$$

Note that the liquid phase contains 53 wt % Cu. Thus, 0.53×62.5 g of liquid $= 33.125$ g of Cu in the liquid phase.

(b) Now we will do the same for the solid phase. The ratio of the length of the line segment from our point of interest A to the liquidus intersection at point E to the total tie line length gives us the fraction of solid.

$$\frac{X-E}{F-E} = \frac{53-47}{63-47} \frac{6}{16} = 0.375 \text{ solid} = 37.5 \text{ g solid}$$

The solid phase contains 37 wt % Cu; thus, 0.37×37.5 g of solid $= 13.9$ g of Cu in the solid phase. *Note*: The grams of liquid and grams of solid $= 100$ g. The grams of Cu in liquid and grams of Cu in solid $= 47$ g, which is the same quantity of Cu that we added to the 53 g of Ni to melt our alloy. We neither create nor destroy material by phase diagrams. The three other points on the diagram are labeled A, B, and C. If the temperature and composition correspond to point A, the alloy is a single-phase solid solution; at B, a two-phase mixture of solid and liquid solutions; and at point C, a single-phase liquid solution.

Many ways of determining phase diagrams are beyond the scope of this book. One method is by the use of *cooling curves*. We will illustrate these curves for the Cu–Ni system, not for the purpose of becoming experts in phase diagram determination, but to better understand the solidified microstructure, which in turn affects the properties of the alloy.

To obtain equilibrium, theoretically we must cool at an infinitely or impractically slow rate; that is, our system must be at equilibrium at each point on the phase diagram. In liquids the diffusion rate is sufficiently fast that equilibrium can be attained or at least approached in a few minutes. In solids it requires a much longer time, and usually cooling curves are not used to detect solid–solid phase transformations. For purposes of illustration, we will assume that the cooling curves in Figure 6.4 were obtained under slow cooling conditions. In this figure our cooling curves are depicted for 100 wt % Cu, 47 wt % Cu–53 wt % Ni, and 100% Ni. Note that the curves have a horizontal constant-temperature section at the melting point of the pure metals. This results from the latent heat of fusion (that is, the heat required to melt the material, which is equal to that released during freezing of the metal) being released as a solidification takes place and that exactly balances the rate of heat removal. Under slow cooling conditions this heat release is sufficient to maintain a constant temperature of the liquid–solid mixture until solidification is complete. There are two phases, liquid and solid, coexisting during solidification.

Now let us consider the alloy cooling curve and at the same time refer to the phase diagram of Figure 6.3. Solidification begins at point T_1 on the phase diagram

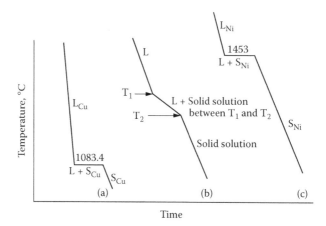

Time

FIGURE 6.4 Cooling curves for (a) 100% Cu; (b) Cu 47%–Ni 53% alloy; and (c) 100% Ni.

and is completed at point T_2, but there is no horizontal segment on the cooling curve. Some heat of fusion is being released, which changes the slope of the cooling-rate curve as the temperature is reduced through the two-phase liquid–solid region of Figure 6.3, but this heat of fusion is not sufficient to maintain a constant temperature. At any point along the vertical line representing the composition Cu 47–Ni 53 we can construct a horizontal tie line and determine the quantities and compositions of the two phases just as we did in Example 6.3. The compositions change during cooling, that of the liquid phase following the liquidus curve and that of the solid phase following the solidus curve. By connecting the solidification points of the pure metals with point E, we can construct the liquidus line, and with point F, the solidus line. The phase diagram has now been completed with only three cooling curves. For greater accuracy, we should use more alloy compositions.

From a practical point of view, we seldom cool under equilibrium or even near-equilibrium conditions. The commercial Cu–Ni alloys (called Monels, an International Nickel trademark) are melted and poured into a mold to produce an ingot that is subsequently fabricated to certain desired forms via the usual rolling or drawing processes. When an ingot is cast, the composition will not be uniform throughout the ingot. Referring again to the 47 wt % Cu–53 wt % Ni alloy of Figure 6.3, when an ingot of this molten alloy is cast, the part to solidify first, that at the mold walls, will have more nickel and less copper than the overall composition (see point G). This part of the ingot will be nickel rich. As solidification proceeds toward the ingot center, the alloy composition changes until at the ingot center the last to solidify (point H) will be copper rich compared to the overall starting composition of 47% Cu–53% Ni.

Usually, columnar grains grow out from the wall of a mold toward the ingot center. Figure 6.5 shows such a grain structure resulting from the pouring of a 70% Cu–30% Zn alloy into a steel mold. These grains form by nucleation and growth. The material near the grain boundaries, the last to solidify, will be copper

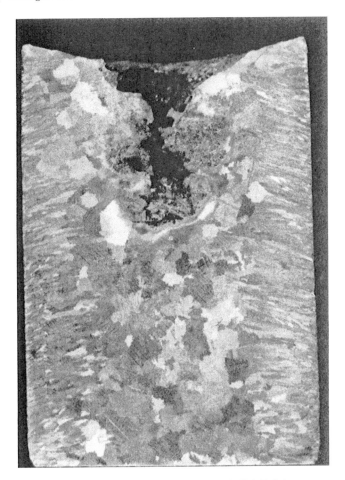

FIGURE 6.5 Cast ingot of brass. (Courtesy of J. Harruf, Cal. Poly.)

rich. Thus, we have segregation throughout a grain as well as an ingot. In many metals *dendrites* are formed during solidification. During the period of columnar growth the solid–liquid interface is not planar, but develops protuberances along the grain boundaries because of compositional changes along the growing front. These shapes are called dendrites. A typical dendritic structure in cast brass is shown in Figure 6.6. These segregated cast structures are undesirable. The composition gradients can be smoothed out by heating the ingot at high temperatures for long periods of time. This treatment is called a *homogenization anneal* by the metallurgist and a *soaking operation* by the plant foreman. Hot working breaks up the columnar grains and dendritic structures, which otherwise would cause anisotropic properties in the ingot, but homogenization prior to hot working is necessary to reduce the alloy segregation to an acceptable level.

Most Monels contain small amounts of other elements. K-Monel contains about 3.0% Al for precipitation-hardening effects. Monels to be used in as-cast

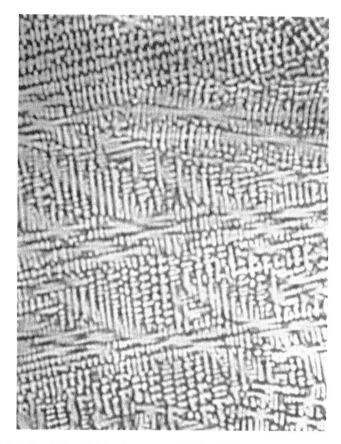

FIGURE 6.6 Dendritic solidification structure in Cu 70%–Zn 30% brass. (Courtesy of J. Haruff, Cal Poly.)

shapes contain small amounts of silicon to promote fluidity, which helps fill the intricate molds designed for final cast shapes. Monels have good corrosion resistance and strength properties (Table 6.1, see pp. 159–160). They are used in chemical industries, marine applications, laundry, and pharmaceutical industries.

Other alloy systems that show complete solid solubility include Ge–Si, Ir–Pt, Mo–W, Mo–V, Ni–Rh, Se–Te, and V–W. Two elements that are completely soluble in each other in the solid state are very similar in chemical behavior and atom size and, of course, must be of the same crystal structure.

6.3.2 Eutectic Systems

The term *eutectic* is derived from the Greek word *eutektos*, which translates to *easily melted*. For simplicity, we divide the eutectic diagrams into the *simple eutectic* systems and those with some solid solubility. The eutectic composition is the lowest melting point of all the possible compositions of alloys in these

TABLE 6.1
Tensile Strength, Yield Strength, and Ductility of Selected Nonferrous Alloys

Alloy	Wt %	Tensile Strength MPa (ksi)	Yield Strength MPa (ksi)	Ductility Elongation	Conditionng[a]	Applications
400 Monel	64 Cu, 31 Ni, 2.5 Fe	550 (80)	240 (35)	40	A	Marine engineering, chemical and hydrocarbon processing equipment, pumps, heat exchangers, valves
K-500 Monel	64 Cu, 31 Ni, 2.5 Al, 2.0 Fe, 1.5 Mn	1100 (160)	790 (115)	20	PH	Pump shafts, marine propeller, shafts, oil well tools
1050 aluminum	99.5 Al min	76 (11)	28 (4)	39	A	Chemical and brewing industries, food containers
		159 (23)	145(21)	7	75% CW	Chemical and brewing industries, food containers
2024 aluminum	93 Al, 4.4 Cu, 2.5 Mg	476 (69)	393 (57)	10	PH T6	Aircraft structures, hardware, truck wheels
7075 aluminum	90 Al, 5.5 Zn, 1.5 Cu, 2.5 Mg	531 (77)	462 (67)	7	PH T6	Aircraft, forged structural parts
C10100 copper	99.95 Cu	220 (32)	69 (10)	45	A	General stamped and drawn products, busbars, lead-in wires, cables, tubes
		380 (55)	345 (50)	4	CW 60%	General stamped and drawn products, busbars, lead-in wires cables, tubes
C26000	70 Cu, 30 Zn	300 (44)	75 (11)	68	A	Radiator cores, socket brass shells, cartridge cases, springs, plumbing fittings, stampings
	70 Cu, 30 Zn	700 (100)	450 (65)	5	CW 50%	Radiator cores, socket brass shells, cartridge cases, springs, plumbing fittings
C17200 Cu-Be	97 Cu, 2.9 Be	1240 (180)	1070 (155)	7	PH	Flexible metal hose, springs, bellows, valves

TABLE 6.1 (Continued)
Tensile Strength, Yield Strength, and Ductility of Selected Nonferrous Alloys

Alloy	Wt %	Tensile Strength MPa (ksi)	Yield Strength MPa (ksi)	Ductility Elongation	Conditionng[a]	Applications
C61000 Al bronze	92 Cu, 8 Al	480 (70)	205 (30)	65	A	Bolts, shafts, tie rods, pump parts
AZ80A magnesium	90 Mg, 8.5 Al, 0.5 Zn	380 (550)	—	7	PH T5	Extruded and forged products, aircraft, automotive, electrical
Ti–6Al–4V	90 Ti, 6 Al, 4 V	1178 (156)	977 (142)	16	Aged	Jet engine fan disks, heat exchangers, nuts, bolts, aircraft structural parts
3	85 Ti, 10 V, 10 V, 2 Fe, 3 Al	1242 (180)	1173 (180)	10	Aged	High-strength air frame components, high toughness

[a] PH, precipitation hardened; CW, cold worked; A, annealed.

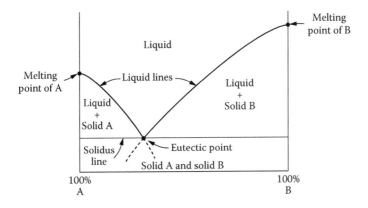

FIGURE 6.7 Schematic of a eutectic phase diagram.

binary systems. The reason that this composition is the lowest melting point of all compositions can be deduced from the schematic of Figure 6.7. The solid curve on the left represents the lowering of the melting point of metal A by the addition of metal B, and vice versa for A added to B. These curves intersect at a point called the *eutectic point*. The continuation of the two curves, represented by the dashed lines, no longer represents melting points. At all temperatures below the point of intersection the system is in the solid state. Thus, the *eutectic point* is the lowest melting point in this alloy system, and since it is a point on our phase diagram map, it is also called the *eutectic temperature* and the *eutectic composition*, all being the same point. Since everything below this temperature is a mixture of two solid phases, we can construct a horizontal line from 100% A to 100% B passing through the eutectic point, completing the phase diagram.

6.3.2.1 Simple Eutectic Systems

The Au–Si system (Figure 6.8) is one of the few simple eutectic systems (i.e., alloys with no solid solubility in each other). The solid solubility of Si in Au is virtually zero, and vice versa. The Au–Si system is also of considerable practical importance in electronic devices because layers of silicon and gold come into direct contact with each other, and at temperatures above the eutectic temperature (363°C) a liquid phase will form.

The melting point of most metals is reduced by the addition of a second metallic element. Referring to Figure 6.8, we see that the melting point of gold is reduced from 1064°C to 363°C by the addition of 19 at % of silicon. Similarly, the melting point of silicon is reduced from 1412°C to 363°C by the addition of 81 at % gold.

Cooling curves for the eutectic composition of the Au–Ni alloy and for the pure metals are constructed along with their corresponding schematic microstructures in Figure 6.9. The cooling curve for the eutectic composition has a horizontal section at the eutectic temperature. Just as for pure metals, the eutectic

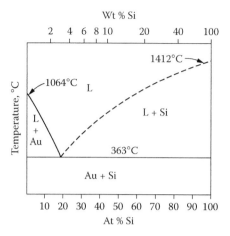

FIGURE 6.8 Au–Si phase diagram (eutectic). (Adapted from R. P. Anantatmula et al., *J. Electron. Mater.*, 4(3), 445, 1975. With permission.)

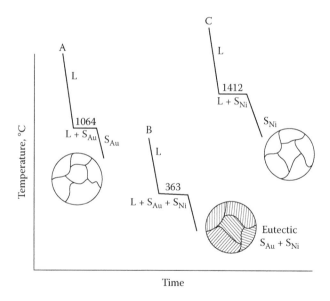

FIGURE 6.9 Cooling curves and microstructure schematics for (A) 100% Au; (B) eutectic composition; and (C) 100% Si.

composition solidifies and melts at a constant temperature. Only pure metals and alloys of eutectic composition have this particular feature. The heat released during solidification is sufficient to maintain a constant temperature until solidification has been completed. Unlike the pure metals, where the pure solid and pure liquid are in equilibrium during solidification, for the eutectic composition we have two solid phases and one liquid phase (in the present case solid Au, solid Si, and a single-phase liquid solution of the two elements), all in equilibrium at the eutectic temperature. This condition can be expressed in the form of the *eutectic reaction* chemical equation as follows:

$$\text{liquid solution of A and B} \rightarrow \text{solid A} + \text{solid B}$$

In the case of the Au–Si alloy, the reaction during solidification is

$$\text{liquid solution of Au and Si} \rightarrow \text{solid Au} + \text{solid Si} \qquad (6.2)$$

Thus, three phases exist together during solidification at the eutectic point. If we held the system at this temperature, there would be a reverse reaction arrow, just as in Equation 6.2. The eutectic reaction is called an *invariant reaction* since it occurs under equilibrium conditions at a fixed temperature and alloy composition that cannot be varied.

When a eutectic composition solidifies at a relatively slow cooling rate, the solid phases are arranged in a parallel platelet form as depicted in Figure 6.9(B), and this is known as the *eutectic microstructure*. In the Au–Si system, if a composition to the left of the eutectic composition (e.g., 1 of 10 at % Si) is slowly cooled from the molten (all liquid) state, pure Au will precipitate out of the liquid when the temperature drops just below the liquidus line (about 760°C). As the liquid continues to cool, more solid Au forms, and thus the liquid will contain a higher percent of Si. By constructing tie lines at successively lower temperatures, it becomes obvious that the liquid composition follows the liquidus line until it reaches the eutectic temperature, where all of the remaining liquid solidifies, just as if the system contained only liquid of eutectic composition. We could strain out all of the solid gold at the eutectic temperature, and then the remaining liquid would solidify with the eutectic microstructure depicted in Figure 6.9; but since we do not strain out the solid gold that exists just above the eutectic temperature, this solid is present and the microstructure is like that depicted in Figure 6.10, along with the corresponding cooling curve. Note that the horizontal section on the cooling curve for the 10 at % Si composition represents the solidification of the eutectic liquid composition — and only that composition. The nonhorizontal part of the cooling curve is that from the precipitation of solid gold as we cool through the liquid–solid two-phase region. Again, some heat is released as the alloy is cooled through the liquid–solid region, but it is insufficient to maintain a constant temperature. The gold that precipitates out in the solid form prior to reaching the eutectic temperature is called the *proeutectic,* or primary, solid (i.e., it forms prior to reaching the eutectic point).

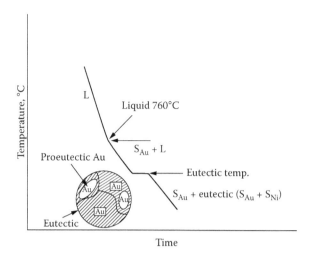

FIGURE 6.10 Cooling curve and microstructure schematic of a Au–Si alloy composition that on solidification results in a combination of a proeutectic Au phase plus the Au–Si eutectic phase arrangement microstructure.

Example 6.2 Convert the atomic percent of the eutectic composition to weight percent (19 at % Si–81 at % Au).

Solution For Si we have 28.1 g/mol/6.02 × 10²³ = 4.67 × 10⁻²³ g/Si atom. For Au we have 197 g/mol/6.02 × 10²³ g/Au atom. Assume 100 atoms of eutectic composition.

19 atoms of Si = 19 × 4.67 × 10⁻²³ g/Si atom = 88.73 × 10⁻²³ g of Si

81 atoms of Au = 81 × 32.724 × 10⁻²³ g/Au atom = 2650.6 × 10⁻²³ g of Au

$$\text{Wt \% Si} = \frac{88.73}{88.734 + 2650.6} \times 100 = 3.2$$

$$\text{Wt\% Au} = \text{balance} = 96.8$$

Example 6.3 For a 60 at % Si–40 at % Au alloy at 800°C, compute the atomic percent of solid Si in equilibrium with the liquid.

Solution Construct a tie line from the liquidus phase boundary (40 at % Si) to the solid 100% Si phase boundary at the 800°C temperature.

Total tie line length = 100 − 40 = 60 units

Since this temperature is closer to the liquidus line than the solidus (the horizontal line at 363°C), it suggests that the bulk of the material is still in liquid form. Thus,

$$\frac{100 - 60}{100 - 40} \times 100 = 66.7 \text{ at } \% \text{ at liquid}$$

or

$$\frac{60 - 40}{100 - 40} \times 100 = 33.3 \text{ at } \% \text{ Si in solid form}$$

The inverse lever rule is sometimes confusing because the length of the segment of the tie line that represents the solid is the segment that intersects the liquidus line. A more logical reasoning, and an easy way to recall which segment goes with which phase, is to follow the vertical composition line downward. Shortly after it intersects the liquidus line, there has not been much time for solid to form. Thus, the shortest segment must represent the quantity of solid. After cooling further we approach the solidus line. Now a lot of time has expired and a lot of solid formed. Therefore, the shortest segment must represent the quantity of liquid remaining.

Another way to remember the inverse relationship is to consider point X in Figure 6.3. The weight of the total alloy at this point is represented by W_{total} at point X, as we are in a 2 phase region.

$$W_{Total} = W_{Solid} + W_{Liquid}$$

If we use the coordinates on the compositional affair to calculate the Ni contact in each phase, we can rewrite

$$W_T \cdot \% \text{ Ni}_T = W_S \cdot \% \text{ Ni}_S + W_L \cdot \% \text{ Ni}_L$$

Substituting and rearranging

$$W_T - W_S = W_L$$

$$W_T \cdot \% \text{ Ni}_T = W_S \cdot \% \text{ Ni}_S + (W_T - W_S) \% \text{ Ni}_L$$

$$W_T (\% \text{ Ni}_T - \% \text{ Ni}_L) = W_S (\% \text{ Ni}_S - \% \text{ Ni}_L)$$

$$\frac{\% \text{Ni}_T - \% \text{Ni}_L}{\% \text{Ni}_S - \% \text{Ni}_L} = \frac{W_S}{W_T} = \text{Solid Fraction}$$

Using Figure 6.3 point X as an example,

$$\text{Solid Fraction} = \frac{XE}{FE}$$

6.3.2.2 Eutectic Systems with Some Solid Solubility

To understand these phase diagrams, we must first define the phrase *terminal solid solubility*. The word *terminal* refers to the end points on the abscissa (composition axis) of the phase diagram. To show all the terminal solid solubility, this solubility line must be extended to the eutectic temperature and then back up along the solidus line until it reaches the melting point of the pure metal of interest. The Ag–Cu phase diagram is a good example of this type of eutectic system (Figure 6.11). In this diagram the liquidus line runs from the melting point of each metal to the eutectic point (779°C and 28.5 wt % Cu). The solidus, indicated as a heavy line, shows that the solid solubility of Cu in Ag decreases with increasing temperature at temperatures above the eutectic temperature. Thus, the α region on the left end of the diagram consists of a solid solution of Cu in Ag, and the β region on the right end is a solid solution of Ag in Cu. A cooling curve of the eutectic composition (28.1 wt % Cu) would appear just as that in the simple eutectic case, and the eutectic microstructure would be the same, except that the layers now would consist of alternate layers of α and β.

Let us construct a cooling curve (Figure 6.12) for a 92.5 wt % Ag–7.5 wt % Cu. First the α solid solution would begin to precipitate out of the liquid solution phase as it crossed the liquidus at about 900°C. As we continue to cool through the α + liquid region, the liquid becomes richer in Cu, but we will never have a eutectic structure in this alloy because we pass into the all-solid α range prior to attaining the eutectic temperature. When we cross the α – α + β line on the phase diagram, some β will precipitate out of the α phase on very slow cooling.

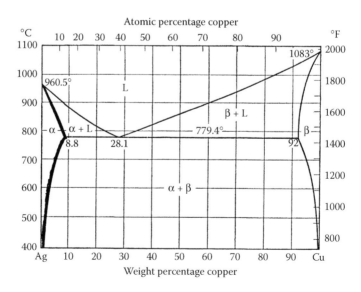

FIGURE 6.11 Ag–Cu phase diagram. (From C. S. Smith, *Metals Handbook*, ASM International, Metals Park, OH, 1948, p. 1148. With permission.)

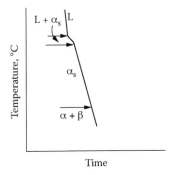

Time

FIGURE 6.12 Cooling curve for a Ag 92.5%–Cu 7.5% alloy (sterling silver). Note that there is no break in the curve at the solid $\alpha - \alpha + \beta$ phase boundary.

This change will never show on typical cooling curves. They are not suitable for detecting solid–solid transformations. This line can be determined by careful examination of the microstructure for the appearance of the second phase. The microstructure would appear as that in a pure metal or a single-phase solid solution (i.e., it would show only grain boundaries) plus some β. This alloy composition is called *sterling silver* and is used for silverware and jewelry. Ag alone is too soft for even these relatively mildly stressed applications.

However, if we cool an 80 wt % Ag–20 wt % Cu alloy from the liquid state, we do cross the eutectic temperature. The cooling and microstructure will be much like that in Figure 6.10, except that a solid solution of proeutectic α silver rather than pure Au is present, and the eutectic structure will consist of alternate layers of the α and β solid solution phases rather than pure metals.

Other alloy systems of this type include Pb–Sn, from which many commercial solders are derived: Bi–Sn, Pb–Sb, and Cd–Zn. The Al–Cu system has a eutectic point at 548°C and at 33.2 wt % Cu (Figure 6.13). The microstructure of a 70 wt % Al–30 wt % Cu showing some proeutectic α in a large amount of the layered eutectic structure is shown in Figure 6.14.

6.3.3 SOME SYSTEMS WITH INTERMEDIATE PHASES

Most alloy systems are not as simple as those portrayed in the preceding section. Many phase diagrams show intermetallic compounds or intermediate phases, which are usually positioned more or less in the central portion of the diagram. The intermetallic compound is one of fixed stoichiometry, such as $CuAl_2$ or Fe_3C, and will show on the phase diagram as a straight vertical line. An intermetallic compound could be considered as a type of intermediate phase, but this term is more often thought of as applying to a compound that has a range of compositions, usually fairly narrow, that is near stoichiometry (e.g., from $A_{0.9}B_{1.1}$ to $A_{1.1}B_{0.9}$). It could also be considered as a solid solution phase of a certain crystal structure that varies in the percent of each atom over the range in which the phase exists.

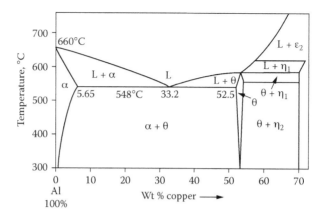

FIGURE 6.13 Al–Cu phase diagram to 72% Cu. (After K. R. Van Horn, Ed., *Aluminum*, Vol. 1, ASM International, Metals Park, OH, 1967, p. 372. With permission.)

FIGURE 6.14 Eutectic microstructure of an Al 70 wt %–Si 30 wt % alloy containing some proeutectic α aluminum (×400). (Courtesy of J. Haruff, Cal Poly.)

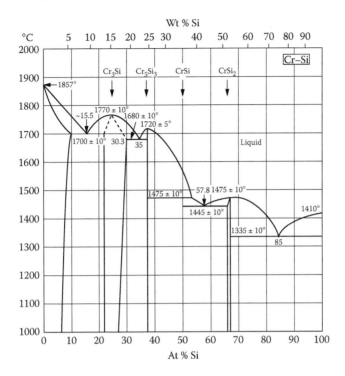

FIGURE 6.15 Cr–Si phase diagram. (From E. A. Brandes, Ed., *Smithell's Metals Reference Book*, Butterworth, Stoneham, MA, 1983, pp. 11–213; as taken from M. Hansen and K. Anderko. With permission.)

Another type of diagram is that where eutectics are formed between intermetallic compounds instead of elements. The Cr–Si phase diagram (Figure 6.15) is an example. Four eutectics and five intermetallic compounds are found in the alloy system. This diagram is of some interest to electronic device manufacturers. Chromium is frequently deposited on silicon wafers because it acts as an adhesive layer. For example, gold does not bond well to silicon; but if a gold layer is desired, often for corrosion protection or high conductivity, chromium is first deposited, which adheres well with silicon, and then a layer of gold is deposited, which adheres to the chromium. If the device happens to experience elevated temperatures, a brittle intermetallic phase could form.

6.4 PRECIPITATION-HARDENING SYSTEMS: ALUMINUM ALLOYS

This is the most important group of alloys in the nonferrous category. They could be classified under the preceding section, intermediate phases, because the precipitates are usually intermetallic compounds, but because of their prominence and distinct processing methods, they warrant separate treatment. They have gained prominence simply because the strength of the base metal, aluminum, for example, can be increased manyfold by small additions of a second element that causes a compound to be formed when the solid solubility limit for this element has been exceeded. Aluminum of commercial purity in the annealed state has a tensile strength of about 28 MPa (10 ksi) and in the cold-worked state on the order of 56 Mpa (20 ksi), whereas in precipitation-hardened aluminum alloys, tensile strengths on the order of 483 MPa (70 ksi) are common (see Table 6.1). This phenomenon was first recognized in aluminum alloys as early as World War I, when attention was focused on Duralium, an Alcoa heat-treatable alloy. The advent of the aircraft industry propelled aluminum to the number one position in the nonferrous alloy category, since strong, lightweight materials were needed to replace parts made of wood and fabrics used in the first aircraft that were developed. The Al–Cu alloys were the first precipitation-hardened alloys developed, have been the most studied, and are the best understood. We still therefore use this classic alloy system to explain the phenomenon of *precipitation hardening*, or as it was known in the 1930s and 1940s, *age hardening*.

Any phase diagram in which the solid solubility of a particular solute in a particular solvent increases with temperature is a potential precipitation-hardening alloy, but it requires more than just the precipitate formation to achieve significant hardening. The hardening comes about in the way the precipitate forms, and maximum hardening is achieved prior to the completion of the formation and appearance of the precipitate compound. Most precipitates are intermetallic compounds, such as $CuAl_2$ in the Al–Cu system. The complete phase diagram of the Al–Cu system can be found in the phase diagram books listed in the Suggested Reading. We are interested only in the aluminum-rich end of the phase diagram and accordingly show that part of the diagram ranging from 100 wt % Al to about 72 wt % Cu (28 wt % Al) in Figure 6.13. This diagram has a eutectic point at 66.8 wt % Al at 548°C, but this point has no bearing on our study of precipitation hardening in this system. The two significant points in the diagram are that of the maximum solid solubility of about 5.7 wt % Cu in Al, which occurs at the eutectic temperature, and the point of appearance of the θ phase, which has the composition of $CuAl_2$ and appears on the phase diagram at about 54 wt % Cu–46 wt % Al. Note that this compound has two atoms of aluminum for every three total atoms; thus, its composition in atomic percent is 66.7 at % Al. If we examine the diagram closely, we notice that the θ phase is not represented by a straight vertical line on the diagram. Therefore, it is an intermediate phase, but the region of the θ phase is very narrow in composition, so we will consider it as being the $CuAl_2$ stoichiometric compound.

One of the earliest developed Al–Cu alloys, and still probably the most widely used, is the 2024 alloy, which contains 4.4 wt % Cu, along with small amounts of Mg and Mn to lower the *solution treating temperature*. This temperature, 493°C, is used to form an all-α solution. When this alloy is quenched from the all-α region, the cooling time is too short to allow the formation of the equilibrium precipitate phase $CuAl_2$. Thus, the α phase containing 4.4 wt % Cu exists at room temperature. The equilibrium diagram shows that such a phase cannot exist at such a low temperature, but the equilibrium diagram deals with equilibrium conditions. Our quenched alloy is thus a supersaturated solid solution of Cu in Al. This is a nonequilibrium metastable condition. The solid solubility of Cu in Al at room temperature is nil. The Cu wants to precipitate out in the form of $CuAl_2$ as dictated by the equilibrium phase diagram. The manner in which this happens and how the properties are affected are the subject of precipitation hardening.

The aging process begins with the formation of clusters of Cu atoms in the aluminum matrix. The term *matrix* as used here means the major part and the continuous phase of the system. In the solid solution state, be it an equilibrium or supersaturated solid solution, the copper solute atoms are distributed more or less randomly on the lattice positions normally occupied by aluminum atoms. In the supersaturated state the compound $CuAl_2$, which contains 33.3 at % Cu atoms, wants to form in order to achieve equilibrium. Rounded off for simplicity, 4.4 wt % Cu translates to about 2 at % Cu. This means that for every 100 atom sites in the supersaturated state, two will be occupied by Cu atoms, but to form $CuAl_2$, we need 33 sites per 100 to be occupied by Cu atoms. Thus, the first stage of the aging process begins with the diffusion of Cu atoms to form regions containing approximately one-third Cu atoms. Since a diffusion process is involved, precipitation is time and temperature dependent. In the Al–Cu system sufficient diffusion can take place at room temperature to obtain significant clustering of atoms in a period of about 5 h. This is enough to cause an increase in hardness and strength. The clusters provide a greater barrier to the passage of dislocations through the matrix than do the same number of atoms distributed randomly, the latter being the normal solid solution hardening discussed earlier. The supersaturated state is stronger than that of pure aluminum, but the cluster state is stronger yet. Remember that anything that restricts dislocation motion strengthens the material. It has been found, by use of x-ray diffraction and electron microscope, that the Cu atoms tend to cluster on the planes of the FCC aluminum lattice. These are frequently called *GP1* zones, named after Guinier and Preston, who pioneered the concepts of the aging process in 1938.

The second stage of aging, often called the GP2 stage, involves the formation of an intermediate lattice (θ'' phase) that has a tetragonal structure. The θ'' phase is coherent (attached) to the planes of the matrix. Thus, it is not a true $CuAl_2$ compound, but the coherency strains between the matrix and the θ'' atom configuration provide tremendous resistance to dislocation motion. The coherency strains are related to the mismatch in atom spacing between the matrix and θ''. This stage coincides with maximum strengthening because of the large coherency strains. The θ'' phase particles are larger than the GP1 zones by about a factor of 10 in both thickness and diameter. A sketch of the distorted lattice surrounding and including the θ'' particle is depicted in Figure 6.16.

The next stage involves the formation of a θ' phase that is incoherent with the matrix; that is, it has broken away from the matrix and has a distinct tetragonal structure. The strength and hardness begin to decrease with the loss of the coherency strains.

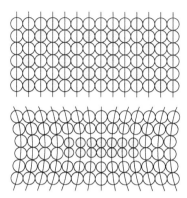

FIGURE 6.16 Schematic of coherency strains created from a supersaturated solid solution (shown above) during the aging of an Al–Cu alloy. (From R. E. Reed-Hill and R. Abbaschian, *Physical Metallurgy Principles*, 3rd ed., PWS-Kent, Boston, 1992, p. 269. With permission.)

The final stage involves the formation of the equilibrium $CuAl_2(\theta)$ phase, which is also tetragonal but of different atom spacing than that in the θ' phase. The strength decreases even more. As these particles grow in size, the alloy softens, even to the point of being in a softer condition than that of the supersaturated solid solution. This is termed the *overaging* stage. Under stress the soft aluminum flows around the large θ particles. A somewhat crude analogy, but one that may assist in understanding the aging process, is to consider marbles (as large $CuAl_2$ precipitate particles) embedded in putty (as the pure aluminum matrix). The putty under stress can flow around the marbles, which provide little resistance to plastic flow. If we take the same quantity of marbles and putty but grind the marbles until they consist of very small glass particles and then mix them with the putty, they provide considerable resistance to the plastic flow of the putty. The fine particles are analogous to the GP2 stage and θ' phase (probably more so like the latter, since there is no coherency between the glass particles and the putty). The point to be made here is that one cannot produce a strong Al–Cu alloy by slow cooling from the liquid or α phases, because such cooling allows large $CuAl_2$ particles to form. To obtain the fine distribution of smaller particles, the alloy must be quenched from the α-phase region and aged to a point that precedes the formation of large particles but instead permits the formation of a coherent intermediate precipitate with the corresponding coherency strains.

A schematic of the various stages of the aging process as revealed by hardness measurements as a function of aging time is shown in Figure 6.17. Not all of these stages will be obvious at all aging temperatures. At room temperatures of around 20°C, significant hardening occurs in about 1 week, but overaging has not been observed to occur at this temperature. If room temperature were that which we would experience in the desert (e.g., 50°C), overaging over a long period of time is possible. Hardness vs. aging time for higher temperatures is shown in Figure 6.18. At temperatures on the order of 250 to 300°C, the process

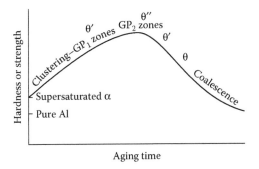

FIGURE 6.17 Schematic of the various stages of the precipitation process as revealed by plotting hardness vs. aging time.

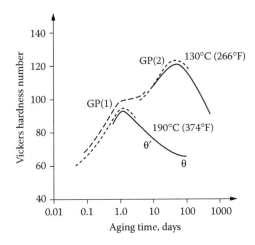

FIGURE 6.18 Hardness vs. aging time in an Al–4% Cu alloy. (From J. M. Silcock, T. J. Head, and H. K. Hardy, *Aluminum*, K. R. Van Horn, Ed., ASM International, Metals Park, OH, 1967, p. 123. With permission.)

occurs very rapidly; some stages may not be readily apparent. In essence the alloy passes through some stages too rapidly to permit detection. Also, the precipitation process is not uniform throughout the alloy. Because of faster diffusion and easier nucleation of new phases along the grain boundaries, precipitation often begins in these regions first.

Often the terms *natural* and *artificial* aging appear in the literature and textbooks, *natural* meaning aging at room temperature and *artificial* signifying aging above room temperature. These words should be avoided. First, room temperature can vary by at least 100°C (e.g., between that in the dead of winter in Alaska vs. the heat of summer in the Sahara Desert [or even in western U.S. deserts such as Arizona]). Second, there is nothing artificial about the aging or precipitation of atoms. It is a diffusion process that varies exponentially with temperature, all the way from 0 K to the solubility limit line.

The processes of precipitation hardening can present some interesting quality control and design issues for the engineer. The quality of the part as measured by mechanical properties must be developed through proper thermal treatments. If treated too long or at the incorrect temperature, the part can overage and when placed in service will change with time, with service temperature causing the part to exhibit declining properties. Conductivity measurements are often used to determine the state of aging. From a design perspective, the service temperature of the part made from a precipitation alloy must be maintained in a range in which the rate of further aging beyond the desired level properties will not overage the part in the designed service life.

Not all precipitation systems show all four stages distinctly. When the lattice parameters (interatomic distances) of the matrix and precipitate are vastly different, coherency strains may be very strong and short lived or absent altogether. Some strengthening occurs because of small precipitates, which restrict dislocation motion. This process is sometimes called *dispersion* hardening rather than precipitation hardening, which involves coherency strains. (Dispersion hardening is common in superalloys and occurs by an aging heat treatment that causes extremely small particles to precipitate.) Many aluminum alloys are strengthened by the precipitation-hardening process. These include the Al–Mg, Al–Zn, Al–Si, Al–Mg–Si, Al–Mn, Al–Li, and Al–Ni alloys. The primary requirement is that the solid solubility increase with temperature. Alloys of the Al–Zn system are the higher-strength aluminum alloys. The Al–Li alloys were recently introduced to reduce the weight of aircraft structures. Li weighs 0.534 g/cm^3 at 20°C compared to 8.96 and 7.14 for copper and zinc, respectively. It has been estimated that replacing the current Al–Cu and Al–Zn alloys in a Boeing 747 aircraft with Al–Li alloys would reduce the weight of the 747 by that equivalent to one large elephant. The chief benefit, however, would be in fuel savings per passenger mile. The strengths of various Al alloys aged to peak hardness are presented in Table 6.1.

6.4.1 OTHER NONFERROUS PRECIPITATION-HARDENING ALLOYS

6.4.1.1 Copper Base

One could search the phase diagrams and come up with a number of potential hardening systems (Problem 6.3). The Cu–Be alloys are the strongest of all Cu alloys. The phase diagram is shown in Figure 6.19. In these alloys the CuBe compound is the final precipitate. Commercial alloys such as C17100, C17200, and C17300 contain about 2 wt % Be and can be aged to strengths on the order of 966 MPa (140 ksi) compared to 345 MPa (50 ksi) for cold-worked copper (see Table 6.1, Figure 6.20). Other copper-based age-hardening systems include Cu–Ti, Cu–Si, and Cu–Cr.

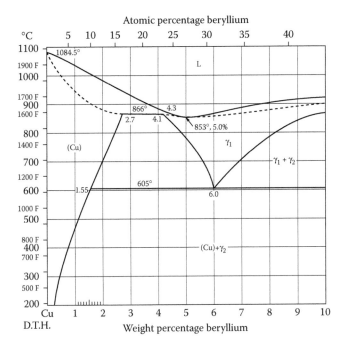

FIGURE 6.19 Cu–Be phase diagram. (From *Metals Handbook*, Vol. 8, 8th ed., ASM International, Metals Park, OH, 1973, p. 271. With permission.)

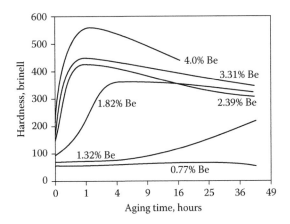

FIGURE 6.20 Precipitation hardening of beryllium.

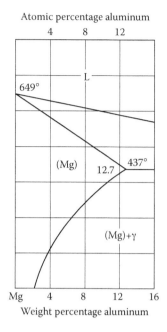

Atomic percentage aluminum

Weight percentage aluminum

FIGURE 6.21 Mg–Al phase diagram. (From *Metals Handbook*, Vol. 8, 8th ed., ASM International, Metals Park, OH, 1973, p. 261. With permission.)

6.4.1.2 Magnesium Base

Several magnesium alloys are precipitation hardenable; the most notable is Mg–Al. These alloys have strengths on the order of 345 MPa (50 ksi). The portion of the Mg-Al phase diagram that pertains to the precipitation-hardening process is shown in Figure 6.21. Mg–Zn alloys are also precipitation hardenable, and many alloys contain both Al and Zn. The American Society for Testing and Materials (ASTM) has a standardized numbering system for magnesium alloys, where the first two letters indicate the alloying elements of major significance, followed by two numbers containing the approximate percentages of these alloying elements. The next letter indicates the series designation for alloys of similar compositions. A letter and a number indicate the heat treatment. For example, AZ61A means the alloy contains about 6 wt % Al and 1 wt % Zn, and the letter A indicates that it was the first alloy of this series.

Mg has a density of 1.74 g/cm^3 compared to Al's density of 2.7 g/cm^3. It is thus lighter than Al, but because of lower strength and ductility, plus higher cost, has not really challenged Al for aircraft usage. Nevertheless, a limited number of aircraft parts are made from magnesium alloys. These alloys are also used in some automotive parts, such as brackets and supports. Small Mg alloy parts can be found in a variety of industrial machinery. Probably the best thing that magnesium has going is its abundance. Seawater contains large quantities of magnesium chloride and carnallite, a magnesium-containing sea deposit.

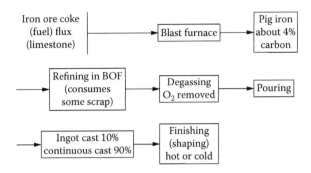

FIGURE 6.22 Steel production.

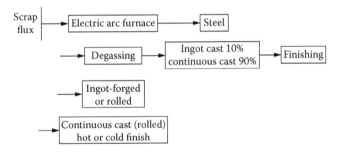

FIGURE 6.23 Mini mill and special steel making.

6.5 EUTECTOID SYSTEMS: IRON–CARBON ALLOYS

Iron and steel products comprise over 90% of the world's metallic materials production. The production for steel is shown in Figures 6.22 and 6.23.

Iron ore is processed at the mine site to produce a Fe_2O_3 ore concentrate. This concentrate is then combined with fuel (coke), limestone, and other fluxing agents in a blast furnace to produce a molten iron called *pig iron*. Pig iron contains about 4% carbon and other elements such as Si, Mn, P, and S. Some of the pig iron is poured into shapes called pigs and is sold directly to iron foundries as melting stock. Most pig iron is further processed in a basic oxygen furnace (Figure 6.24). In the basic oxygen furnace scrap steel is added to the molten bath and O_2 is blown in under high pressure to reduce the carbon content, producing *steel*.

In some regions of the world, steel is produced at mini mills that use recycled materials such as ground-up auto bodies for melting stock. Melting is done in an electric furnace, and the batch sizes are somewhat smaller than those from a blast furnace. These mills generally focus on regional product markets.

Once the carbon content of the steel has been reduced to the appropriate level, the molten steel must be converted into a useful shape. About 90% of the steel is converted to a primary shape using a continuous casting process. From

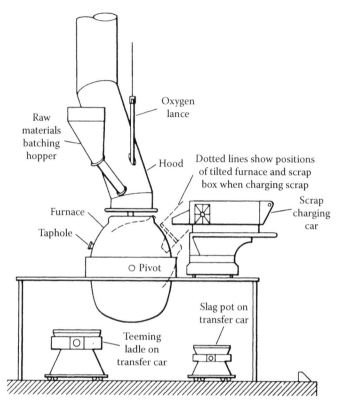

FIGURE 6.24 Basic oxygen process for converting pig iron to steel (From *Metals Handbook*, Desk ed., ASM International, Metals Park, OH, 1985, pp. 22–26. With permission.)

the primary as-cast shape, the steel is processed to its final salable configuration through a series of rolling processes. Steel can be rolled into shape and left to cool. This product is commonly known as *hot-rolled* steel and is characterized by a surface scale, wider property variation, and less accurate dimensions. Commonly this is steel that would receive further thermal and mechanical processing in the manufacturing of the final part. The steel shapes can receive additional processing by removal of the surface scale and final rolling at room temperature. This product is known as *cold-rolled* steel. This steel has a clean surface, less property variation, and better dimensional accuracy. It is commonly used in applications in which no further thermal processing is conducted in the manufacturing process.

Steel is purchased to a specification that is designated in the United States by a four- or five-digit code:

$$AA \times BB$$

where

AA = the alloy family (major added alloying element)
BB = the carbon content in tenths of a percent
X = usually B or Pb has been added or other special additions

Examples of common steels are

1010: A carbon steel with 0.1% carbon. In addition, Mn, S, and P are
present in some amount.
2320: A Ni bearing steel with 0.2% carbon. Ni content is nominally 3.5%.
4140: A Cr Mo steel with 0.4% carbon.

6.5.1 IRON–IRON CARBIDE PHASE DIAGRAM

In the iron–carbon phase diagram we use only that part of the diagram from 100
wt % Fe to 93.3 wt % Fe–6.7 wt % C. The compound Fe_3C contains 6.67 wt %
C (25 at % C), so it is more convenient to use the Fe–Fe_3C diagram than the
Fe–C diagram. The difference between the two is insignificant, as can be seen in
Figure 6.25 where both diagrams are displayed; the Fe–Fe_3C diagram is shown
by the solid lines and the Fe–C diagram by dashed lines. We usually round off
the 6.67 wt % C content of the Fe_3C compound to 6.7 wt % C for convenience
in our calculations. For all but the cast irons, we can use the abbreviated diagram
of Figure 6.26 for the heat treatment processes to be described here. Also note
that in this diagram the 0.77 wt % C point has been rounded to 0.8 wt % C, and
the 0.022 wt % C point to zero in order to facilitate some computations to be
illustrated later. To further simplify the discussion, we will concentrate on the
regions of the diagram below 1300°C, as we will concentrate on the property
management of steel and solid-state transformation processes centered about the
eutectoid reaction.

6.5.2 DEFINITIONS

In the Fe–Fe_3C diagram (Figure 6.25) it is important to gain an understanding of
the various symbols and their significance to the process of property management
in steels:

Ferrite: This phase is called the α phase since it is on the extreme left of
the diagram. It is an interstitial solid solution of carbon in body-centered
cubic (BCC) iron, where the carbon resides in the largest interstitial
sites. We will refer to this phase hereafter as α-iron in an attempt to
reduce the number of *ites* one must remember. The α phase is a relatively
weak (soft) iron since it contains a maximum of 0.022 wt % C only at
727°C (eutectoid temperature).
Austenite: This is the γ phase, which consists of an interstitial solid
solution of carbon in FCC iron, where the carbon atom resides in the

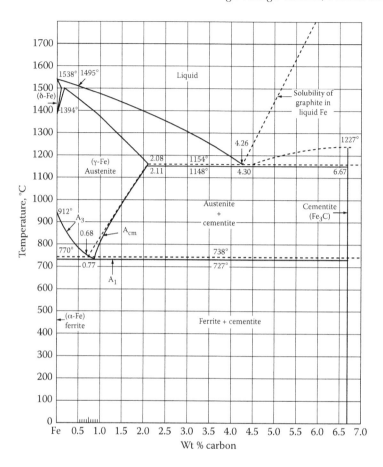

FIGURE 6.25 Fe–Fe$_3$C (solid lines) and Fe–C (dashed lines) phase diagram. (From G. Krauss, *Metals Handbook*, Desk ed., ASM International, Metals Park, OH, 1985, p. 28–32. With permission.)

    largest interstitial site. The FCC lattice has a larger interstitial site than does the BCC lattice and therefore can contain more carbon, up to 2.11 wt % C at 1148°C and up to 0.8 wt % C at 727°C. Under equilibrium conditions this phase cannot exist below 727°C.

Cementite: This phase is Fe$_3$C, the iron carbide compound, and will be so called hereafter. It has an orthorhombic crystal structure (which you should forget) and, like intermetallic compounds, is hard and brittle (which you should remember).

Martensite: This is a metastable phase that results from the rapid cooling (*quenching*) of a steel from the γ region to room temperature. It cannot be found on the equilibrium diagram. It has a tetragonal structure, with the carbon atoms residing in a body-centered-tetragonal crystal structure, in a similar position as the BCC unit cell. The body-centered-tetragonal

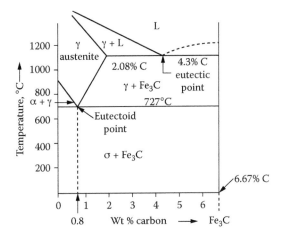

FIGURE 6.26 Abbreviated Fe–Fe$_3$C phase diagram to 6.7% carbon.

cell is merely an elongated BCC cell, and at low carbon contents the martensite is a BCC structure (i.e., the elongation is barely if at all detectable). Martensite is considered to be supersaturated with carbon because it would be more stable if the carbon atoms were not present. This metastable phase is very hard and brittle and most often is not present in a finished product. It plays the role of an intermediate step in the heat treatment process, where the objective is to obtain a more desirable microstructure, somewhat analogous to the supersaturated phase in the quenched Cu–Al alloy, although the latter is not hard and brittle. We do not have a Greek symbol for martensite, so we will use the letter *M*.

6.5.3 Eutectoid Steel

The *eutectoid point* on our abbreviated Fe–Fe$_3$C map in Figure 6.26 is the lowest temperature and composition at which the γ phase can exist. It is the point corresponding to a temperature of 727°C and a composition of 99.2 wt % Fe–0.8 wt % C. This is the *eutectoid point* of the Fe–iron carbide system. This point resembles the eutectic point described in Section 6.3.2. When a liquid of eutectic composition is slowly cooled through the eutectic point, it is transformed into two solid phases that exist in the parallel-plate formation known as the eutectic microstructure. At the eutectoid point, however, we have the eutectoid reaction,

$$\text{solid A} \rightarrow \text{solid B} + \text{solid C}$$

Students often confuse the terms *eutectic* and *eutectoid*. One memory gimmick that is worth a try is to spell the word *solid*. Notice that it contains the letters "oid," which you can associate with the eutectoid reaction, which deals with a solid-to-solid transformation, whereas the eutectic reaction begins with a

Austenite	→	Fe$_3$C	+	α
maximum		carbide		Ferrite
carbon		6.7% C		0.03% C
solubility 2%				

FIGURE 6.27 Equilibrium reaction when steel (austenite) cools below 727°C.

liquid. At the eutectoid point, the *eutectoid reaction* takes place on cooling a 0.8 wt % C alloy composition slowly through the eutectoid temperature. At this temperature the eutectoid reaction is (and see Figure 6.27)

$$\gamma \rightarrow \alpha + Fe_3C$$

This reaction is one of a solid transforming to two different solids and forms a parallel-plate microstructure (just like the eutectic) of the two phases called *pearlite* (Figure 6.28). This reaction takes place by nucleation and growth of the Fe$_3$C platelets, which usually nucleate at the γ grain boundaries (Figure 6.29) but sometimes grow out from previously nucleated colonies. The strength and hardness vary somewhat with cooling rate, the faster cooling rates (e.g., air cooling)

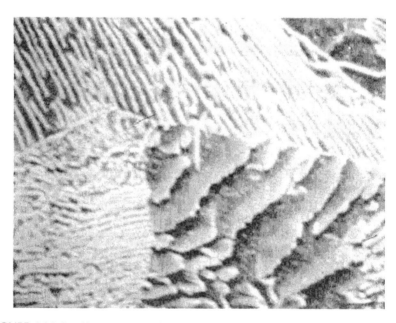

FIGURE 6.28 Pearlite composed of Fe$_3$C platelets (white) separated by ferrite platelets. Plates of Fe$_3$C in the lower right were cut almost parallel to the surface (×7000). (Courtesy of Rama Wallingford, Cal. Poly.)

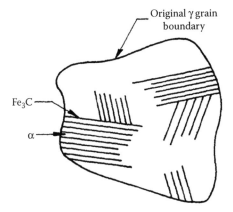

FIGURE 6.29 Schematic of pearlite formation from γ-iron (austenite). (Courtesy of Rama Wallingford, Cal. Poly.)

forming thinner plates, but more than are formed at a slower cooling rate. The finer pearlite is the stronger of the two, but in general the hardness of pearlite is in the vicinity of Rockwell C (HCR) 20 and has corresponding tensile strength of about 773 MPa (112 ksi). A plain carbon steel of eutectoid composition is a 1080 steel

Example 6.4   Compute the percentages on a weight basis of the α and Fe_3C phases in a 100% pearlite microstructure. (Refer to Figure 6.26.)

Solution   Using the 0.8% C vertical line and 20°C as our point of interest, construct a tie line from the left boundary of 0% C to the right boundary of 6.7% C. We can now use our inverse lever rule as follows:

$$\% \, \alpha \frac{6.7 - 0.8}{6.7 - 0} = \times 100 = 88\%$$

$$\% \, Fe_3C = \frac{0.8 - 0}{6.7 - 0} \times 100 = 12\%$$

Pearlite will always consist of 88% α–12% Fe_3C no matter how fine or coarse the platelets and no matter in which plain carbon steel it exists.

6.5.3.1 Hypoeutectoid Steels

Steels with carbon contents below 0.8 wt % C are known as hypoeutectoid steel (*hypo* meaning "less than"). Let us examine the slow cooling of a 1040 steel (i.e., one that has a 99.6% Fe–0.4% C composition) from the γ-phase region (refer to Figure 6.26). When the alloy has been cooled to the line that separates the γ from the α + γ region (approximately 800°C), the α phase begins to form. As cooling continues, one can construct tie lines from the 100% Fe boundary (since the maximum carbon content is only 0.022% in the α phase, we will assume that it is zero for computational purposes) to the boundary running from pure iron to the eutectoid point and thereby determine the composition of the remaining γ phase at each temperature and corresponding composition along this boundary. Of course, as low-carbon α-iron precipitates out of the γ phase, the γ contains more and more carbon until at 727°C and 0.8% C it has the eutectoid composition.

On further cooling, the remaining γ transforms to pearlite at the eutectoid point. The microstructure obtained on cooling at various points along the 0.4% C vertical line is sketched in Figure 6.30. This 1040 slowly cooled steel is somewhat softer than the 1080, since it contains some of the softer proeutectoid α phase. The phase that precipitates out prior to the eutectoid temperature is called a *proeutectoid* or primary phase, just as in the case of the eutectic system the first phase precipitating out of the liquid was called the proeutectic phase. A 1040 microstructure of the type sketched after the pearlite (eutectoid) formation will have an Rockwell C hardness less than 10 (at this HRC level it is advisable to use a different scale), or a DPH hardness number of about 190 and a tensile strength of around 518 MPa (75 ksi). These mechanical properties will vary with pearlite spacing and α grain size, but it gives one an idea of how the strength increases with the percent pearlite. Figure 6.31 shows the final 1040 slowly cooled microstructure.

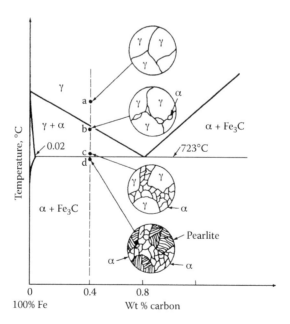

FIGURE 6.30 Transformation of a 1040 (0.4% C) hypoeutectoid steel on slow cooling. (From W. F. Smith, *Principles of Materials Science and Engineering*, McGraw-Hill, New York, 1986, p. 447. With permission.)

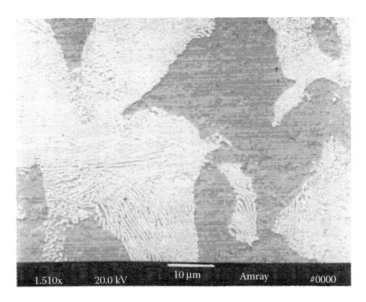

FIGURE 6.31 1040 steel, slowly cooled microstructure showing the proeutectoid α phase plus the eutectoid pearlite at room temperature. (Courtesy of W. Morris, Cal. Poly.)

Example 6.5 Compute the weight percent of the proeutectoid α phases and the weight percent of pearlite for a 1040 steel slowly cooled from the γ region. Also compute the total α.

Solution We must construct a tie line at a temperature just above the eutectoid temperature (e.g., 728°C). The percent γ at this point is obtained as follows:

$$\% \, \gamma = \frac{0.4 - 0}{0.8 - 0} \times 100 = 50\%$$

The proeutectoid α is the balance, or 50%. This can be checked by the inverse lever rule:

$$\% \text{ proeutectoid } \alpha = \frac{0.8 - 0.4}{0.8 - 0} \times 100 = 50\%$$

Assume 100 g of starting γ. At 728°C only 50 g of γ remains, which transforms to pearlite via the $\gamma \rightarrow \alpha + Fe_3C$ reaction. The pearlite contains 88% eutectoid α. Thus, $50 \times 0.88 = 44$ g of eutectoid α. Total $\alpha = 50$ g of proeutectoid $\alpha + 44$ g of eutectoid $\alpha = 94$ g. Thus, the total $\alpha = 94\%$. We can check this by constructing a tie line at 20°C, still using the vertical 0.4% C line as our composition of interest. Using the inverse lever rule, we have

$$\% \text{ total } \alpha = \frac{6.7 - 0.4}{6.7 - 0} \times 100 = 94\%$$

Below the eutectoid temperature only the α and Fe_3C phases exist, but the computation does not tell us how the α phase is distributed. To determine the distribution, we must use a tie line immediately above the eutectoid temperature as we did in the first part of the computation above to determine the % of primary phase.

6.5.3.2 Hypereutectoid Steels

Steels of carbon content greater than the 0.8% C eutectoid composition, but less than the 2.1% C content, where the cast irons begin, are termed *hypereutectoid* steels (*hyper* meaning "more than"). In Figure 6.32 we can follow the slow cooling of a 1.2% C steel and examine the resulting microstructures. In this case proeutectoid Fe_3C first precipitates out of the γ when the line proceeding from the eutectoid point upward to the right is crossed. Between this line and the horizontal eutectoid temperature line, the phases γ and Fe_3C exist. The proeutectoid Fe_3C atoms form at the γ grain boundaries. As the alloy is cooled through this two-phase region, the remaining γ is continually reduced in carbon content until it reaches the 0.8% C eutectoid composition, at which point it transforms to pearlite. This structure is difficult to deform plastically. The Fe_3C is almost as brittle as

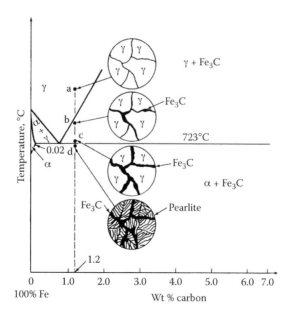

FIGURE 6.32 Transformation of a 1.2% C hypereutectoid steel on slow cooling. (From W. F. Smith, *Principles of Materials Science and Engineering,* McGraw-Hill, New York, 1986, p. 449.)

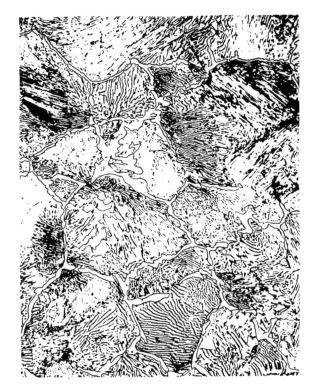

FIGURE 6.33 Microstructure of a slowly cooled high-carbon (1.5% C hypereutectoid) steel. Note the white proeutectoid Fe₃C surrounding the pearlite colonies (×1000). (Courtesy of W. F. Forgeng, Cal. Poly.)

glass and cracks under the slightest shock. If a piece of this alloy in this condition were dropped from a height of several feet onto a concrete floor a fracture would result. The final microstructure after slow cooling is shown in Figure 6.33.

Example 6.6 (a) Compute the percent proeutectoid Fe₃C that would result from slow-cooling a 1095 steel from the all-γ range. (b) Compute the total Fe₃C.

Solution
(a) Again using our inverse lever rule,

$$\% \text{ proeutectoid Fe}_3\text{C} = \frac{0.95 - 0.8}{6.7 - 0} \times 100 = 2.5\%$$

(b)

$$\% \text{ total Fe}_3\text{C} = \frac{0.95 - 0}{6.7 - 0} \times 100 = 14.2\%$$

The equilibrium structures and properties of the plain carbon steels have a variety of applications. Low-carbon 1010 steel sheet would be used where good formability is required (e.g., in forming automobile fenders). Sheet steel of even lower carbon (e.g., 1005 to 1007) is used in most auto bodies. The low-carbon steels can be cold-worked to increase the strength levels where both low carbon and higher strength are required. Cold-rolled sheet steel, which is subsequently tin plated and formed into cans, has a carbon content of about 0.08%. For years, 1080 steels have been used for massive parts such as railroad rails. The hyper-eutectoid plain carbon steels are used in the heat condition (quenched and tempered) for high-strength applications such as springs, cutting tools, and files.

6.6 PROPERTY MANAGEMENT OF STEEL (NONEQUILIBRIUM PROCESSING)

Common shapes (rounds, squares, billets, plate sheets, etc.) of steel are used as delivered from the mill or supply sources. These products have a microstructure that is the result of the cooling from the final processing step, and the properties of these steels are sufficient for many applications. Steel properties can be managed over a wide range for more specific applications by manipulating the time–temperature relationships. Depending on the type of steel, one can produce a file with a hard cutting edge or a common nail with high ductility. The key to managing the properties is the control of energy during the eutectoid transformation.

The phase diagram allows the prediction of what will happen to the steel at equilibrium conditions. In the manufacture of parts, equilibrium processing is rarely an option. The final microstructure is dependent on the rate at which the eutectoid process happened. Hence, a new type of map relating temperature (relative energy content) and time is needed to describe the transformation process. This is called an *isothermal transformation* (IT) diagram. This diagram depicts the eutectoid transformation with time.

$$\gamma \rightarrow \alpha + Fe_3C$$

In very simple terms the eutectoid reaction is a three-step process:

Break the bonds: When the austenite is taken below the eutectoid temperature, there is not enough energy to maintain the FCC crystal structure, and hence the Austenite becomes unstable. The iron atoms cannot remain in that crystalline format, so the existing bonds are broken and the atoms start to reorganize.

Move the atoms: The atoms move by diffusion, which is a process whose rate is exponentially dependent on temperature.

Create a new structure: Formation of the new structure, $\alpha + Fe_3C$, must be nucleated from the former FCC austenite and grow.

Usually the break-the-bonds step is not too difficult. The true property-controlling mechanism lies in the next two steps. Movements and nucleation are interdependent. If the temperature (relative energy) is high, the atoms can move about easily, but the process of nucleation is difficult. The number of atoms needed to form a critical nucleus must be relatively large to be stable in the high energy state. If the temperature is low, movement (transportation) is slow but nucleation is easier since the critical size is smaller at the lower energy levels.

6.6.1 ISOTHERMAL TRANSFORMATION (IT) DIAGRAMS

Let us first consider a 1080 IT diagram. Because 1080 is a eutectoid steel, there is no pretransformation of alpha or carbide upon cooling from the austenitizing temperature to below the eutectoid horizontal. If we take a number of small samples of this steel (small so they will react quickly to temperature change), austenitizing them (raising the temperature to above the eutectoid temperature 727°C), and hold them until they are uniformly heated. Next, let us quickly cool several samples to a temperature below the eutectoid horizontal and allow them to remain at that temperature (experimentally, this is done in a liquid salt or metal bath) for various times and then quench them in water. The water quench removes the energy necessary for diffusion, and the reaction quickly stops capturing the degree of transformation to that point.

At the temperature shown in sample A of Figure 6.34, transportation (diffusion) is fast, but nucleation is slow. Between A and A_s an unstable austenite exists for a while until the first nucleus is formed. Between A_s and A_f the transformation proceeds until at A_f all the unstable austenite is gone and only ferrite and carbide remain.

In sample B, between B and B_s the same unstable austenite exists, but note that the time to start the reaction to ferrite and carbide is quicker because at this

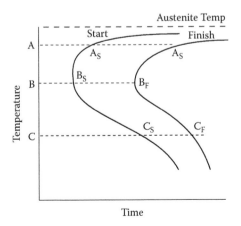

FIGURE 6.34 Schematic isothermal diagram.

level of energy the transportation is slower but the critical nucleus size is smaller. At this point we have the closest approach of the curve to the temperature axis.

At the temperature shown in sample C, the transportation mechanism is much slower than in A, and movement of the atoms to form a first nucleus takes considerably longer. The nucleus size is much smaller than in either A or B and consequently the structure formed is finer.

The microstructures formed during each of the above transformation paths represent significantly different properties but with the same materials: α and Fe_3C. Names have been given to these different patterns to identify them. The name pearlite is credited to H. C. Sorby, who invented the reflecting light microscope in 1870 and in 1886 published a paper describing a "pearl-like" constituent in railroad rail steel. Bainite was named for E. C. Bain for his work in the 1930s on the transformation studies in the range from 550 to 250°C. Spheroidite is the near equilibrium structure and names for its appearance. The descriptions of these materials are as follows:

> *Pearlite:* A microstructure consisting of parallel platelets of α and Fe_3C. This product results when the γ phase of 0.8% C is slowly cooled (e.g., in ambient air or in the furnace after the power has been switched off). It has a higher hardness than the α phase but much lower than the hard martensite phase.
>
> *Spheroidite:* A microstructure consisting of spherically shaped Fe_3C particles in an α matrix. This is a very soft product — even softer than pearlite of the same carbon content.
>
> *Bainite:* A microstructure consisting of small particles of Fe_3C in an α matrix. It is obtained by cooling the γ phase at moderately slow rates. This microstructure has very desirable mechanical properties. Its hardness is between that of pearlite and the hard but brittle martensite, yet it has good ductility and toughness.

The regions where these microstructures are formed are shown in Figure 6.35.

In each case these microstructural patterns are mixtures of two phases as determined by the energy available during the transformation. Another alternative is to remove the available energy by quenching. Quenching at a fast enough rate to completely suppress the normal diffusion transformation mechanism can create a phase called *martensite.*

When the austenite is quenched, the immediate instability occurs in the FCC structure when the eutectoid horizontal is crossed, but because of the rapid heat transfer rate experienced by quenching, no energy is available for diffusion. The structure is "quick-frozen," trapping the carbon atoms in interlattice positions and creating a great deal of internal strain. In order for this treatment to be effective, the cooling path must miss the nose of the IT curve. This cooling path is called the *critical cooling rate line* (Figure 6.35A).

$$\gamma \rightarrow \text{martensite}$$

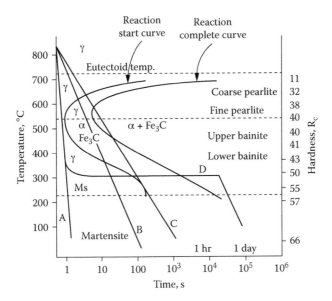

FIGURE 6.35 Isothermal transformation (IT) curves for a 1080 steel, listing the micro-structures obtained by a variety of cooling rates from the γ region of the phase diagram.

This reaction begins at the horizontal line marked Ms for martensite start, about 200°C for a 1080 steel, and is completed by the time it reaches room temperature. Martensite is a very hard (≈Rc 65) brittle phase with a distinct microstructure (Figure 6.36). At carbon contents less than 0.6%, it consists of domains of laths of slightly different orientations. At higher carbon contents it has more of a platelet appearance. The martensite will remain in this metastable state for infinite periods at room temperature. Martensite has been found in meteorites that are millions to billions of years old.

The maximum hardness of the martensite is determined by the carbon content as shown in Figure 6.37, which represents the maximum potential hardness for a given carbon content assuming that a cooling rate can be achieved that will completely miss the nose of the IT curve for that steel and complete the martensite reaction. This is not easy to accomplish. As we will see later, the physical size of the part, the cooling media, and the other elements added to the steel have a bearing on defining the critical cooling rate for a given steel.

In the example above, a 1080 plain carbon steel was used because it is of eutectoid composition, the transformation to α + Fe₃C is 100%, and the shape of the IT curve is not complicated. Grain size, carbon content, and alloying elements change the basic shape of the curve and its position on the time axis.

6.6.1.1 Grain Size

As previously discussed, grain size has a marked effect on the properties of a metal because of the grains' ability to disrupt dislocation movement. Grain

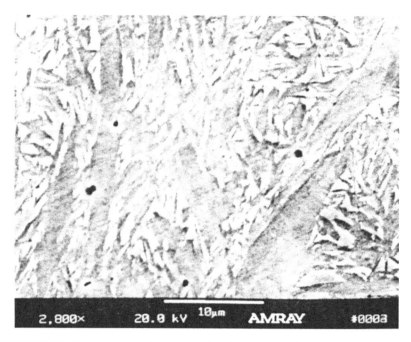

FIGURE 6.36 Microstructure of martensite in a background of untransformed γ.

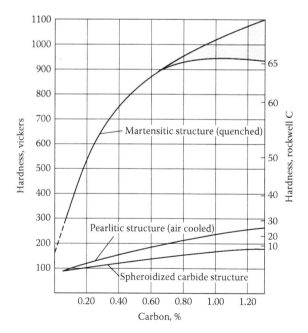

FIGURE 6.37 Maximum hardness vs. percent C. (E. C. Bain and H. W. Paxton, *Alloying Elements in Steel*, ASM International Metals Park, OH, 1966. p. 37. With permission.)

boundary areas are not as well organized as the bulk of the grain and consequently offer an easier path way for the diffusion of atoms. Since the eutectoid reaction is diffusion based, small grains (large grain boundary area) will promote faster diffusion and therefore make the transportation phase of the reaction faster. The net effect is to move the position of the nose of the curve to the left.

6.6.1.2 Carbon Content

The carbon content alters the shape of the curve by adding a ferrite or carbide pretransformation reaction on top of the curve. For example, a 1040 steel, when cooled from the austenite region, passes through a ferrite zone on the phase diagram. This preeutectoid transformation reduces the amount of the eutectoid reaction by producing a proeutectoid (primary) ferrite or, in a 1090 steel, a carbide microconstituent. Both of these prereactions happen before the eutectoid and effectively move the curve left or right on the time axis using the 1080 as benchmark:

0 to 8 % C: Curve moves right.
0.8 + % C: Curves moves left.

6.6.1.3 Alloying Elements

These have the greatest influence on the position and shape of the curve. Alloys are added to increase strength and impart special properties to the steel. Strength is gained by interfering with the movement of dislocations in the metal. This also interferes with the diffusion mechanism. Alloys restrict the ability of the atoms to move in the lattice and consequently increase the time required for the eutectoid reaction to initiate and complete. Also, adding alloying elements to the steel increases the number of elements in the phase diagram, adding dimensions to the phase diagram (ternary or more) that are reflected in the transformation to ferrite and carbide.

An example of this is shown in Figure 6.38a,b. In the 4150 IT diagram (a) one sees the progression of the eutectoid reaction. Small samples are lowered to the temperature of transformation and quenched in water after varying times. In the commercially available diagram (b), note the time frame, the reaction to pearlite does not start for almost 40 seconds, and the bainite nose (lower curve) does not start for 9 seconds. If this is compared to the 1080 IT diagram Figure 6.35, it can be easily seen that the cooling rate necessary to miss the nose of the curve for the alloyed steel is slower. This slower cooling rate would allow larger parts to be through-hardened, or a lesser cooling media (oil) might be used, reducing the possible distortion in the part as a result of the martensite reaction.

6.6.1.4 Continuous Cooling Curves

In normal industrial practice, except for special circumstances, steel is never transformed isothermally, yet the curves in Figure 6.35 were obtained via the isothermal cooling of samples. Continuous cooling curves have been developed

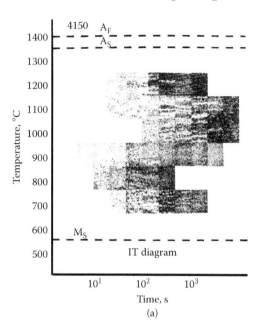

(a)

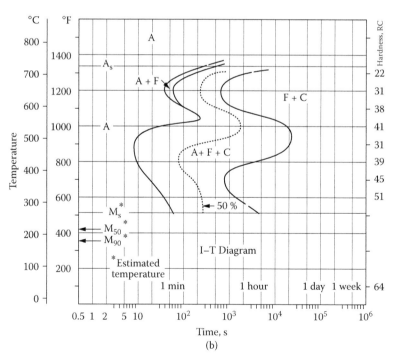

(b)

FIGURE 6.38 IT diagram of a 4150 steel as derived from the microstructures obtained after quenching from successive increases in holding times at constant temperatures. (Courtesy of Robin Churchill, Cal. Poly.)

for many of the more common alloys. In Figure 6.39 a continuous cooling curve for a eutectoid 1080 steel is superimposed on the IT curves discussed above. In the continuous cooling diagram the start and finish lines are shifted to slightly longer times and slightly lower temperatures. The dashed lines represent various cooling rates, and the microstructures resulting from these cooling rates are listed at the end of each dashed line. IT curves for most commercial steels are available in handbooks, many published by the steel manufacturers. Continuous curves are often difficult to obtain commercially.

6.6.2 HARDENABILITY

In common industrial practice, steel properties are managed by the way the steel is cooled from the austenite region to room temperature. Hardness is controlled by the amount of carbon in the steel and the ability to maximize the hardness is controlled by the alloy content of the steel. The term *hardenability* is used to refer to this sensitivity of the steel to the rate of cooling to room temperature.

Low alloy steels have a better hardenability than do plain carbon steels. Whereas the plain carbon steel (example 1010) can be hardened to a depth approaching 1/16 in, the low-alloy steel (alloy additions of 0.5 to 3% of elements such as Mn, Cr, and Mo one can obtain martensite at a slower cooling rate, even as slow as air-cooling in some steels, and can be hardened to depths of 1 in. Some alloys and their compositions are listed in Table 6.2.

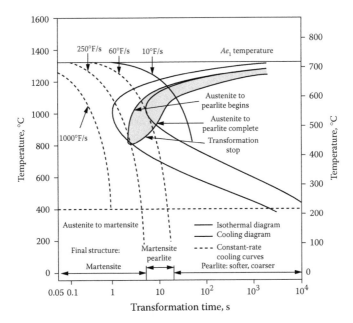

FIGURE 6.39 Continuous cooling diagram for a 1080 steel. (After R. A. Grange and J. M. Kiefer as adapted in E. C. Bain and H. W. Paxton, *Alloying Elements in Steel*, 2nd ed., ASM International, Metals Park, OH, 1966, p. 254. With permission.)

TABLE 6.2
Composition of Selected Alloy Steels

Steel	C	Mn	Ni	Cr	Mo	Other
Low-Carbon Quenched and Tempered Steels						
A514	0.15–0.21	0.80–1.10	—	0.50–0.80	0.18–0.28	—
HY 80	0.12–0.18	0.10–0.40	2.00–3.25	1.00–1.80	0.20–0.60	0.25 Cu
Medium-Carbon Ultrahigh-Strength Steels						
4130	0.20–0.33	0.40–0.60	—	0.8–1.10	0.15–0.25	—
4340	0.38–0.43	0.60–0.80	1.65–2.00	0.70–0.90	—	—
Carburizing Bearing Steels						
3310	0.08–0.13	0.45–0.60	3.25–3.75	1.40–1.75	—	—
5120	0.17–0.22	0.70–0.90	0.55	0.70–0.90	—	—
8620	0.20	0.80	—	0.50	0.20	
Heat-Resistant Chromium-Molybdenum Steels						
UNSK12122	0.10–0.20	0.30–0.80	—	0.50–0.80	0.45–0.54	—
UNSK31545	0.15	0.30–0.060	—	2.65–3.35	0.80–1.06	—

6.7 HARDENABILITY MEASUREMENTS: THE JOMINY END-QUENCH TEST

One could obtain a good idea of the hardenability by quenching a round bar or plate, then cutting a cross-section sample some distance away from the ends of the piece and making hardness measurements across the diameter or plate thickness. Such hardness traverses that might be obtained on steels of both low and high hardenability are depicted schematically in Figure 6.40. The best hardenability was obtained in the alloy containing 1.0% Cr plus 3% Ni. Good-hardenability steel shows a nearly constant hardness traverse, decreasing only slightly in the center. It is nearly all martensite throughout its cross section. Plain carbon steel contains a martensitic shell with pearlite in the remaining cross section. The latter microstructure is, of course, a result of the decrease in cooling rate as we proceed inward toward the center of the bar. In low-hardenability steel, pearlite forms at rather shallow depths and throughout the central sections — thus the large dip observed in the hardness traverse.

W. E. Jominy and A. L. Boegehold in 1939 proposed a convenient test for hardenability, now known as the *Jominy end-quench test*, in which a bar is cooled by water quenching on one end only (Figure 6.41). Thus, there exists a spectrum of cooling rates along the bar. This can be visually observed by the change in color along the length of the bar as it is being cooled. The quenched end experiences a very high cooling rate, while at the opposite end the cooling rate approaches that of air cooling (i.e., the water has little or no effect on the extraction of heat on this end). Figure 6.42 shows the cooling rates along the specimen.

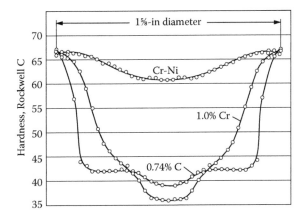

FIGURE 6.40 Hardness traverses across cross sections of low- and high-hardenability steels. (From E. C., Bain and H. W. Paxton, *Alloying Elements in Steel*, 2nd ed., ASM International, Metals Park, OH, 1966, p. 156. With permission.)

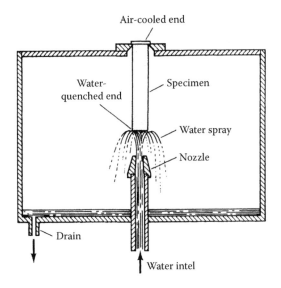

FIGURE 6.41 Jominy end-quench test. (From *Metals Handbook*, Vol. 1, 10th ed., ASM International, Metals Park, OH, 1990, p. 466. With permission.)

Specimen dimensions and other details are given in the ASTM A255 specification. The cooling rates along the Jominy bar represent different paths through the IT or continuous cooling (CC) diagrams. With each path comes a different set of microstructures (and therefore properties) depending on the steel type. If we combine the cooling rate data from Figure 6.42 and the cooling path illustration from Figure 6.39, we get the combined diagram shown in Figure 6.43. The cooling rates of the positions on the Jominy are independent of the steel type. The results

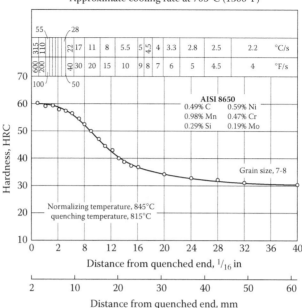

FIGURE 6.42 Cooling rates along the length of a Jominy end-quench specimen. (From *Metals Handbook*, Vol. 1, 10th ed., ASM International, Metals Park, OH, 1990, p. 466. With permission.)

of that cooling rate when imposed on the IT diagram give rise to different properties for a specific steel. Hardenability curves for 1050 and 4150 steels are shown in Figure 6.44. Some steels have the letter H at the end of the numerals (e.g., 4340H). The H specifies that the hardness values must fall within certain maximum and minimum values along the hardenability curve. The ASM *Metals Handbook*, Vol. 1, 10th ed., shows these H-bands' hardenability curves (pp. 485–570) for 85 carbon and low-alloy steels.

6.7.1 COMMERCIAL HEAT TREATMENTS

In industry common heat treatments are given to steel parts to create the desired end-use properties. The details of treatment for a specific steel and part geometry vary considerably, but the types of treatment fall into some general categories.

6.7.7.1 Annealing

This term is used to describe a number of processes generally used to soften the steel:

> *Full anneal:* This is an austenitizing treatment followed by slow cooling, usually in a furnace or other controlled environment. The purpose of

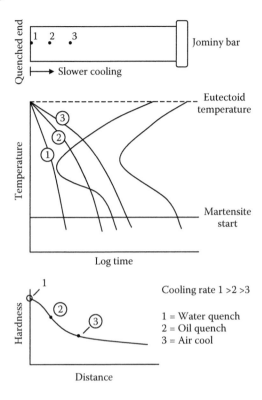

FIGURE 6.43 Combined Jominy and cooling path illustration.

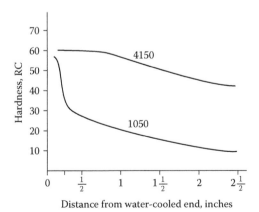

FIGURE 6.44 Hardenability curve comparisons for 1050 and 4150 steels. (Adapted from E. C. Bain and H. W. Paxton, *Alloying Elements in Steel*, ASM International, Metals Park, OH, 1966, p. 136. With permission.)

this treatment is to create the lowest hardness and strength for a particular steel in preparation for a manufacturing process such as machining or mechanical forming. For example, differential ring gear forgings are commonly annealed after forging to create a structure suitable for cutting gear teeth.

Spherodizing anneal: This is similar to the full anneal, but in this process a specific microstructure *spheroidite* (Figure 6.45) is called out as an end result. In this process the parts are held just below the eutectoid (about 700°C) for varying lengths of time as is needed to change the structure. Again, this process is intended to create a low-hardness, low-strength product but is used for tool steels and alloy steels where slow cooling in a furnace is not sufficient to create the desired softness level for the next manufacturing process. These parts are subsequently re-treated to develop the structures necessary to satisfy the service conditions of the part.

Process anneal and recrystallization anneal: As the name suggests, this is done during the processing of a part and is intended to remove the effects of a cold working process. The part or material is held at a temperature above the recrystallization temperature for an appropriate length of time for the size of the part.

Stress relief anneal: The parts are held at a temperature below the eutectoid for a short period of time to allow for diffusion of atoms and the removal of residual stresses resulting from fabrication processes such as welding. Generally this process is carried out to stabilize the parts or structure prior to machining.

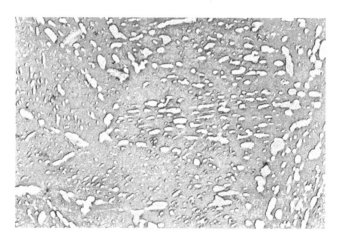

FIGURE 6.45 Spheroidite microstructure in an 8620 steel obtained by heating for 20 h at 600°C (×2500). (Courtesy of Allen Chien, Cal. Poly.)

6.7.7.2 Normalizing

The parts are austenitized and air cooled to room temperature. The intention is to create a uniformly distributed pearlitic structure throughout the part. Depending on the steel type, this can be a hardening or softening process. For low-carbon steel it is generally a softening or no-change process, but for higher-alloy parts it can be a process that can increase the hardness of the part. Caution must be exercised in air cooling to guarantee part uniformity during summer and winter. Often still air at 22°C is recognized as the standard air cool. Normalizing is often carried out on cast parts that have experienced uncontrolled cooling after shake-out (see Figure 6.35, Path C).

6.7.7.3 Quenching

The intent of quenching is to remove heat from the part and create a uniform martensitic structure. Depending on the steel, this can be accomplished using various media. A water quench has a quench severity of 1, with salt brine solutions 1.2 more severe (faster heat transfer) and oil about 0.3 less severe. A part is transferred to the quench tank as quickly as possible and submerged in the circulating media until it reaches the desired temperature (Figure 6.35, Path A [water], Path B [oil].

6.7.7.4 Tempering

The temperature of the parts is raised to a temperature below the eutectoid, and after a period of time (depending on the size of the part) the parts are cooled to room temperature. The intention of this process is to achieve a controlled softening of the parts (Figure 6.46). In the quenching process a uniformly hard martensite is formed. At the as-quenched hardness level for a given steel, maximum strength and wear resistance are achieved, but in higher carbon steels this often results in a hard brittle material. For example, steel used for a hammer needs to be strong and have the ability to resist abrasion, but if the as-quenched material is used, the potential for shattering when a hard object is struck is high. A compromise is reached in this product by tempering the hammer to a slightly lower hardness, which imparts a degree of toughness to the steel and makes it a safer tool to use.

Carbon atoms are trapped in the interstitial positions in the BCC lattice. In the tempering process energy is added through heating the part, allowing diffusion to take place and the eutectoid reaction to continue. Carbon atoms are allowed to move from the interstitial positions to form Fe_3C and ferrite. This structure is called *tempered martensite* and is a mixture of two phases.

6.7.7.5 Austempering

This process is used to create a bainite structure by quenching the parts to just above the martensite start temperature and isothermally transforming them to form bainite. Bainite (Figure 6.47) is a very desirable structure since it has a high hardness

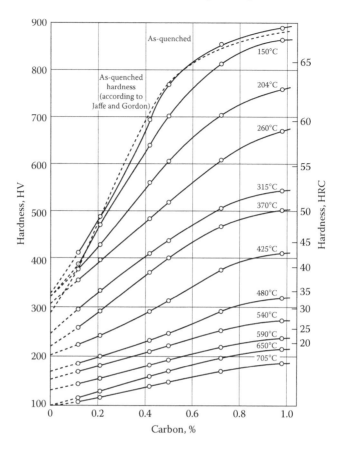

FIGURE 6.46 Tempered hardness of plain carbon steels vs. carbon content and tempering temperature. (From R. A. Grange, C. R. Hibral, and L. F. Porter, *Met. Trans. A.*, p. 1775, 1977. With permission.)

and high toughness. Commercial applications of this treatment are limited because it requires precise control of the quenching and holding process. This process has been used successfully in creating austempered ductile iron, a commercial cast product offering exceptional wear resistance (Figure 6.35, Path D).

6.7.7.6 Marquenghing

This is a process in which the parts are cooled to just above the martensite start temperature and held to a time point just before the bainite start point and quenched to form martensite. This process is used to minimize the temperature differential between the center and surface of a part, thereby reducing the potential for cracking and distortion during the martensite formation. This process is not commonly used because of the need for precise control of the quenching and holding times.

FIGURE 6.47 Bainite microstructure — A151 8620 steel austenitized at 1575°F for 30 minutes and quenched at 600°F and held for 240 minutes. (Courtesy of The American Iron and Steel Institute.)

DEFINITIONS

Age hardening: Heat-treating process that increases hardness via precipitation of some type of phase, often a compound.

Austempering: Isothermal heat treatment of steel below the nose of an IT curve but above the Ms (Martensite start) temperature that results in an all-bainitic structure.

Bainite: Microstructure that consist of a mixture of α-iron with small Fe_3C particles that has a hardness in the Rc range 40 to 55.

Binary system or Binary phase diagram: Alloy or phase diagram in which more than 80% of the composition is made up of two elements.

Cementite: Fe_3C (iron carbide).

Dendrites: Grain structure in alloys where growth is favored in certain directions, giving a Christmas tree–like appearance. Snowflakes also have a dendritic appearance.

Equilibrium: That condition whereby phases can exist in the same system with zero net change among the phases.

Equilibrium diagram (also phase diagram): Map that plots the conditions under which phases can exist in equilibrium, the conditions being temperature, composition, and pressure.

Eutectic microstructure: Microstructure consisting of parallel platelets of the α and β phases that are formed when the eutectic structure solidifies.

Eutectic point: Lowest melting point and composition in a binary alloy.

Eutectic reaction: Liquid → A_s + B_s (s = solid).

Eutectoid composition: Point and a temperature on a phase diagram where one solid decomposes to form two different solid phases.

Ferrite: α phase on the Fe–C diagram, which is a solid solution of carbon in iron (BCC).

Gamma (γ): In steel, the solid solution phase, consisting of carbon in FCC iron.

Gamma prime (γ'): Phase, usually an intermetallic compound, that exists in superalloys and provides high-temperature strength.

Hardenability: Capability to harden steel. The depth to which a steel bar can be hardened.

Isothermal transformation (IT) diagram: Diagram that shows the phases of steel (also used in other alloys) that exist during cooling as a function of time and temperature.

Jominy end-quench test: Method of measuring hardenability whereby a specimen is quenched from one end only.

Liquidus: Line or lines on a phase diagram above which the system is completely liquid.

Martensite: Metastable phase formed by quenching.

Matrix: In an alloy, the major and continuous phase.

Metastable: Condition whereby a phase exists even though thermodynamic equilibrium does not exist.

Microalloyed steels: Type of HSLA steel that results from a small dispersion of carbides combined with a fine grain size.

Monel: International Nickel Corp. trademark for a series of Cu–Ni alloys.

Overaging: Aging beyond the point of maximum strength in the precipitation-hardening process. The loss of coherency strains and coalescence of the precipitate occurs.

Pearlite: Eutectoid microstructure in Fe–C alloys that is formed from alternate platelets of α-iron (ferrite) and Fe_3C (cementite).

Phase diagram: See Equilibrium diagram.

Pig iron: Product that results from the reduction of iron oxide to iron in a blast furnace.

Precipitation hardening: See Age hardening.

Proeutectic phase: Solid phase that forms in solidification of a liquid solution of two elements that form a eutectic system. It appears at temperatures below the liquidus and above the eutectic temperature.

Proeutectoid phase: Solid phase that precipitates out of a single-phase solid solution as the solid solution phase enters a two-phase solid range (on the phase diagram) above the eutectoid temperature. Occurs only in eutectoid alloy systems.

Segregation: Nonequilibrium separation of elements that occurs during alloy solidification.

Solidus: Line or lines on a phase diagram below which all phases are in the solid state.

Spheroidite: Small Fe_3C particles in ferrite that result from a spheroidizing anneal for long periods at temperatures just below the eutectoid temperature.

Supersaturated solid solution: Solid solution that has solute in excess of that dictated by the phase diagram. It is a metastable phase usually formed by quenching from a temperature where the solute is not in excess of the solid solubility limit.

Tempering: Process of annealing martensite whereby it decomposes into α + Fe_3C.

Terminal solid solubility: Solid solubility permitted by the equilibrium diagram near the pure base metal position on the diagram.

QUESTIONS AND PROBLEMS

6.1   In Example 6.3 the liquid quantity in a composition of 60 at % Si–40 at % Au at 800°C was 66.7% of the total on an atom percent basis. Convert this to weight percent silicon precisely by interpolating from the weight percent figures given on the diagram.

6.2   For the composition in Problem 6.1, sketch the resulting microstructure if this alloy were slowly cooled to room temperature.

6.3   List three nonferrous and one ferrous precipitation-hardening alloy along with their strengths and ductilities in the aged condition (internet search).

6.4   Why must elements that have complete solid solubility in the solid state possess the same crystal structure?

6.5   For a 70% Cu–30% Ni alloy at equilibrium at a temperature of 1175°C, compute the amount of the solid phase. What is the composition for all phases within this alloy system at this equilibrium temperature?

6.6   If the alloy in Problem 6.6 became a solid by being poured into a mold, sketch or describe how the composition gradient might exist across the ingot. Sketch the grain structure that would exist in this casting if it solidified slowly but yet too fast for equilibrium to be attained, and show which regions have a nickel content greater than 30% Ni.

6.7   Assume that a gold foil was placed on a silicon wafer (assume the foil is one-tenth the weight of the silicon) and heated to 365°C for a period of several hours and then suddenly cooled. Sketch the microstructure and identify the phases that would be present in the solidified compact.

6.8   For a 60 wt % Cu–40 wt % Ag alloy that was melted and cooled to 770°C, where equilibrium was established, compute the quantity of proeutectic α, eutectic α, and total α in the resulting microstructure. Sketch this microstructure.

6.9 Estimate the hardness that one would obtain on a 2024 Al alloy (a) quenched from the all-α region, (b) quenched and aged at 190°C for 24 h, and (c) furnace cooled from the all-α region.

6.10 Consider pure copper and a Cu–2 wt % Be alloy. Which has the higher strength and which has the higher electrical conductivity? Which would be the better material for a power line and which would be more suitable for electrical contacts where impact is involved?

6.11 For a 1020 steel cooled to room temperature under near-equilibrium conditions, compute the total of (a) the proeutectoid α phase and (b) the total Fe_3C phase. Sketch the resulting microstructure and estimate the alloy's strength.

6.12 Describe two methods of obtaining a very desirable microstructure of around 40 to 45 RC hardness in a 1080 steel.

6.13 What would be the benefits of adding small amounts (e.g., 0.5 wt %) of Cr and Mo to a 1080 steel? What would be the difference in the IT curves?

6.14 What heat treatment would you use on a 10100 steel slowly solidified alloy to make it useful? What would the strength be after your treatment?

6.15 Are there any advantages to oil-quenching a 1080 steel vs. water-quenching? List the phases and the microstructures that would result in each case.

SUGGESTED READING

American Society for Metals, *Metals Handbook*, 9th ed., Vol. 6, *Welding, Brazing and Soldering*, ASM International, Metals Park, OH, 1983.

American Society for Metals, *Metals Handbook*, Desk ed., ASM International, Metals Park, OH, 1985.

American Society for Metals, *Metals Handbook*, 10th ed., Vols. 1 and 2, ASM International, Metals Park, OH, 1990.

Brandes, Eric A., Ed., *Smithells' Metals Reference Book*, 6th ed., Butterworth, Stoneham, MA, 1983.

Dieter, G. E., *Mechanical Metallurgy*, 3rd ed., McGraw-Hill, New York, 1983.

Guy, A. G., *Elements of Physical Metallurgy*, 2nd ed., Addison-Wesley, Reading, MA, 1959.

Massalski, T. B., Ed., *Binary Alloy Phase Diagrams*, ASM International, Metals Park, OH, 1986.

Smith, W. F., *Principles of Materials Science and Engineering*, McGraw-Hill, New York, 1990.

7 Ceramics

7.1 INTRODUCTION

In this chapter an overview of properties of ceramics and their applications is provided. It is more or less a summary of books that gives a more comprehensive picture of ceramics, such as *Modern Ceramic Engineering,* by David Richardson. A majority of the pictures and explanations are taken from *Breviary Technical Ceramics,* a recommendable paperback about ceramic materials from the Association of the German Ceramics Industry (see http://www.keramverband.de). Refer also to the American Ceramic Society (http://www.ceramics.org) for much more specific information including a list of manufacturers.

The word "ceramics" originated from the Greek word *keramos*, which translates to "potter's clay." Ceramics have a long history. Many centuries ago, dating back to about 6000 B.C., clay products, made by burning naturally occurring clays, were probably the first manufactured ceramics. The silicate glasses date back to a few thousand years B.C. Many centuries later, porcelain ceramic products were introduced that were made by mixing 50 to 60% of the mullite-type clays ($3Al_2O_3$ $2SiO_2$) with flint (ground SiO_2) and feldspar, an anhydrous alumina silicate containing K^+, Na^+, or Ca^{2+} ions. When fired, this mixture resulted in crystalline phases bonded by a complex glassy phase. Those ceramics are traditional ceramics that are still known today from many applications such as tableware, decorative ceramics, vitreous white ware, and wall and floor tiles.

During the past 40 years or so, newer, more scientifically developed ceramics have emerged that have been labeled with a number of different titles. We may call those ceramics high-performance ceramics, structural or engineering ceramics, or advanced ceramics. Advanced ceramic materials have many advantageous and often unique properties that make them suitable for many applications. We find a great variety of ceramics based on Al_2O_3, ZrO_2, Si_3N_4, and SiC, just to name some basic types of oxides, carbides, and nitrides.

Ceramic materials are usually electrically and thermally insulating, have high hardness, and may have very low thermal expansion coefficients. Their shape is extremely stable because of the absence of capacity of plastic deformation and high Young's modulus. Compression strengths 10 times greater than the bending or tensile strengths can be achieved. In comparison with metals, ceramics are particularly suitable for applications at high temperatures, since the characteristics of ceramic materials are far less strongly influenced by temperature than metals. Ceramics are corrosion and wear resistant.

Because of these advantages, we find ceramics wherever we go. Without ceramic insulators, many household devices would not function, and without safety devices made of ceramic, a reliable electricity supply would be unthinkable.

Ceramic substrates and parts are the basis for components in all areas of electronics. Owing to high wear resistance and low corrosion, ceramics are used for sliding and bearing elements in machine and plant construction. Ceramics are needed for industrial furnaces and high-temperature technology.

An important point always to bear in mind when applying ceramics is this: *ceramics are brittle.* Metals are able to forgive small errors of construction such as incorrect tolerances, because they are able to disperse local stress peaks through plastic deformation. Ceramics cannot do this because they have no ability of plastic deformation. This leads, for example, to a different approach of strength. General processing difficulties also arise because of their brittle nature. We will come back to those issues later.

7.2 CERAMIC MATERIALS

7.2.1 DEFINITIONS

Ceramic materials are inorganic and nonmetallic. They are generally molded from a mass of raw materials at room temperature and gain their typical physical properties through a high-temperature sintering process. We also regard glass, enamel, glass–ceramic, and inorganic cementitious materials (cement, plaster, and lime) as ceramic materials.

Except for glass, ceramics have a crystalline structure; the crystal structures are more complex than metals. Another difference from metals is the bonding behavior. Oxides show mainly ionic bonding, and carbides and nitrides are mainly covalently bonded.

Glasses are based on SiO_2, with various further additions, mainly Na_2O for various reasons.

Technical ceramics are ceramic products for engineering applications (e.g., turbo charger rotors) and for nonengineering applications (e.g., clay products for tableware).

High-performance ceramic might be a better term for ceramics for engineering applications. High-performance ceramic is defined as a highly developed, high-strength ceramic material that is primarily nonmetallic and inorganic and possesses specific functional attributes. We can subdivide these ceramics according to their properties and applications into the following groups:

- *Structural or construction ceramics*, sometimes also called industrial or engineering ceramics, are materials that must withstand, for example, mechanical stresses, bending, or pressure.
- *Functional ceramics* are ceramic parts that posses specific electric, magnetic, dielectric, or optical properties. An example is a capacitor dielectric made from barium titanate that shows extremely high permittivity.

- *Electrical ceramics* are used for specific electric and electronic applications owing to, for example, their excellent insulating properties in combination with good mechanical strength. The electronics industry also takes advantage of characteristics such as ferroelectric behavior, semiconductivity, nonlinear resistance, ionic conduction, and superconductivity. Examples are insulators made from Al_2O_3 and an oxygen sensor made of ionic conducting ZrO_2.
- *Cutting tool ceramics* are applied in machining processes (lathe bits, drilling, milling) because of their excellent wear and high temperature resistance. Si_3N_4 is a ceramic used for high-speed cutting of break wheels made form Al alloys.
- *Medical ceramics* are used for applications within the human body, for example, products for replacing bones, teeth, or other hard tissue. A famous application is the implant made of Al_2O_3 for hip bones owing to its good wear resistance and good compatibility with the human body.

The definitions of ceramics might be a bit confusing because the categories may overlap. However, for engineering application we regard all the above-mentioned ceramic groups as high-performance ceramics. The high-performance concept is primarily used to distinguish them from technical ceramics based on clay used for tableware applications and other nonengineering applications.

7.2.2 CERAMIC MATERIAL TYPES

In addition to grouping ceramics by application, we can group them according to their chemical composition or microstructure. Table 7.1 lists the most important technical ceramics.

TABLE 7.1
Types of Important Technical Ceramics

Silicate Ceramics	Oxide Ceramics	Nonoxide Ceramics	
		Carbides	Nitrides
Porcelain	Aluminum oxide, Al_2O_3	Silicon carbide, SiC	Silicon nitride, Si_3N_4
Steatite	Zirconium oxide, ZrO_2	Boron carbide, B_4C	Aluminum nitride, AlN
Cordierite	Magnesium oxide, MgO		Boron nitride, BN
Mullite	Titanium oxide, TiO_2		
	Aluminum titanate, Al_2TiO_4		
	Lead zirconium titanate		
	Dispersion ceramic		
	(Al_2O_3/ZrO_2)		

7.2.2.1 Silicate (SiO$_2$-Based) Ceramics

Silicate ceramics, as the oldest group among all the ceramics, represent the largest proportion of technical ceramics. The major components of these multiphase materials are clay, kaolin, feldspar, and soapstone as silicate sources. Additionally such components as alumina and zircon are used to achieve special properties such as higher strength. After sintering, the material exists in a crystalline phase and a >20% glassy phase based on SiO$_2$.

Owing to their relatively low sintering temperatures and the ready availability of the natural raw materials, silicate ceramics are much cheaper than the oxide or nonoxide ceramics, which will be described below. The latter require expensive synthetic powders and high sintering temperatures.

Silicate ceramics are used in a wide range of applications in the electrical equipment industry. Typical uses are insulators, fuse cartridges, catalysts, and enclosures in high- and low-voltage applications. Figure 7.1 shows ternary phase diagrams, which help in understanding the compositions of various silicate ceramics explained in the following chapters.

7.2.2.1.1 Porcelain

Technical porcelains belong to the alkaline alumina silicate porcelain group. The various porcelain types are quartz porcelain, alumina porcelain, and up to lithium porcelain. As electricity began to be used in homes and industry, porcelain types already offered excellent properties such as high mechanical strength, excellent electrical insulating properties, and outstanding resistance to chemical attack. As energy consumption rapidly increased, so did the needs of electrical technology, and the properties of porcelain were continually improved. This led to today's alumina porcelain, which exhibits noticeably greater strength and reliability even under extreme thermal stress or thermal shock conditions. However, raw materials for

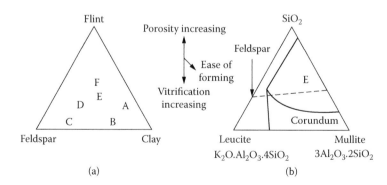

FIGURE 7.1 Compositions of raw materials used to produce certain ceramic products: (A) wall tiles; (B) vitreous whiteware; (C) floor tiles; (D) electrical porcelain; (E) hard porcelain; (F) semivitreous porcelain. (From I. J. McColm, *Ceramic Science for Materials Technologies*, Blackie & Son Ltd., 1983, p. 1. With permission.)

FIGURE 7.2 Sockets and housings made from steatite. (From *Breviary Technical Ceramics*, ed. Informationszentrum Technische Keramik in cooperation with member firms of the Technical Ceramics Section. With permission.)

alumina porcelain are more expensive than raw materials for quartz porcelain, so quartz porcelain represents an economic alternative for moderate load applications.

7.2.2.1.2 Steatite

Steatite is a ceramic based on natural raw materials and consists mainly of soapstone [$Mg(Si_4O_{10})(OH)_2$], a natural magnesium silicate, with the addition of clay and feldspar or barium carbonate. Various steatite types exist: normal steatite and special steatite. The latter is known as high-frequency steatite and has a low loss factor. However, special steatite is not only used for high-frequency parts. Because of its excellent workability, it is also used for thin parts, requiring a constant thickness. This allows thermally induced mechanical stress to be controlled. Typical applications are sockets, control housings, insulating beads, low-voltage power fuses, and base plates (see Figure 7.2).

7.2.2.1.3 Cordierite

A simplified approximation of the composition of cordierite ceramic is 14% MgO, 35% Al_2O_3, and 51% SiO_2. Cordierite is fabricated by sintering soapstone or talcum with the addition of clay, kaolin, fireclay, corundum, and mullite. Cordierite materials have a low coefficient of thermal expansion. This is the reason for their outstanding thermal shock resistance combined with good mechanical strength. Cordierite, moreover, behaves inertly toward heating element alloys at temperatures up to 1000 or even 1200°C. Therefore, cordierites are often found in electric heating applications. Applications include insulators for continuous-flow electric water heaters, heating element pipes, heating element supports in furnaces, link heaters, heating cartridges for soldering irons, and so on. Cordierite is also used as a catalyst carrier in automobiles.

7.2.2.1.4 Mullite

Pure mullite (3 Al_2O_3 2 SiO_2) consists of 82.7% Al_2O_3 and 17.3% SiO_2 by mass. By varying the composition of the Al_2O_3–SiO_2 system, specific modifications of the mullite material's properties can be achieved. Mullite ceramics have a microstructure whose mineral phases consist of pure mullite, corundum (Al_2O_3), and glass (SiO_2). Mullite usually still has a porosity of about 10% after sintering.

Therefore, it is highly resistant to thermal shock and it has a low coefficient of thermal expansion. Applications include kiln furniture for temperatures up to 1700°C, even in oxidizing atmospheres, and carrier rollers in high-temperature furnaces. Because of its low thermal conductivity and high corrosion resistance, it is also used as an industrial refractory material.

7.2.2.2 Oxide Ceramics

7.2.2.2.1 Aluminum Oxide

Aluminum oxide, Al_2O_3, is the most important technical oxide ceramic and has the widest range of applications. Millions of parts made of alumina, such as spark plugs, insulators, and many others, are produced per day. Densely sintered Al_2O_3 shows:

- High strength and hardness
- Temperature stability
- High wear resistance
- High corrosion resistance

As can be seen from Table 7.2, different types of aluminum oxide have Al_2O_3-contents from 80 to more than 99%. The nonpure alumina materials contain mainly a glassy phase as a further ingredient. This influences processing properties, for example, lower sintering temperatures, and physical properties of the alumina types. Different powder formulations, for example, powders with various particle sizes, lead to materials with different properties.

Aluminum oxide materials satisfy all the requirements of insulation materials for applications in electrical engineering. Outstanding values for bending strength — higher than the strength values of the silicate ceramics, high wear resistance, and excellent high-temperature properties — make the material suitable for a lot of mechanical applications. An overview of applications is listed in Table 7.3 and a few examples of applications in Figure 7.3.

7.2.2.2.2 Magnesium Oxide (MgO)

MgO is mostly manufactured in high purity as a single-component ceramic. It shows good electrical insulation and good thermal conductivity. MgO is used in heat engineering primarily for small tubes as heating elements, for instance, in immersion heaters, heating cartridges, and thermocouples.

7.2.2.2.3 Zirconium Oxide (ZrO_2)

Owing to its oxygen ion conductivity, zirconium oxide is used, for instance, as an oxygen sensor. This application will be explained in more detail later. ZrO_2 has gained importance because of its high fracture toughness, good strength, wear resistance, corrosion behavior, very good tribological properties (makes it very well suited for slide rings), and low thermal conductivity. Figure 7.4 shows, for example, turn round rolls for wire drawing made of high-wear-resistant ZrO_2. Quite unique is also a comparatively high thermal expansion coefficient, which

FIGURE 7.3 Examples of applications of alumina. (Left) Thyristor housing in which FeNi alloy is joined to the alumina component by brazing. (Right) Tubes for ash transport in power plants. (From *Breviary Technical Ceramics*, Informationszentrum Technische Keramik in cooperation with member firms of the Technical Ceramics Section. With permission.)

FIGURE 7.4 Turn round rolls for wire drawing made of high-wear-resistant ZrO_2 clamped together by steel wheels. (From *Breviary Technical Ceramics*, Informationszentrum Technische Keramik in cooperation with member firms of the Technical Ceramics Section. With permission.)

is similar to cast iron. All other ceramics usually show lower coefficients of thermal expansion that cause mismatches by joining them to cast iron or steel.

If we look at data sheets like the one given in Table 7.2, we find various types of ZrO_2:

ZrO_2-FSZ: fully stabilized zirconia
ZrO_2-PSZ: partially stabilized zirconia
ZrO_2-TZP: tetragonal zirconia polycrystal

TABLE 7.2
Properties of Oxide Ceramics

	Symbol	Unit	Al_2O_3 Aluminum oxide <90%	Al_2O_3 Aluminum oxide 92–96%	Al_2O_3 Aluminum oxide 99%	Al_2O_3 Aluminum oxide >99%	ZTA Aluminum oxide/ZrO_2-toughened	PSZ Partially stabilized ZrO_2
Mechanical:								
Open porosity	—	[Vol %]	0	0	0	0	0	0
Density, min	ρ	[g/cm³]	>3.2	3.4–3.8	3.5–3.9	3.75–3.98	4.0–4.1	5–6
Four-point bending strength	σ_B	[MPa]	>200	230–400	280–400	300–580	400–480	500–1000
Elastic modulus	E	[GPa]	>200	220–340	220–350	300–380	380	200–210
Hardness	HV_{10}	[GPa]	12–15	12–15	12–20	17–23	16–17	11–12.5
Stress intensity factor	K_{IC}	[M_{Pa} Ohm]	3.5–4.5	4–4.2	4–5.5	4–5.5	4.4–5	5.8–10.5
Weibull modulus	m	[--]	10–15	10–15	10–15	10–15	10–15	20–25
Electrical:								
Breakdown strength	E_d	[kV/mm]	10	15	15	17	—	—
Withstand voltage	U	[kV]	15	18	18	20	—	—
Permittivity 48–62 Hz	ε_r	[--]	9	9	9	9	—	22
Loss factor at 20°C, 48–62 Hz	$\tan \delta_{pf}$	[10^{-3}]	0.5–1.0	0.3–0.5	0.2–0.5	0.2–0.5	—	—
Loss factor at 200°C, 1 MHz	$\tan \delta_{1M}$	[10^{-3}]	1	1	1	1	—	—
Specific resistance at 200°C	$\rho v_{>20}$	[Wm]	10^{12}–10^{13}	10^{12}–10^{14}	10^{12}–10^{15}	10^{12}–10^{15}	10^{14}	10^{8}–10^{13}
Specific resistance at 600°C	$\rho v_{>600}$	[Wm]	10^{6}	10^{6}	10^{6}	10^{6}	10^{6}	10^{3}–10^{6}

Thermal:

Mean coeff. of lin. expansion at 30–1000°C	$\alpha 30{-}1000$	$[10^{-6}K^{-1}]$	6–8	6–8	6–8	7–8	9–11	10–12.5
Specific heat capacity at 30–1000°C	$C_{p,30-1000}$	$[Jkg^{-1}K^{-1}]$	850–1050	850–1050	850–1050	850–1050	800	400–550
Thermal conductivity	$\lambda 30{-}100$	$[Wm^{-1}K^{-1}]$	10–16	14–24	16–28	19–30	15	1.53
Resistance to thermal shock		Assessed	Good	Good	Good	Good	Good	Good
Maximum temperature of use	T	$[°C]$	1200–1400	1400–1500	1400–1500	1400–1700	1500	800–1600

Source: From *Breviary Technical Ceramics*, Informationszentrum Technische Keramik in cooperation with member firms of the Technical Ceramics Section. (With permission.)

TABLE 7.3
Applications of Aluminum Oxide

Area of Application	Products
Sanitary industry	Sealing element
Electrical engineering	Insulation parts
Electronics	Substrate
Machine and plant construction	Wear protection (coating of surfaces with wear-resistant tiles)
Chemical industry	Corrosion protection coatings, coatings against melts and slags owing to the resistance of Al_2O_3 against vapors, melt, and slags even at high temperatures and filters
Measurement technology	Protective tube for thermocouples, substrates for capacitive pressure measuring devices
Human medicine	Implants
High-temperature applications	Burner nozzles, support tubes for heat conductors

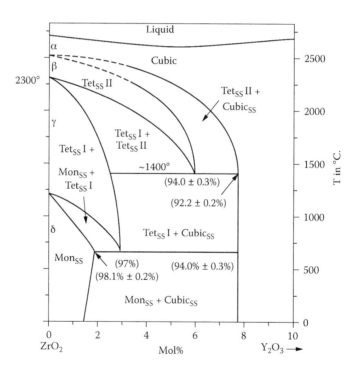

FIGURE 7.5 Phase diagram of a binary ZrO_2–Y_2O_3 system. (From various references.)

Zirconia occurs in various crystal structures: monoclinic, tetragonal, and cubic. Figure 7.5 shows a phase diagram of a ZrO_2–Y_2O_3 system. In pure ZrO_2 the change in crystal structure from tetragonal to monoclinic and vice versa occurs at ca. 1200°C. This causes a volume change and therefore stresses in the sintered component owing to phase transformations. Those stresses will lead to cracks. Consequently, this volume change is undesired in technical applications, so densely sintered parts must be manufactured in a stabilized crystal form as cubic or tetragonal ZrO_2 types. Those phases must be stabilized by adding, for example, Y_2O_3 (see Figure 7.5). At about 8 mol% Y_2O_3, no phase transformations of the cubic phase occur up to the melting temperature. Other important stabilizers are MgO, CaO, and occasionally CeO_2, Sc_2O_3, and Yb_2O_3.

7.2.2.2.4 ZrO₂-FSZ

The cubic structure is preserved owing to the addition of other oxides as described above. The volume change caused by a phase transformation, undesirable for technical applications, does not take place in FSZ.

7.2.2.2.5 ZrO₂-PSZ

This zirconia type exists of a matrix of cubic phases with tetragonal regions. The tetragonal crystal form is metastable, and its transformation into the monoclinic phase can be prevented through appropriate process-control and annealing techniques. This leads to a *prestressed* microstructure that is associated with an increase in strength and toughness.

7.2.2.2.6 ZrO₂-TZP

Polycrystalline tetragonal zirconium oxide is now made up completely of tetragonal grains with an extremely fine-grained microstructure. This is achieved by using extremely fine initial powders during the manufacturing process. Owing to the fine microstructure and metastable tetragonal structure, this material has extraordinarily high mechanical strength, possibly even exceeding 1.500 MPa.

In PSZ and TZP ZrO_2 types the metastable tetragonal structure leads to an increase in strength because the tetragonal metastable grains slow or deflect crack growth. Once a crack has been initiated and starts to grow, it will intersect a tetragonal phase. Because of the energy release caused by crack growth, the tetragonal phase transforms into the monoclinic phase and the volume increases. This volume increase closes the crack, slowing or deflecting its growth. This mechanism is called the transformation toughening method.

7.2.2.2.7 Zirconium Oxide–Toughened Alumina

The fracture toughening method is not restricted to zirconia ceramics. By adding tetragonal zirconia to aluminum oxide, a so-called ZTA, zirconium oxide–toughened aluminum oxide, can be produced with improved bending strength compared to conventional alumina types.

7.2.2.2.8 Aluminum Titanate

Aluminum titanate is the stoichiometric solid solution of Al_2O_3 and TiO_2 (Al_2O_3 • TiO_2). It has some remarkable properties different from most other ceramics. It has a very low coefficient of thermal expansion, a low Young's modulus, and a low thermal conductivity. In combination with a high residual porosity and formation of microcracks, according to anisotropic thermal expansion of the individual aluminum titanate crystals, it reveals excellent thermoshock resistance. The material is used, for instance, for portliners, cylinder liners, spacing rings of catalytic converters and in foundry engineering. Mechanical strength values are low, but, for example, in portliners and cylinder liners the liquid metal is poured around the ceramic; and during cooling, because of high thermal expansion of the metal, a compressive stress is applied to the ceramic. The strength of ceramics under compressive loads is much higher than under tension loads and hence aluminum titanate can be used as a thermal barrier for this application owing to its low thermal conductivity.

7.2.2.2.9 Titanium Dioxide (TiO$_2$)

Titanium dioxide materials consist of TiO$_2$, or titanates, and are used for capacitors in high-frequency electronics.

7.2.2.2.10 Barium Titanate

Barium titanates are also used as functional ceramics, as capacitors because of high permittivity, as temperature sensors in instrumentation and control technology, and as limit sensors in motor and machine protection. The latter application is due to a positive temperature coefficient (PTC) of the ohmic resistance. Those ceramics are called PTC ceramics. This effect is characterized by a very sharp rise in electrical resistance — several powers of 10 — starting at a specific temperature.

7.2.2.2.11 Lead Zirconate Titanate [Pb(Zr$_x$Ti$_{(1-x)}$)O$_3$]

Lead zirconate titanate is a functional ceramic and is the most important piezoelectric ceramic material. The piezoelectric effect links both electrical and mechanical properties. The direct piezoelectric effect refers to an electrical charge, detectable as a voltage drop, created in proportion to the mechanical deformation of the lead zirconate titanate crystals. Instead of applying a load and measuring a voltage drop, an electrical voltage can be applied to this ceramic, which leads in a deformation. The latter effect is called a reciprocal, or inverse, piezoelectric effect. Piezoceramics are used in a wide range of applications such as transducers in telecommunications, acoustics, hydroacoustics, materials testing, ultrasonic processing, flow measurement, distance measurement, and medical technology. Furthermore, they are used as actuators for micropumps in optical systems, gas valves, ink jet printers, and many further applications.

7.2.2.2.12 Sintered Fused Silica

Silicon oxide ceramics, SiO$_2$, or fused silica ceramics or quartzware, are sintered from amorphous silicon dioxide powders. The goal in manufacturing is to obtain the typical properties of SiO$_2$, with its thermal expansion coefficient close to zero. This leads to a ceramic with extremely high thermal shock resistance but with fairly low mechanical strength compared to other oxide ceramics. The area of application lies in areas where extreme thermal shocks are experienced and in thermal treatment of glass products. In the latter case, the high purity and the compatibility with glass prevent any contamination such as discoloration.

7.2.2.3 Nonoxide Ceramics

Nonoxide ceramics are:

- Carbides
- Nitrides
- Oxynitrides

The high portion of covalent bonds is responsible for quite unique properties, such as very high hardness and high temperature strength. The nonoxide ceramics are made exclusively from synthetic raw materials and extremely fine-grained powders, and the sintering process requires an oxygen-free atmosphere. This makes the process more costly compared to oxide ceramics because this implies sintering in vacuum or inert gas at temperatures that can reach well over 2000°C.

7.2.2.3.1 Carbides

7.2.2.3.1.1 Silicon Carbide

Silicon carbide is by far the most important carbide ceramic. It is characterized by a variety of unique properties:

- Very high hardness
- Corrosion resistance even at high temperatures
- Oxidation resistance even at high temperatures
- High strength even at high temperatures
- Good tribological properties
- Good thermal shock resistance
- Very high thermal conductivity
- Low thermal expansion

The high temperature properties are mentioned quite often in the list above and therefore SiC is, for instance, an excellent material for burning chambers (see Figure 7.6). Bearings are made of SiC because of good tribological properties

FIGURE 7.6 Burner made from SISIC. (From *Breviary Technical Ceramics*, Information-szentrum Technische Keramik in cooperation with member firms of the Technical Ceramics Section. With permission.)

and high hardness. It has recently been used as a brake disk in automotive applications, which will be explained in more detail later.

When we speak of SiC, we must be aware that there are several different varieties of the material, depending on the manufacturing process. Here we want to mention some dense silicon carbide types. The corresponding manufacturing processes will be explained in the following chapter.

RBSIC: Reaction-bonded silicon carbide
SISIC: Silicon infiltrated silicon carbide
SSIC: Sintered silicon carbide
HPSIC or HIPSIC: Hot (isostatic) pressed silicon carbide
LPSIC: Liquid-phase sintered silicon carbide

Reaction-bonded SISIC is composed of about 85 to 94% SiC and 15 to 6% metallic silicon. The specific formed foam-like SiC and carbon are infiltrated by metallic silicon. The reaction between metallic Si and carbon leads to bondings between the SiC grains. An advantage of this process is, for example, that no shrinkage takes place, so large parts can be manufactured with precise dimensions.

Pressureless SSIC is produced by sintering fine-grained SiC powder under atmospheric pressure. An advantage of this material is the high-temperature strength maintained up to temperatures of 1600°C.

Hot-pressed and hot isostatic-pressed SiC exhibit even better mechanical specifications than SSIC because of the applied high pressure during the manufacturing process, which forms a nearly pore-free SiC product. The process is limited to parts of relatively simple or small geometry, but the parts are used for the most challenging applications.

Liquid-phase sintered SiC is a mixed oxynitride material. Oxide ceramic powders are added to the SiC powder. The oxide additions melt during sintering, which leads to increased sintering activity and hence to dense products.

7.2.2.3.1.2   Boron Carbide (B_4C)

Boron carbide ceramics are manufactured from submicron B_4C powder. As for SiC, various types (sintered boron carbide [SBC], hot-pressed boron carbide [HPBC], and hot-isostatic-pressed [HIP] boron carbide [HIPBC]) are available. B_4C has very high wear resistance in combination with low density (2.52 g/cm^3), high mechanical strength, and high Young's modulus. It is applied in the field of ballistic protection. Only cubic boron carbide and diamond exceed the hardness of B_4C.

7.2.2.3.2   Nitrides

7.2.2.3.2.1   Silicon Nitride (Si_3N_4)

Silicon nitride is one of the most interesting nitride ceramics. It achieves a combination of outstanding material properties not yet reached by other ceramics:

FIGURE 7.7 Si$_3$N$_4$ balls for bearings. (From *Breviary Technical Ceramics*, Information-szentrum Technische Keramik in cooperation with member firms of the Technical Ceramics Section. With permission.)

- High toughness
- High strength even at high temperatures
- Outstanding thermal shock resistance
- Remarkable resistance to wear
- Low thermal expansion
- Good corrosion resistance

Silicon nitride therefore is appropriate for the toughest application conditions and is used for machine components with very high dynamic stresses and reliability requirements. Examples of high-temperature applications are valve and turbo charger rotors for passenger cars. However, these applications have not yet entered the market. Other successful examples are balls for bearings (shown in Figure 7.7) and tool tips for high-speed cutting.

Various Si$_3$N$_4$ types are used according to the manufacturing process:

SSN: Low-pressure-sintered Si$_3$N$_4$.

GPSSN: Gas-pressure-sintered silicon nitride. This smart process applies a N$_2$ gas pressure of up to 100 bar during sintering, which leads to high-performance materials.

HPSN: Hot-pressed silicon nitride.

RBSN: Reaction-bonded silicon nitride. Relatively economical metallic Si powder is used as a raw material, and after forming the sintering takes place under a N$_2$ atmosphere and the Si reacts with nitrogen during the sintering process with no shrinkage.

7.2.2.3.2.2   Sialons

Silicon aluminum oxynitride (Si-Al-O-N → SIALON) is blended from silicon nitride with alumina-type ceramics. The advantage of this process is a lower sintering temperature. SIALONs have relatively high fracture toughness and therefore are frequently used for cutting tools. The low wetting by liquid Al and other molten nonferrous metals has made SIALONs more or less the standard material for thermocouple protection tubes in the foundry industry.

7.2.2.3.2.3   Aluminum Nitride (AlN)

Aluminum nitride is an interesting functional ceramic used for applications in electrical engineering because of a unique combination of properties:

- AlN has high thermal conductivity of up to 220 W/(m·K), which is comparable to the thermal conductivity of metallic aluminum.
- AlN is an isolator.

This remarkable combination and others not mentioned here have advantages in processing that make AlN a suitable material for substrates for semiconductors, high power electronic parts, and heat sinks.

7.3   MANUFACTURING OF CERAMIC PRODUCTS

The microstructure governs the properties of a material, and the manufacturing process strongly influences the microstructure. It seems to be easy to produce ceramic materials if we just list the manufacturing steps, but a thorough knowledge of technology as well as a good materials science and technology background is necessary to manufacture high-quality products. It is important to monitor a lot of parameters during the manufacturing process in order to "produce quality" and not to "prove quality." This is valid especially for ceramics because, for example, sintering processes may take a long time at very high temperatures. This is a cost-effective process that must be controlled carefully. In this chapter we will not be able to explain all manufacturing steps in detail. The objective of this chapter is to give an overview of manufacturing steps and to convey basic knowledge.

Ceramic materials are produced by powder metallurgical processes. The individual manufacturing steps are listed in Figure 7.8. Starting with the ceramic powder, it might be alloyed with additives and then pressed, sintered, and finished. Let us explain some important parameters during each manufacturing step.

7.3.1   RAW MATERIALS AND ADDITIVES

Quality starts with the availability of a specified raw powder. It must have:

- A specific grain size
- A specific surface area
- High purity

Mistakes at this early state, for example, too many impurities, cannot be corrected during the further manufacturing steps. Thus, all parameters must be

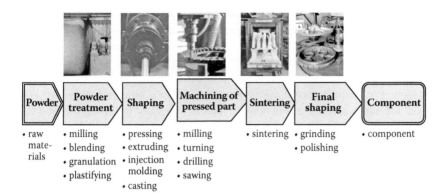

Powder	Powder treatment	Shaping	Machining of pressed part	Sintering	Final shaping	Component
• raw materials	• milling • blending • granulation • plastifying	• pressing • extruding • injection molding • casting	• milling • turning • drilling • sawing	• sintering	• grinding • polishing	• component

FIGURE 7.8 Manufacturing steps from powder to component. (From *Breviary Technical Ceramics*, Informationszentrum Technische Keramik in cooperation with member firms of the Technical Ceramics Section. With permission.)

5 μ 2 μ

FIGURE 7.9 Coarse and fine Al_2O_3 powder. (From *Breviary Technical Ceramics*, Informationszentrum Technische Keramik in cooperation with member firms of the Technical Ceramics Section, ISBN 3-924158-57-6. With permission.)

carefully specified and controlled. For example, grain size and surface area have a remarkable effect on the sintering activity and hence on the density and mechanical properties of the final product.

Figure 7.9 shows Al_2O_3 powder of different grain sizes. The finer powder will lead to higher-strength alumina qualities.

For improving pressing, sintering conditions, and so on, further additives are necessary that are as significant as the raw powder itself. Additives are:

- Sintering aids (inorganic)
- Usually organic forming aids:
  - Liquefaction agents
  - Plasticizers
  - Binders

Sintering additives are mainly used for lowering the sintering temperature or increasing the sinter activity. In alumina ceramics glassy phases are common sintering additives.

Why do we add organic additives? If we want to produce a substrate this will be done by slip casting, and therefore a liquid suspension of ceramic powder is needed. For getting good flowing powders, organic binders are needed for a specified agglomeration of the powder. This is important for filling forming tools automatically with powder. A very fine nonagglomerated powder would stick together, which hinders the powder from good flowing behavior. Thermocouple rods, for example, are extruded. In order to extrude a ceramic powder, an organic binder is needed to hold the powder particles together. Of course, all organic ingredients have to be removed carefully during the firing process.

7.3.2 FORMING

The powder must be compacted to form a more or less solid body, called a green body, at least sufficient for subsequent handling. If necessary, this green body can be machined economically before firing. Because of the extreme hardness of ceramics after sintering, their machining would be very expensive. When applying various forming processes, care must be taken to avoid significant density gradients and textures. Those issues can be amplified by sintering, leading to distortions and internal stresses. The choice of a suitable forming process is usually determined by economic factors. Methods of shaping ceramic parts can be divided into the following types:

Pressing: 0 to 10% moisture
Plastic forming: 15 to 25% moisture
Casting: >25% moisture

Let us explain now the most important forming processes.

7.3.2.1 Dry Pressing

This is the most economic process for large production runs and is suitable for both simple and complex geometries. Nonclumped granulates are compressed in steel dies, sometimes cemented carbide dies, designed appropriately for the part to be manufactured. Figure 7.10 shows schematically the single- and double-ended dry pressing, and Figure 7.11 shows a dry-pressing process in which three cutting tool tips are manufactured in one step.

7.3.2.2 Wet Pressing and Moist Pressing

This forming method allows the manufacturing of complexly shaped geometries such as screw threads, side holes, recesses, and undercuts. The moisture usually lies in the range of 10 to 15%. Compared to dry pressing, the compression is

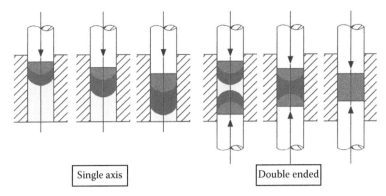

Single axis   Double ended

FIGURE 7.10 Single-axis dry pressing and double-ended pressing with regions of different compression (gray level). (From *Breviary Technical Ceramics*, Informationszentrum Technische Keramik in cooperation with member firms of the Technical Ceramics Section. With permission.)

FIGURE 7.11 Dry pressing. (From *Breviary Technical Ceramics*, Informationszentrum Technische Keramik in cooperation with member firms of the Technical Ceramics Section. With permission.)

limited because the materials allow only small compressive strains. Sometimes it is necessary to dry the pressed parts before sintering.

7.3.2.3 Hot Pressing

This method is used to manufacture components with a density close to the theoretical maximum. It uses with hot dies and hence forming and sintering are combined in one manufacturing step.

7.3.2.4 Cold Isostatic Pressing

Simple rubber molds determine the initial form. The rubber mold is filled with the ceramic powder, and the mold must be carefully closed. The mold is next placed in a pressure chamber filled with liquid (water with some additives). A

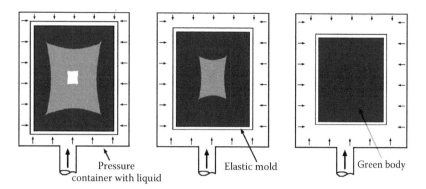

FIGURE 7.12 Isostatic pressing with regions of different compression (gray level). (From *Breviary Technical Ceramics*, Informationszentrum Technische Keramik in cooperation with member firms of the Technical Ceramics Section. With permission.)

pressure of up to 5000 bar is applied, and owing to the isostatic loading, a green body of homogeneous density will be achieved. The principle is shown in Figure 7.12. This method is suitable for uniformly pressed blanks and large parts that may be machined in the green state.

7.3.2.5 Hot Isostatic Pressing

The isostatic pressing will be carried out at high temperatures. This requires an autoclave-like furnace in which isostatic gas pressures (usually N_2 gas) of up to 3000 bar at temperatures up to approximately 2000°C are applied. The ceramic usually is covered with a silicate glass envelope. This process leads to the highest density of all manufacturing processes.

7.3.2.6 Extrusion

This process is comparable to the extrusion of plastics. However, the machines are more complicated because of higher wear of the feed screws, dies, and so on. Extrusion is carried out using piston extruders or vacuum screw processes. The process is shown schematically in Figure 7.13. The homogenized mass of higher moisture compared to former processes is pressed through a nozzle thus forming endless products, such as axles and pipes.

7.3.2.7 Injection Molding

Like extrusion, this process is a plastic-forming process. It is comparable to the injection molding of plastics, but only the basic principle is the same. All parts of the machine have to accommodate the high wear of the ceramic mass. The powder contains a fairly high amount of binder to allow the injection molding. Injection molding is principally used for complexly shaped products. A typical component would be a turbo charger rotor made of Si_3N_4. The process is limited

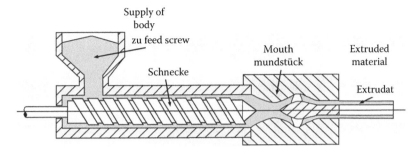

FIGURE 7.13 Extrusion process. (From *Breviary Technical Ceramics*, Informationszentrum Technische Keramik in cooperation with member firms of the Technical Ceramics Section. With permission.)

by the relatively high cost of the die. Furthermore, the binder has to be burned off before sintering, a quite complex process.

7.3.2.8 Casting

7.3.2.8.1 Slip Casting

This is a simple method for the manufacture of prototypes and parts with complex geometries. It involves a stable suspension, referred to as the slip, that is poured into a porous absorbent plaster mold. Extraction of the suspending liquid forms a layer of ceramic particles on the mold wall.

7.3.2.8.2 Tape Casting

A ceramic slip containing various organic additives is poured onto an endless steel strip carried by rollers (a tape-casting machine is shown in Figure 7.14). The slip flows continuously from a reservoir through an adjustable slot onto the strip. Hot air is blown over the strip in the opposite direction to dry it, and by use of organic binders, a flexible tape of green ceramic is obtained. Typically, tapes from 0.25 to 1.0 mm are formed. The tapes are suited for the manufacture of substrates, housing, capacitors, and multilayer transducers.

7.3.3 FIRING PROCESS (DRYING, BURNING OUT, SINTERING)

The ceramic green body contains, and this is valid for all forming processes, a specific amount of moisture and often organic additives acting as binders or plastifyers. This means that before the actual sintering process can start, the moisture or organic binders have to be removed. The moisture and organic binders are volatile or decompose at temperatures of about 400°C, which is still far below sintering temperature. However, those processes are not trivial. Drying and burning out are carefully adapted to a temperature–pressure–atmosphere–time profile in order to achieve nondestructive and reproducible removal of these additives.

After drying and burning out, the compacted powder is less closely bonded together, so careful handling of the products during subsequent steps is required.

FIGURE 7.14 Tape-casting machine. (From *Breviary Technical Ceramics*, Information-szentrum Technische Keramik in cooperation with member firms of the Technical Ceramics Section. With permission.)

This is the why, whenever possible, drying, burning out, and sintering are integrated in one process cycle.

Whereas the density of the green body lies in the range of 70% (percent of theoretical density), depending on forming processes, the ceramic part will achieve densities well above 90% after sintering. The shrinkage is strongly dependent on the material. Shrinkages of various ceramics are given in Table 7.4.

The sintering reactions lead to a densification of the material, resulting in a reduction of porosity. A product of high strength will be achieved after sintering. What are the mechanisms that lead to such a large shrinkage?

TABLE 7.4
Longitudinal Shrinkage of Some Materials

Material	Shrinkage in Percent
Silicon-infiltrated silicon carbide	Approximately 0
Sintered silicon carbide	18–20
Cordierite	3–8
Alumina porcelain	13–16
Aluminum oxide	17–20
Zirconium oxide	25–32

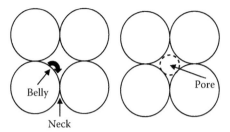

FIGURE 7.15 First sintering step: material transport (dark arrow) to the necks caused by capillary forces. (Right) Pore formation after filling up the necks.

The processes occurring during the sintering of ceramics are very complex. The sintering rate is dependent on numerous parameters such as grain size, surface area, purity, compaction, and sintering atmosphere. The main sintering steps are:

- Formation of round pores. If we regard the pressed particles as balls, they form a gap of specific caro-type shape between the particles (see Figure 7.15). At the contact areas between the round particles the particles form "necks," and between the necks the particle shows a "belly." A capillary force motivates the atoms to move to the neck and fill it up so that a round pore is formed. The mechanisms of atom movements are diffusion (surface and volume), evaporation of atoms at the bellies, and condensation at the necks.
- Reduction of porosity. In the second step the system reduces its porosity, which causes shrinkage. The material transport occurs by volume and grain boundary diffusion. The driving force for the diffusion is the interfacial energy of the system. Decreasing pore size leads to decreasing interfacial energy and allows the system to lower its energy. Controlling the diffusion rates is very important for achieving high densities. The pores should be as long as possible and located at grain boundaries. Then the shrinkage can carry on. If the pores lose their grain boundary, they will stay as residual pores in the inner grain, resulting in a residual porosity.

Sintering is a diffusion-controlled process, and we are now at least familiar with its basic mechanisms. However, the sintering process is quite complex. The main goal in developing ceramic manufacturing processes is the possibility of reducing sintering temperatures or, generally speaking, increasing the sintering activity. Some basic parameters for increasing sintering activity can be concluded:

- Small particle sizes and optimum particle size distribution increase sintering activity. Bimodular particle size distributions, for example, allow and increase in density after pressing.

TABLE 7.5
Typical Sintering Temperatures of Various Ceramics

Ceramic	Sintering Temperature in °C/°F
Alumina porcelain	Ca. 1250/ca. 2282
Quartz porcelain	Ca. 1300/ca. 2372
Steatite	Ca. 1300/ca. 2372
Cordierite	1250–1350/2282–2462
Al_2O_3	1600–1800/2912–3272
Sintered SiC	Ca. 1900/ca. 3450
Si3N4	Ca. 1700/ca. 3090

- Applying high external pressure increases the sintering velocity.
- Liquid phase sintering leads to increased sintering activity owing to high diffusion rates in the liquid between solid particles.

The sintering temperature therefore is dependent on the material itself and on the manufacturing parameters. Some typical sintering temperatures are given in Table 7.5.

7.3.4 MACHINING

In Figure 7.8 various processes are classified. We distinguish between green machining after drying, white machining after burning out, and prefiring and hard machining after sintering. We will not go into detail on the surface characteristics in this book, and we refer the interested reader to the individual standards just to say that sometimes even lapped or polished surfaces are required, for example, for extremely flat surfaces such as for slip rings and sealings.

7.3.5 SURFACE TREATMENTS

For specific applications the surfaces of ceramics are metallized or glazed or enameled (*enbobed*). Metallization, for example, is necessary if the ceramic must be joined to a metal by brazing or soft soldering. Sometimes a thin gold layer may be deposited on the ceramic surfaces.

Applying a glaze makes the surfaces smooth and visually more attractive. However, sometimes glazing is also necessary for technical reasons. A thin nonvitreous enamel layer is applied by dipping, rolling, spraying, or brushing. It is a porous glass-free surface layer used, for example, in furnace engineering for protecting ceramic surfaces against corrosive or mechanical attack.

7.3.6 CERAMIC COATING

In this chapter we should mention the coating of metallic surfaces with ceramic thick film and thin film techniques. The coatings offer excellent properties such as high wear resistance, electrical insulation, and low thermal conductivity. Thick films (>20 μm) are applied, for example, by thermal spraying. Plasma spraying has shown excellent results in this application. During plasma spraying, oxide-type ceramic powders are melted in a plasma flame at approximately 10,000°C and sprayed on a previously sandblasted metal surface. Also further pretreatments of the metallic surface are common. The ceramic coating often requires an adhesive layer on the metal surface. An example for this application is the thermal barrier coating of turbine blades with Al_2O_3 or ZrO_2. Other examples are plasma-sprayed tools in textile and wire-drawing machines or the coating of heating elements with Al_2O_3.

Thin films (<20 μm) are applied by physical vapor deposition or chemical vapor deposition by galvanic or sol-gel processes. A well-known example is the gold-colored TiN coating achieved by physical vapor deposition on various tool applications.

7.4 PROPERTIES OF TECHNICAL CERAMICS

7.4.1 GENERAL REMARKS

Ceramics have unique properties that in many respects cannot be achieved by other materials. We already mentioned many properties while describing the individual ceramic types. The variety of applications of ceramic materials arises from their specific properties. Let us highlight some positive properties again:

* Low density
* High hardness
* High mechanical strength
* High stiffness
* High wear resistance
* High working temperatures
* Low thermal conductivity or even high thermal conductivity
* Good electrical insulation
* Dielectric and ferroelectric properties

The functional properties such as specific dielectric and ferroelectric properties, piezoelectric behavior, electrical insulating properties, thermal barrier behavior, and wear resistance have led to numerous applications.

In this chapter we want to focus our attention on the mechanical properties of ceramics because the strength behavior of ceramics is different from the strength behavior of metals. Ceramics have no ability of plastic deformation. Even though ceramic materials can achieve very high strength values, they will

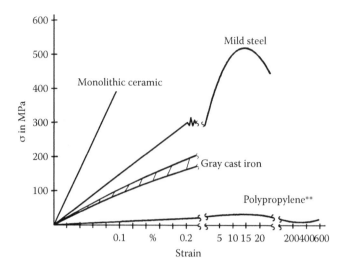

FIGURE 7.16 Stress–strain relationship for different materials. (From *Breviary Technical Ceramics*, Informationszentrum Technische Keramik in cooperation with member firms of the Technical Ceramics Section. With permission.)

break because they are brittle. In a batch of 100 specimens we have a certain probability that some specimen may break at fairly low stresses owing to crack propagation of inherent crack-like defects. This requires a statistical approach of the strength values, which we want to explain.

7.4.2 MECHANICAL PROPERTIES

In Figure 7.16 the stress–strain relationship of a ceramic (alumina) is compared with other technical materials, mild steel, and plastics. Some mechanical properties for various ceramics are given in Table 7.6. The ceramic shows the highest stiffness because of high Young's modulus, but the ceramic is brittle, which limits its structural applications. However, in cases where components are exposed to large compressive loads, ceramic materials are widely used. In Table 7.6 we find values for the bending strength and the fracture toughness of K_{Ic} and the Weibull modulus. We will explain these strength values below.

7.4.2.1 Bending Strength

Tensile tests of ceramic materials are extremely difficult to perform because failure occurs predominately in the gripping. Therefore, the strength of ceramics is mainly measured by three-point or four-point bending tests. Figure 7.17 shows a schematic illustration of the four-point bending test. We assume that the reader is familiar with the test procedure that is standardized in the German DIN EN 60-672 and DIN EN 843-1 and in the ISO standards.

TABLE 7.6
Overview of Mechanical Properties of Various Ceramics

Material Properties		Porcelain	Steatite	Al$_2$O$_3$ (>99%)	ZrO^2 PSZ	Silicon Nitride				Silicon Carbide			Mild Steel	Cast Iron
						SSN	HPSN	RBSN	Sialon	SSIC	RSIC	SISIC		
Density	g/cm³	2.3	2.7	3.94	6	3.3	3.4	2.5	3	3.15	2.8	3.12	7.85	7.3
Bonding strength (25°C)	MPa	110	140	520	1000	1000	900	330	355	600	120	450	300–450	95–170
Bending strength (1000°C)	Mpa	–	–	–	–	–	–	300	355	650	130	450	–	–
Elastic modulus	GPa	70	110	360	210	330	800	180	231	450	280	350	200–210	70–130
Fracture toughness	MNm$^{-3/2}$	–	–	5.5	> 8	8.5	8.5	4	2.2	5	3	5	140	–
Thermal expansion	10^{-6}K^{-1}	8	8.5	8	5	3.5	3.28	3	3	4.8	4.8	4.8	10	12
Weibull modulus	m	–	–	>10	>15	>10	>10	>10	15	>10	4.8	>10	–	–

Source: From *Breviary Technical Ceramics*, Informationszentrum Technische Keramik in cooperation with member firms of the Technical Ceramics Section.

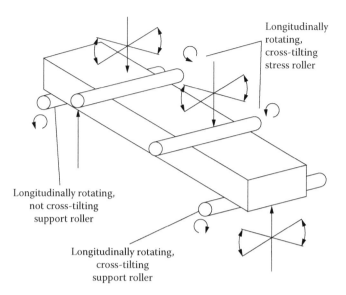

FIGURE 7.17 Schematic illustration of the four-point bending test. (From *Breviary Technical Ceramics*, Informationszentrum Technische Keramik in cooperation with member firms of the Technical Ceramics Section. With permission.)

The test piece must be made by the process from which the final product is expected to be manufactured. Any surface treatment such as grinding or polishing should be noted and be representative for the surface of the component, since the fracture strength of a ceramic can be affected by the surface quality. The minimum number of test species for each test is 10.

7.4.2.2 Statistical Approach to Strength: Weibull Distribution

There is a significant difference regarding strength of metals and ceramics. Let us first consider the behavior of crack propagation under tensile and compressive loads. A crack can propagate if crack-tip opening occurs. In metals the crack-tip radius may increase by overloads because of plastic deformation. This is not possible for ceramic materials. Therefore, the internal stress ahead of the crack tip will increase and lead to crack propagation. For ceramics we have to apply the weakest link model. This means the longest available crack in a structure will propagate after achieving a critical load and after brittle failure occurs. The most critical defect determines the reliability of the product.

Applying compressive stresses leads to a different mechanism of crack growth. Damage occurs ahead of the crack tip, which can be explained by a kind of wear ahead of each individual crack. Not only one (long and sharp) crack will propagate, but all cracks will propagate. Obviously this mechanism requires much more energy, and the compressive strength of ceramic is 5 to 10 times higher

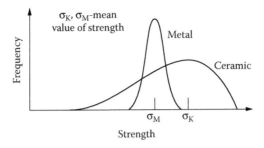

σ_K, σ_M-mean value of strength

Metal

Ceramic

Frequency

σ_M σ_K

Strength

FIGURE 7.18 Strength distribution within batches. (From *Breviary Technical Ceramics*, Informationszentrum Technische Keramik in cooperation with member firms of the Technical Ceramics Section. With permission.)

than the strength under tensile loads. It can be concluded that tensile stresses are much more critical.

Let us get back to the weakest-link model. Figure 7.18 shows schematically a result of a strength test series of at least 10 specimens. Metals show a Gauss distribution of strength values, so we are able to measure the yield strength or fracture strength showing a comparatively low scatter. For ceramics a completely different strength distribution occurs. The scatter of the strength values is much bigger, and with a certain probability we find specimens that fail at fairly low stresses. Those parts contain the most statistically critical defects (long, sharp inherent cracks). To avoid failure in service life, we need a statistical approach for the strength behavior of the ceramic.

Weibull derived a mathematical expression for the distribution of strength values of ceramic materials:

$$P = 1 - \exp\left[-\left(\frac{\sigma - \sigma_u}{\sigma_o}\right)^m\right] \tag{7.1}$$

where

$\exp(x) = e^x$

P = probability of failure

σ = applied stress

σ_u = minimum strength

σ_o = parameter of the Weibull distribution (the strength at a failure probability of 63%)

m = Weibull modulus

The Weibull modulus is the most important parameter in this equation because it is a value that stands for the scatter of the strength results. A low value for m means high scatter, and a high value of m means low scatter. Some typical values are:

- m ~ 5: Concrete
- m ~ 10–20: Typically achieved values for ceramics today
- m ~ 100: Steel

The value for steel is included to give a feeling for those parameters. For steel, it does not make sense to measure a Weibull modulus.

Let us try to understand the parameters in Equation 7.1. The equation can be simplified if we set $\sigma_u = 0$. This would be the most conservative approach for the strength distribution.

$$P = 1 - \exp [-(\sigma/\sigma_o)^m] \qquad (7.2)$$

If the applied load is zero, we expect a probability of failure of zero; $P = 0$. If we take $\sigma = 0$ and calculate P, we achieve $P = 0$ no matter what the value of m might be.

$$P = 1 - \exp (0) = 0$$

The meaning of the parameter σ_o can be understood if we assume that the applied stress is equal to σ_o; hence, $\sigma = \sigma_o$. Then Equation 7.2 results in

$$P = 1 - \exp [-(\sigma_o/\sigma_o)^m] = 1 - \exp [-(1)^m] = P = 1 - \exp (-1)$$

$$P = 0.63$$

As shown in Figure 7.19, σ_o is a parameter of the Weibull distribution for the strength at a failure probability of 63%. This parameter may vary from batch to batch.

If m is such an important value, we have to calculate this value from a test series. For this purpose, we have to transpose Equation 7.2 into a different form. First we want to have the exponential form just on one side:

$$1 - P = \exp [-(\sigma/\sigma_o)^m] \qquad (7.3)$$

Now we want to get rid of the minus in the exponent:

$$1/(1 - P) = \exp [(\sigma/\sigma_o)^m] \qquad (7.4)$$

Applying the natural logarithm leads to

$$\ln [1/(1 - P)] = (\sigma/\sigma_o)^m \qquad (7.5)$$

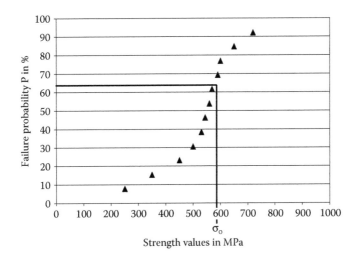

FIGURE 7.19 S-type strength distribution on a linear scale.

If we now apply a natural logarithm again, we achieve a linear equation of the type $y = m\,x + b$.

$$\ln \ln [1/(1 - P)] = m \ln \sigma - C \qquad (7.6)$$

In a diagram where $\ln \ln [1/(1 - P)]$ is plotted vs. $\ln \sigma$, the slope of this curve represents the Weibull modulus m.

Example In a bending test 12 specimens were tested and achieved the following bending strength values: 825, 870, 900, 915, 935, 970, 940, 930, 910, 890, 850, and 800 MPa. Now we put the measured value in a sequence with increasing strength. According to Table 7.7 we calculate the failure probability by $P = N/N_i + 1$.

The calculated Weibull modulus is $m = 18$. This means that for this batch we measure a comparatively high Weibull modulus. Today m values of >15 should be achieved. Figure 7.20 shows the results graphically. In Figure 7.20 another strength distribution with a Weibull modulus of 9.3 is given. This ceramic batch has a lower strength combined with a higher scatter.

It is obvious that the amount of defects statistically will increase with increasing component volume. This means that the above-calculated strength behavior is only valid for a standard-sized specimen. In service the volume under load of the product must be taken into account. The strength of larger parts is thus less than what is measured on test samples. Using Weibull statistics, the following component strength results from the volume relationship:

$$\sigma_{part} = \sigma_{test\;piece} \; \times \; (V_{test\;piece}/V_{part})^{1/m} \qquad (7.7)$$

TABLE 7.7
Calculation of Weibull Modulus

N	σ in MPa	P = N/(N$_i$ + 1)	ln σ	1/(1 − P)	ln ln [1/(1 −P)]
	800	0.08	6.68	1.08	−2.53
2	825	0.15	6.72	1.18	−1.79
50	00	0.23	6.75	1.30	−1.34
70	00	0.31	6.77	1.44	−1.00
90	00	0.38	6.79	1.63	−0.72
00	00	0.46	6.80	1.86	−0.48
10	00	0.54	6.81	2.17	−0.26
15	05	0.62	6.82	2.60	−0.05
30	00	0.69	6.84	3.25	0.16
10	935	0.77	6.84	4.33	0.38
11	940	0.85	6.85	6.50	0.63
12	970	0.92	6.88	13.00	0.94

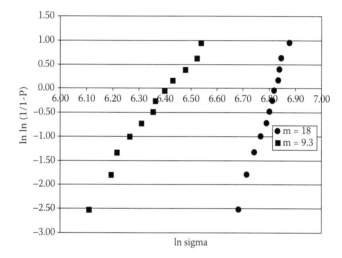

FIGURE 7.20 Example of strength behavior with high and low Weibull moduli. The higher the Weibull modulus is, the more consistent the material and also the narrower the probability curve of the strength distribution. This means that inherent defects are more evenly distributed throughout the entire volume. Possible defects are pores, impurities, and cracks.

The volume under stress is to be inserted here for the component volume, and the stress distribution must be identical ($\sigma_{tensile}$ or $\sigma_{4\,pt\,bending}$). This relationship does not apply to compressive stress.

7.4.2.3 Stress Intensity Factor

Ceramic materials are subjected to brittle fracture. The stress intensity factor K_I (mode I stands for tensile loading) has been adopted from fracture mechanics to determine the behavior of brittle materials with respect to crack growth. K_I is defined as follows:

$$K_I = \sigma \sqrt{\pi a} \cdot Y \ (a/W) \tag{7.8}$$

where

K_I = stress intensity factor in MPa $\sqrt{m}$
σ = applied stress
a = crack length
Y = correction factor depending on the geometry (crack length–to–specimen width ratio)

Increasing load and longer inherent cracks will lead to higher stress intensities. If the specimen or component reaches a critical crack length, a_c, it will fail. Hence, a critical stress intensity factor K_{Ic} can be concluded:

$$K_{IC} = \sigma_c \sqrt{\pi a_c} \cdot Y \ (a/W) \tag{7.9}$$

K_{Ic} is given in manufacturers' data sheets. In combination with the knowledge of the crack length, this allows us to calculate a critical applied stress. Ceramics with high-stress intensity factors such as silicon nitride or tetragonal zirconia polycrystalline (TZP) have a high resistance to crack propagation.

7.5 DESIGN ASPECTS

The ceramic product must be reliable and cost must be optimized. Ceramic manufacturing requires a complex firing process combined with large shrinkage rates. Every type of forming and treatment process entails certain demands that must be considered in design. The physical properties of the ceramic and metal components must also be considered. We are not able to give a comprehensive picture of the design of ceramic parts in this chapter, but we will explain some important design guidelines, mainly regarding the mechanical behavior of ceramics.

Ceramic materials are brittle and their strength is influenced by external stresses, whereas tensile stresses are particularly critical and by the amount size distribution of defects. This leads to some important design guidelines:

- Avoid large tensile stresses.
- Avoid sharp edges or notches.
- Apply a large loading area (homogeneous stress distribution).
- Avoid point loads (equivalent to previous guideline).

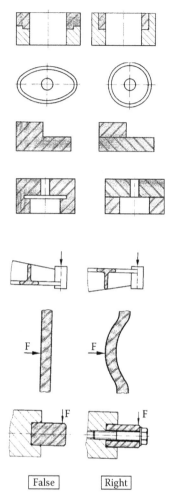

FIGURE 7.21 Design aspects. (Left) Simple shapes. (Right) Minimizing tensile stresses. (From *Breviary Technical Ceramics*, Informationszentrum Technische Keramik in cooperation with member firms of the Technical Ceramics Section. With permission.)

- Avoid large temperature gradients.
- Decrease component volume if possible; use a modular design.

We will give a few examples in order to discuss some issues of the design guidelines. Keep shapes simple and minimize tensile stresses. Figure 7.21 shows some examples of simple design guidelines. In each picture the right row of examples shows the suggested design. The left shows that:

- Steps and offsets should be avoided because of stress concentration and for easier manufacturing.
- Oval parts should be avoided. In the case of a circular form a more homogeneous stress distribution will occur during loading.
- Modular structures are preferred to lower component volume and improve reliability. The probability of having defects in a small volume is lower than in a large volume.

On the right of Figure 7.21 examples for minimizing tensile stresses are shown:

- Profiles should be chosen that minimize tensile stresses. In the right example the ribs are under compressive stress.
- Choosing an arch instead of a beam converts tensile into compressive stresses in this case.
- The ceramic component is prestressed with a compressive load, and hence tensile stresses are minimized.
- Avoid material accumulations and stress concentrations.

Figure 7.22 shows some design guidelines for avoiding material accumulations and stress concentrations. Again, the right side of each picture is the recommended design. The left side of Figure 7.22 suggests that you:

- Aim for even wall thickness in extrusions.
- Separate the nodes.
- Avoid thick ends on moldings.

All three examples are recommended because the component volume under stress would be reduced, and according to the Weibull weakest-link concept, this would lead to a lower probability of failure. The right side of Figure 7.22 suggests that you:

- Avoid sudden changes in cross-sectional area.
- Minimize notch-like structures.
- Provide large contact areas.
- Avoid corners and sharp edges, round inner edges, and cut-outs.
- Avoid long, sharp edges (risk of breakage).

These few examples show the strong dependence of the strength of a ceramic component on the design. In order to use ceramic materials in mechanical engineering, many more issues must be considered. The final finishing should be minimized, and it is important that manufacturing details be considered. This requires a close cooperation between manufacturing and design engineers in order to get a good ceramic product.

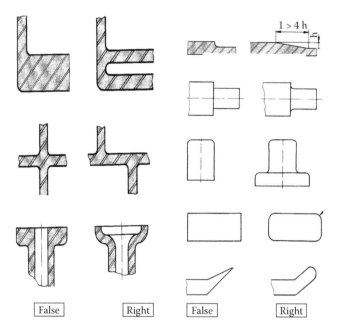

FIGURE 7.22 (Left) Avoid material accumulations. (Right) Avoid stress concentrations. (From *Breviary Technical Ceramics*, Informationszentrum Technische Keramik in cooperation with member firms of the Technical Ceramics Section. With permission.)

7.6 CASE STUDIES

So far, a number of examples where the unique ceramic properties are beneficial for the application have been mentioned. Below, we have chosen some specific case studies to explain in more detail.

7.6.1 CERAMIC BRAKE DISK

Usually brake disks are made of gray cast iron. Ceramic brake disks offer a step into the future for this application. In 1998 Porsche started the development of a brake disk made of a carbon-reinforced SiC ceramic, C/SiC, for the Porsche 911 GT2 (see Figure 7.23). It was introduced in the market in 2000.

The ceramic brake disk has many advantages such as:

- Higher temperature strength than gray cast iron
- Constant brake performance even at high temperatures
- Long lifetime, low wear
- Less vibration
- Corrosion resistance
- Up to 70% reduced weight

FIGURE 7.23 Ceramic brake wheel (www.sglcarbon.com).

For example, the ceramic disk saves 5 kg per wheel. This reduces the weight of the suspension and means a reduction in unsprung masses, with a further improvement of shock absorber response, acceleration, and steering behavior.

The disadvantage of this high tech material is the high cost. The manufacturing of this material is a challenging job, and quality control leads to high costs because each part has to be controlled by a sound check of first eigenfrequency.

Let us briefly review the manufacturing steps to get a better understanding of this ceramic product. The main processing steps are:

- Mixing of carbon short fibers plus carbon powder plus resin.
- Pressing.
- Carbonizing to a porous body (900 to 1100°C).
- Infiltration with liquid silicon (1500 to 1600°C) reacting with matrix carbon but not reacting with carbon fibers. This process has to be controlled very carefully, and the metallurgical reactions must be fully understood.

Finally, the product of SiC + C-fibers (rest Si) exists with a residual porosity of about 2%. Further applications for this ceramic are brake disks for high-speed trains and a ceramic clutch.

7.6.2 Oxygen Sensor

ZrO_2 ceramic is the heart of an oxygen sensor. This oxygen sensor is used, for example, for controlling gasoline injection to minimize emission by the catalyst in an automotive application. This is a controlled catalyst, and Bosch Company has manufactured more than 250 million oxygen sensors worldwide. The control principle is shown in Figure 7.24. The oxygen sensor (A) measures the oxygen content in the exhaust gas, which leads to information about the gasoline-to-air ratio in the combustion chamber. This information, more or less an mV-signal, is forwarded to the control unit (B). This unit now controls the manifold pressure and calculates the amount of exhaust gas returning to the combustion chamber as well as the injection initiation point and the quantity of the injected fuel.

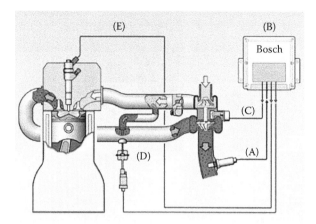

FIGURE 7.24 Principle measurement system of controlling gas injection (www.Bosch.de or www.Bosch.com).

FIGURE 7.25 Bosch oxygen sensor (www.Bosch.de or www.Bosch.com).

This control system is necessary to achieve the high efficiency of the catalyst and minimize the exhaust gases. An oxygen sensor is shown in Figure 7.25. The heart of the sensor actually is a sensor element made of zirconia ceramic. According to the sketch in Figure 7.26, the ceramic is exposed to the exhaust gas on one side, which has a low oxygen content and hence a low oxygen partial pressure Po_2. The opposite side of the ceramic is exposed to the environmental air, which has a high oxygen partial pressure. This gradient of the oxygen partial pressures leads to a driving force for the diffusion of O^{2-} ions in the ceramic from one side to the other, resulting in a voltage drop that can be measured between the two surfaces. In order to measure this voltage drop, the ceramic surface must be coated with a metal electrode — platinum in this case because of the extremely corrosive environmental condition of the sensor. The ZrO_2 measuring cell acts like a solid electrolyte, and the voltage drop can be calculated by the Nernst equation:

$$U_{th} = RT/2F * \ln (\sqrt{Po_2^{air}}/\sqrt{Po_2^{exhaust\ gas}})$$

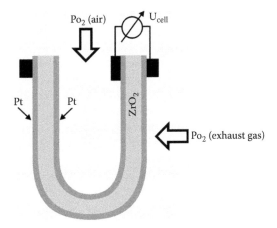

FIGURE 7.26 Sketch of the ZrO_2 sensor element.

The reaction at the surface where the ceramic is exposed to the air is

$$O_2 + 4\ e^- \rightarrow 2\ O^{2-}$$

which represents the anode. At the exhaust gas surface the reaction will be

$$2\ O^{2-} \rightarrow O_2 + 4\ e^-$$

which is the cathode. Consequently, O^{2-} ions must be able to diffuse through the ZrO_2 lattice, which is possible in this ceramic owing to its ion conductivity properties.

A metallurgical problem arises from manufacturing of the ZrO_2 ceramic. As can be seen form the $ZrO_2–Y_2O_3$ phase diagram in Figure 7.5, there is a phase transformation from tetragonal to monoclinic during cooling at about 1200°C that causes a volume change. A solid ceramic is not able to accommodate this volume change, and hence cracking will occur. As sintering temperatures lie above this temperature, a pure ZrO_2 cannot be sintered crack free to a final product. Therefore, it is necessary to stabilize the ZrO_2 ceramic. As described in Section 7.2.2.2.3, by adding Y_2O_3, a cubic phase can be achieved at more than 8 mol% Y_2O_3. The cubic phase is stable without any phase transformations up to the melting point. This is one example of fully stabilizing ZrO_2. Other additions such as CaO are also common.

DEFINITIONS

Aluminum oxide: Densely sintered Al_2O_3 shows high strength and hardness, temperature stability, high wear resistance, and high corrosion resistance. The most often used ceramic materials, mainly for electric applications.

Bending strength: Because of the brittleness of ceramics, it is very difficult to perform tensile tests. Three- and four-point bending tests are the most common strength tests.

Organic forming aids: Liquefaction agents, plasticizers, and binders are additives used to agglomerate very fine powder or hold particles together for further handling, such as slip casting, extruding, and pressing. They must be removed before sintering.

Silicate ceramics: After sintering, the material exists in crystalline phases and >20% glassy phase based on SiO_2. Porcelain, steatite, cordierite, and mullite are silicate ceramics.

Silicon carbide: SiC ceramic available in various types according to manufacturing. A carbon-reinforced SiC is a future material for break disks.

Silicon nitride: Si_3N_4; high-performance ceramic with very good high-temperature properties and high hardness.

Sintered fused silica: Silicon oxide ceramic, SiO_2, or fused silica ceramic or quartzware sintered from amorphous silicon dioxide powders.

Sintering additives: Mainly used for increasing the sinter activity. In alumina ceramics glassy phases are common sintering additives.

Titanates: $BaTiO_3$ types of compounds that often exhibit piezoelectric behavior.

Weibull modulus: A parameter that stands for the scatter of strength values in one batch.

Zirconia: High-strength ceramic with ion conductivity. For metallurgical reasons, technical ZrO_2 ceramics contain further additions to stabilize a cubic or tetragonal phase.

QUESTIONS AND PROBLEMS

7.1 What are the advantages of hot pressing over the cold pressing and sintering combination?

7.2 What are the advantages of cold isostatic pressing compared to dry pressing?

7.3 List three possibilities for improving the sintering activity and justify your choice.

7.4 List four factors that affect the density of pressed products.

7.5 What is the reason for the low electrical conductivity of alumina?

7.6 How can you improve the thermoshock resistance of ceramics?

7.7 What is the reason that zirconia cannot be used as pure ZrO_2?

7.8 What is the mechanism of increasing fracture toughness in PSZ and TZP zirconia?

7.9 List three reasons why ceramic materials are more brittle than metals.

7.10 What are the purposes of sintering aids?

7.11 What is the difference between slip casting and compression molding? What processing method is analogous to these processes in metals?

7.12 What is meant by the term *reaction sintering* and how does it differ from ordinary sintering?

7.13 Why do we need a statistical approach for describing the strength behavior of ceramics?

7.14 What is the Weibull modulus and what does it stands for?

7.15 Give some typical examples of oxide, carbide, and nitride ceramics.

SUGGESTED READING

American Society for Metals, *Engineered Materials Handbook*, Vol. 4, ASM International, Metals Park, OH, 1992.

Breviary Technical Ceramics, ed. Informationszentrum Technische Keramik in cooperation with member firms of the Technical Ceramics Section.

Doremus, R. H., *Glass Science*, Wiley, New York, 1973.

Kingery, W. D., Bowen, H. K., and Uhlmann, D. R., *Introduction to Ceramics*, 2nd ed., Wiley, New York, 1976.

McColm, I. J., *Ceramic Science for Materials Technologies*, Chapman & Hall, London, 1983.

Rice, R. W., The compressive strength of ceramics, in *Materials Science Research*, Vol. 5, W. W. Kriegel and H. Palmour, Eds., Plenum Press, New York, 1971, p. 195.

Richerson, D. W., *Modern Ceramic Engineering*, Marcel Dekker, New York, 1982.

Stokes, R. J., *Fracture*, Vol. 7, H. Liebowitz, Ed., Academic Press, New York, 1972, Chapter 4, p. 159.

8 The Structure and Properties of Polymers

8.1 INTRODUCTION

Common polymers are long chain molecules based on carbon. Like metals and ceramics, a single polymer molecule may contain millions of individual atoms, all bonded together. It is not possible to determine exactly how many carbon (C) and hydrogen (H) atoms are linked together in a single molecule of polyethylene ($[CH_2CH_2]_n^-$), for example, the average molecular weight is often on the order of 100,000 or more, indicating that there are approximately 8,000 carbon atoms and 16,000 hydrogen atoms in a single molecule of polyethylene. Properties are therefore related to the average molecular weight and vary from short chain waxes to long chain engineering materials.

8.2 BASIC CHEMISTRY REVIEW*

Polymers, like metals, solidify. Metals are usually solidified by cooling from the melt; polymers typically solidify through a (or a series of) chemical reactions. Because of this, it is important to understand the basic chemistry responsible for polymers, which is the chemistry of carbon.

Carbon has an atomic number of 12 and is the smallest atom with 4 valence electrons. Carbon will form covalent bonds with itself, oxygen, hydrogen, nitrogen, and the halogens (F, Cl, Br, and I). Covalent bonds arise when a pair of electrons is shared between two atoms. This is shown for a single molecule of ClF in Figure 8.1. Note the two electrons occupy a region of space between the centers of the two atoms. This space, or orbital, is well defined for each bond. The bond shown in Figure 8.1 is a single bond. In some cases, two pairs of electrons are shared between two atoms. This is called a double bond. Each pair of electrons occupies a distinct region of space. Thus, there are two separate bonds in a double bond. Triple bonds can also form.

FIGURE 8.1 Covalent bond between chlorine and fluorine.

---

* The chemistry presented in this review is intended to be a brief review of material presented in a fundamental chemistry course as it applies to polymers. Nothing presented in this section should be taken outside this context. The information presented in most chemistry texts will be more complete.

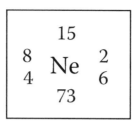

FIGURE 8.2 Lewis dot structure for Ne, showing the placement of electrons.

In order to form a covalent bond, it must be possible to form an electron pair between two atoms. In order to predict the structure of an organic (organic means based on carbon) molecule, one needs to first determine how many unpaired electrons exist in each atom. These unpaired electrons are available for bonding. To draw the Lewis structure for an element, one counts the number of valence electrons in an element and places them around the elemental symbol in the order shown in Figure 8.2. Note that the first four electrons are unpaired (1, 2, 3, 4), the next four will form pairs (1–5, 2–6, 3–7, 4–8). The actual placement of electrons is not critical; however, what is critical is that the first four electrons are unpaired and only the latter four are paired. Thus, oxygen, with six valence electrons, would have two sets of paired electrons and two unpaired electrons. The Lewis structures for the most important elements in polymers are shown below:

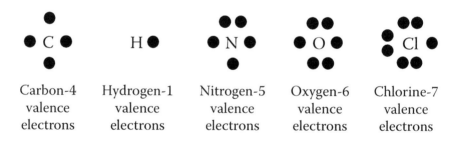

| Carbon-4 valence electrons | Hydrogen-1 valence electrons | Nitrogen-5 valence electrons | Oxygen-6 valence electrons | Chlorine-7 valence electrons |

Hydrogen therefore can only form one covalent bond, meaning that hydrogen can only bond to one other atom. Chlorine also has only one unpaired electron, so chlorine can only bond to one other atom. Oxygen has two unpaired electrons, thus it can form two single bonds, bonding to two other atoms, or it can form a double bond with a single atom. Nitrogen has three unpaired electrons, which means it can form three single bonds, a single bond and a double bond, or a triple bond. Carbon has four unpaired electrons. Carbon can therefore form four single bonds, two double bonds, or a single and a triple bond.

When determining the structure of an organic molecule, one needs to be aware of the number of bonds. For example, the structure of CH_3COOH is shown in Figure 8.3. Note in each case:

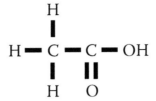

FIGURE 8.3 Lewis structure of acetic acid.

- Hydrogen (H) is bonded to only one atom. Three hydrogen atoms are bonded to a carbon (C) atom, and one to an oxygen (O) atom.
- Oxygen (O) forms two bonds. In one case an oxygen atom forms a double bond with a single carbon atom. In the other case, oxygen is bonded to two other atoms — carbon and hydrogen.
- Carbon (C) forms four bonds. In one case, carbon forms four single bonds, bonding with three hydrogen atoms and one other carbon atom. In the other case, carbon forms one double bond and two single bonds. It forms a double bond with an oxygen atom and a single bond with another carbon atom and another oxygen atom.

Also note that the structure shown in Figure 8.3 resembles the chemical formula CH_3COOH.

The molecule shown in Figure 8.3 has a functional group — COOH. This is an acid group. Functional groups contain the atoms that participate in chemical reactions. We will limit our consideration to three types of functional groups. These are shown below:

$$R-OH \qquad R-NH_2 \qquad R-\underset{O}{\overset{\|}{C}}-OH$$

Alcohol   Amine   Acid

Note that R is used to indicate an *organic root*. This can be anything, for example in Figure 8.3, R refers to the CH_3 group.

Organic molecules react and form new organic molecules, that is the focus of organic chemistry. The functional groups shown above can react with each other forming water and a *linkage*. Two of these reactions are shown in Figures 8.4 and 8.5. The formation of an amide is shown in Figure 8.4 and that of an ester is shown in Figure 8.5. Note that water is a by-product of both reactions.

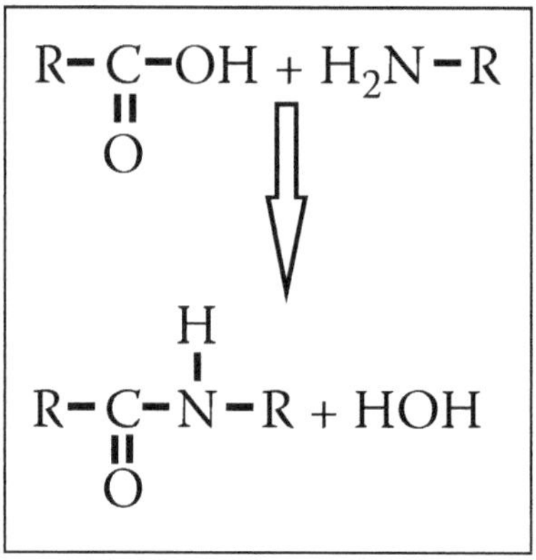

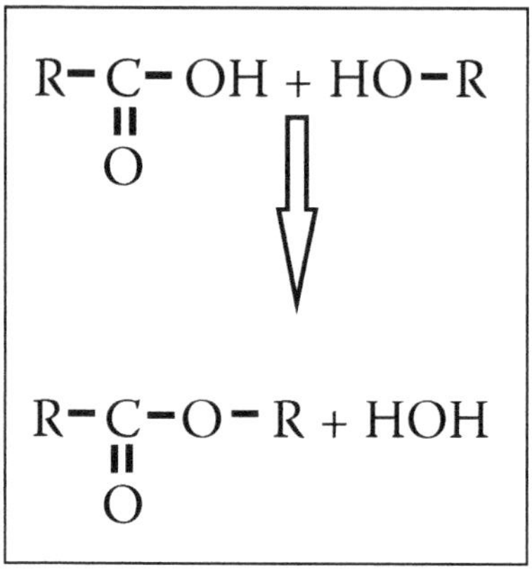

$$R-\underset{\overset{||}{O}}{C}-OH + H_2N-R$$

$$\Downarrow$$

$$R-\underset{\overset{||}{O}}{C}-\overset{\overset{H}{|}}{N}-R + HOH$$

$$R-\underset{\overset{||}{O}}{C}-OH + HO-R$$

$$\Downarrow$$

$$R-\underset{\overset{||}{O}}{C}-O-R + HOH$$

FIGURE 8.4 Reaction between an amine and an acid to form a polyamide.
FIGURE 8.5 Reaction between an alcohol and an acid to form an ester.

8.3 POLYMERIZATION REACTIONS

The reactions shown in Figure 8.4 and Figure 8.5 would not form polymers. The product would be a new molecule and water. However, in the cases shown in these figures, each reaction has only one functional group.

8.3.1 CONDENSATION

If each reactant has two functional groups, then it is possible to form a polymer. This type of reaction, like the reactions shown in Figure 8.4 and Figure 8.5, are consideration reactions. Water is a product of the chemical reaction. The same is true for condensation polymerization. Water is a product of the reaction. While in industrial practice it is common for the by-product to be a chemical other than water, we are going to restrict our discussion to reactions in which water is a by-product.

Consider the potential product of reacting $HOOC(CH_2)_6COOH$ and $H_2N(CH_2)_6NH_2$ as shown in Figure 8.6. This reaction is very similar to the one shown in Figure 8.4. The amine functional group (NH_2) and the acid functional group ($COOH$) will react and form an amide *linkage* — $NHCO-$ and water. This is shown in Figure 8.7. The reaction will not cease in this case. There is no possibility that one and only one $HOOC(CH_2)_6COOH$ molecule reacted with one and only one $H_2N(CH_2)NH_2$ molecule. Therefore, more than one of the product molecules formed. The product molecule has an amine functional group (NH_2) and an acid functional group ($COOH$). These can react and form an amide linkage — $NHCO-$ and water. This is shown in Figure 8.8. The reaction will continue, and a molecule of "n" units will form. The single unit enclosed in [S] is called a *mer*, and because the molecule has many *mers*, it is called a polymer. The $[S_n]$ is important; it clearly denotes that the material is a polymer.

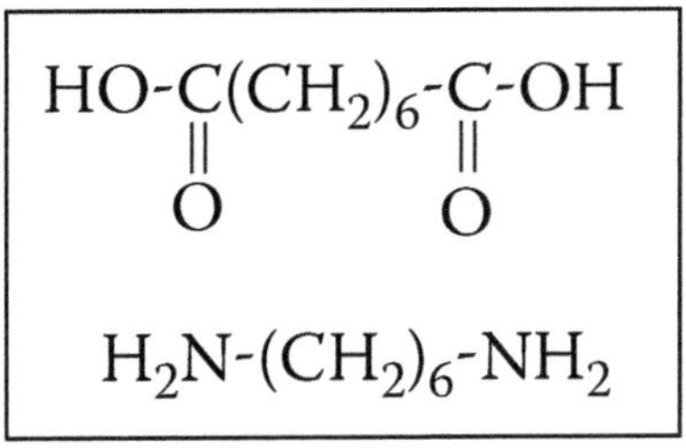

$$HO-\underset{\underset{O}{\|}}{C}(CH_2)_6-\underset{\underset{O}{\|}}{C}-OH$$

$$H_2N-(CH_2)_6-NH_2$$

FIGURE 8.6 Two reactants that could form a polyamide.

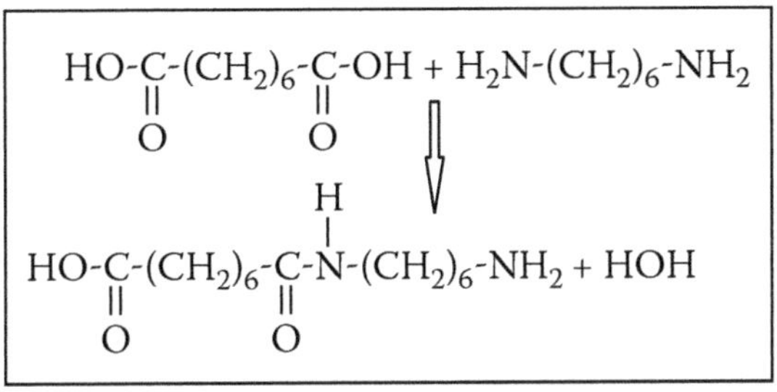

$$HO-\underset{\underset{O}{\|}}{C}-(CH_2)_6-\underset{\underset{O}{\|}}{C}-OH + H_2N-(CH_2)_6-NH_2$$

$$HO-\underset{\underset{O}{\|}}{C}-(CH_2)_6-\underset{\underset{O}{\|}}{C}-\overset{\overset{H}{|}}{N}-(CH_2)_6-NH_2 + HOH$$

FIGURE 8.7 The reaction of a difunctional alcohol and a difunctional amine.

$$n\left[HO-\underset{\underset{O}{\|}}{C}-(CH_2)_6-\underset{\underset{O}{\|}}{C}-\overset{\overset{H}{|}}{N}-(CH_2)_6-NH_2\right]$$

$$HO\left[\underset{\underset{O}{\|}}{C}-(CH_2)_6-\underset{\underset{O}{\|}}{C}-\overset{\overset{H}{|}}{N}-(CH_2)_6\right]_n-NH_2 + n\text{-}1 \ HOH$$

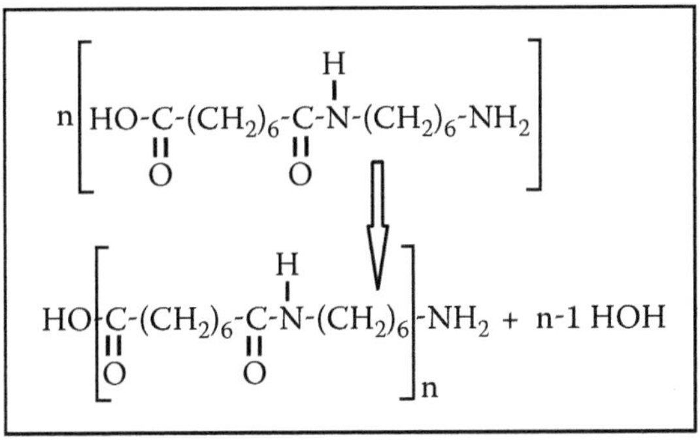

FIGURE 8.8 Formation of a polyamide from the product of a difunctional alcohol and a difunctional amine.

8.3.2 ADDITION

Some polymers, called addition polymers, are formed without producing a by-product. Before discussing addition polymers, or the reactions by which they are formed, two facts must be made clear.

First, a carbon–carbon double bond is not two carbon–carbon single bonds. The second bond is not as strong as the first bond. It is called a π-bond and

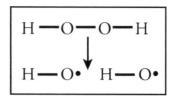

FIGURE 8.9 Formation of two free radicals from hydrogen peroxide.

overlaps the first bond that forms. Therefore, energy is reduced if a carbon–carbon double bond is replaced by two carbon–carbon single bonds.

Second, it is possible for a molecule to exist with an unpaired electron, although only for a short time. Such a molecule is electrically neutral; it has no charge. These molecules are extremely reactive.

Molecules containing carbon–carbon double bonds can be attacked by free radicals and form polymers. This is because the product of the reaction is a free radical and can react with another molecule containing another carbon–carbon double bond. Just like the product of the condensation reaction had two functional groups and could react with another molecule, so can the free radical.

An addition polymer is formed by adding a small (maybe one part per million) amount of initiator and monomer (a molecule with a carbon–carbon double bond). In the first step the initiator forms two free radicals. This is shown in Figure 8.9, where the O–O bond in H_2O_2 splits and forms two HO free radicals. Note that there is no charge on the HO free radical; it is not an OH$^-$ ion. This occurs because the O–O bond is weak, requiring only 142 kJ/mol of energy to break. In contrast, C–H single bonds require 347 kJ/mol of energy to break. In the organic reactions we will discuss, C–H, C–C, and C=O bonds will not break.

The free radical will react with the second bond in the C=C bond. The product that forms is a free radical as shown in Figure 8.10. The free radical (the HO molecule with an unpaired electron) will form a C–O bond, leaving an unpaired electron on the molecule. Note that this free radical is also unstable. The reaction can thus continue, and a polymer will form as shown in Figure 8.11. Because only a small amount of initiator is used, it is unlikely that the same double-bonded molecule will be attacked by two free radicals. If this happened, the reaction would terminate and no polymer would form. Because the product of the reaction shown in Figure 8.10 and Figure 8.11 is reactive (just like the product shown for condensation polymerization), the chemical reaction will continue and a polymer will form. The polymer which forms is shown in Figure 8.12. Note that the HO groups are not shown as part of the polymer. This is because they are present in very small amounts.

8.3.3 Cross-Linking

Cross-linked polymers are polymers in which the chains are physically connected by other polymer chains. Such molecules are really three-dimensional

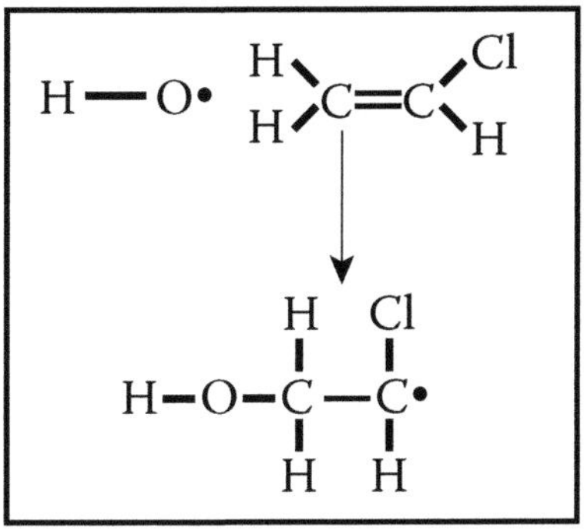

FIGURE 8.10 Reaction of free radical and molecule containing a C=C bond.

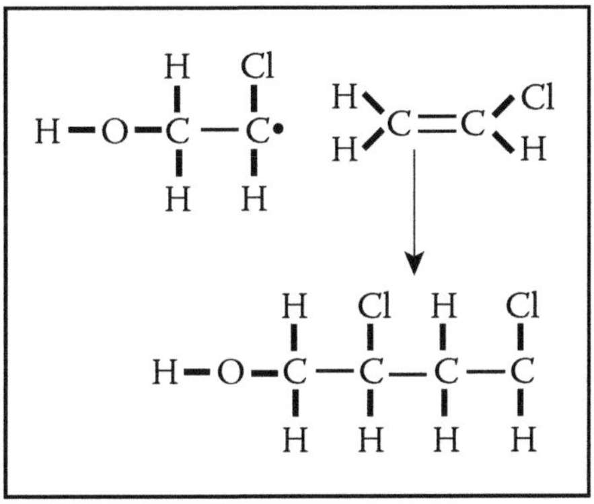

FIGURE 8.11 Continuation of reaction shown in Figure 8.10.

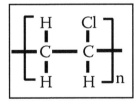

FIGURE 8.12 Polymer that forms from the reaction shown in Figure 8.11.

networks. Cross-linked polymers can form through either addition or condensation polymerization.

Condensation polymers form because the product of a chemical reaction has two functional groups that can keep reacting with each other. If the product contains more than two functional groups, then it is possible to connect the chains that form. This is shown schematically in Figure 8.13. The polymer that forms will have extra functional groups as shown in Figure 8.14

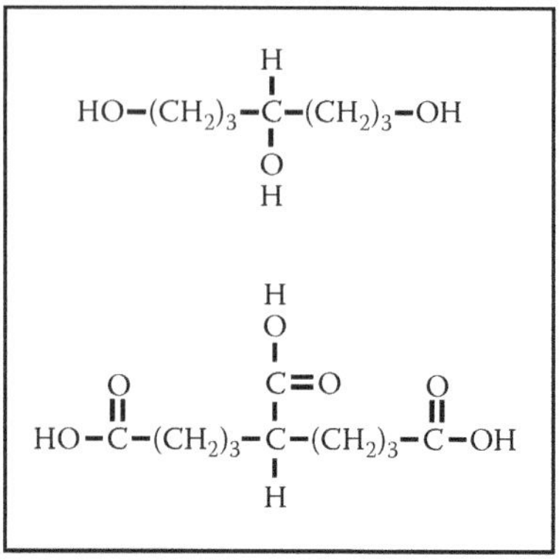

FIGURE 8.13 Reactants to form a cross-linked polymer through condensation polymerization.

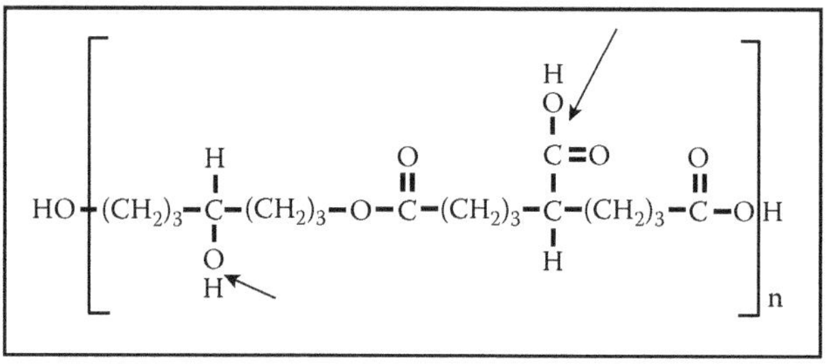

FIGURE 8.14 Polymer formed from reactants shown in Figure 8.13.

The arrows in Figure 8.14 show the extra functional groups that could continue to react. This reaction is shown in Figure 8.15.

The extra OH group shown in Figure 8.15 can react with the extra COOH group in Figure 8.15. The result is shown in Figure 8.16. Note that the polymer chains are physically connected; it is essentially one molecule.

While cross-linking polymerization as shown in Figure 8.13 through Figure 8.16 is easy to visualize, in most cases cross-linking occurs through addition polymerization. To form a cross-linked polymer through addition polymerization, it is required that the polymer have an extra reactive group, just like an extra functional group was required to form a cross-linked polymer through condensation polymerization. Thus, a polymer with a C=C bond in the chain can cross-link. The most common example is polybutadiene, shown in Figure 8.17. The monomer $CH_2CHCHCH_2$ reacts with a small amount of H_2O_2. Addition polymerization occurs, and the polymer forms as shown. As was shown for cross-linking through condensation polymerization in Figure 8.15 and Figure 8.16, when the polymer contains reactive groups, cross-linking can occur.

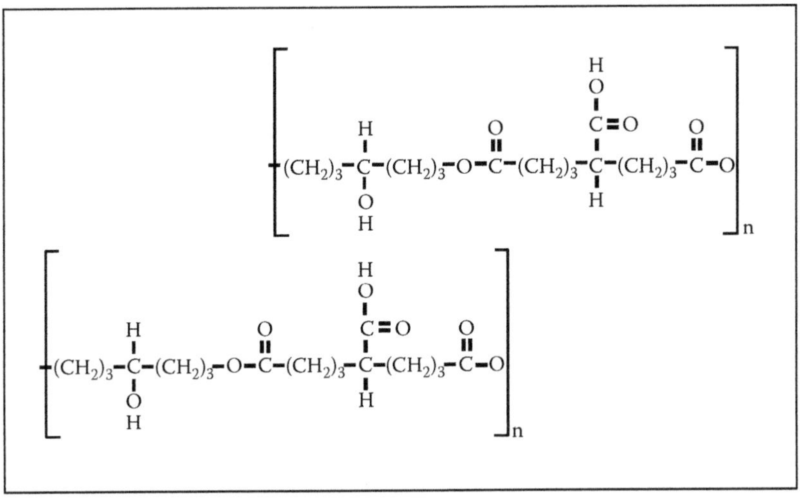

FIGURE 8.15 Reaction of extra functional groups prior to the formation of a cross-link.

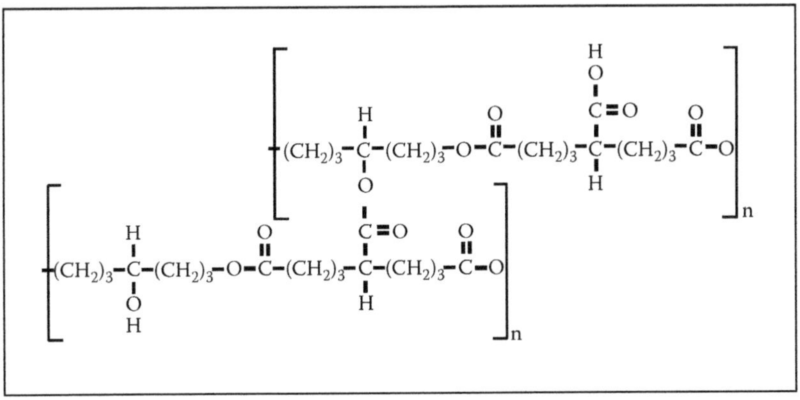

FIGURE 8.16 Cross-linked polymer formed by condensation polymerization. This is the product of the reaction shown in Figure 8.15.

The cross-linking of polybutadiene can also occur with the addition of sulfur. This is called vulcanization. Sulfur has two unbonded electrons, as shown in Figure 8.18. The reaction is more complicated than S_2 (which does not exist in nature) reacting with the polymer shown in Figure 8.17. However, it is clear in Figure 8.18 that two double bonds in adjacent polymer molecules have been replaced by a series of single bonds, which connect the two molecules. It should be noted that many epoxies are also cross-linked. The epoxy group is shown in Figure 8.19. The bonds in the COC group have a bond angle of 60°. The equilibrium angle is 109.5°. Thus, there is a lot of excess energy in the molecule,

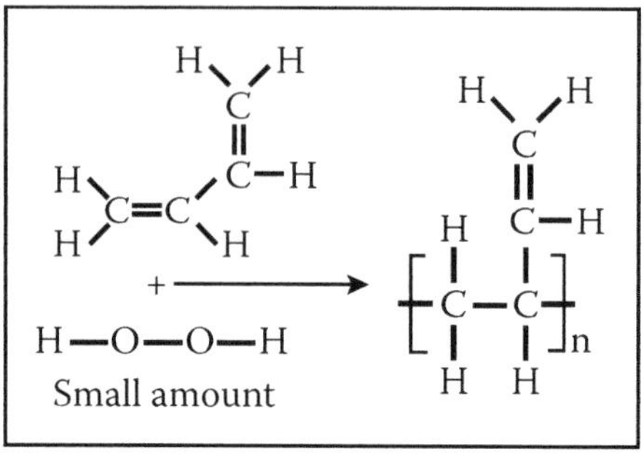

FIGURE 8.17 Formation of polybutadiene.

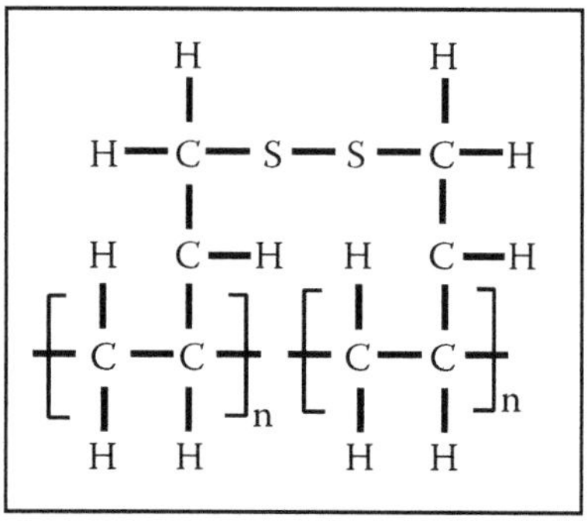

FIGURE 8.18 Cross-linking of polybutadiene.

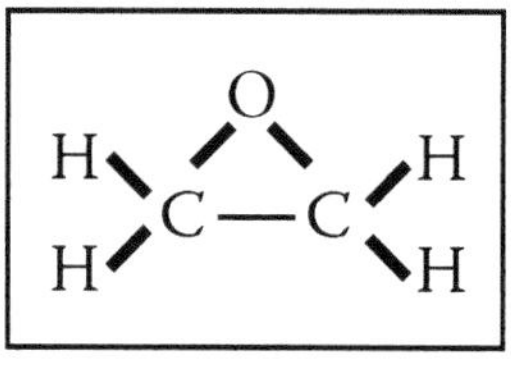

FIGURE 8.19 Epoxy group.

and it is very reactive. Polymers with epoxy groups in the chain can be cross-linked.

8.4 MECHANICAL PROPERTIES OF POLYMERS

As for all materials, the mechanical strength of polymers depends on structure. The strength of polymers can vary greatly, and the origin of this variation is the structure of the molecule.

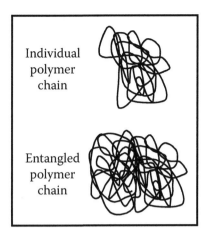

Individual polymer chain

Entangled polymer chain

FIGURE 8.20 Arrangement of polymer chains.

8.4.1 ORIGIN OF STRENGTH: INHIBITION OF CHAIN SLIDING

Consider that a polymeric material (the size of which you could hold in your hand) contains millions of long chain molecules. Each of these long chain molecules could be a polymer in which the degree of polymerization, "n," is 10^5 to 10^6. This means that the molecule contains several hundred thousand C–C bonds in the chain. There is no way (except in very exceptional cases) that the chain will be one straight line. Rather, each individual chain will be jumbled, and the combination of chains will be entangled. This is shown in Figure 8.20.

When the polymer is subjected to a mechanical load, the chains shown in Figure 8.20 disentangle. If this disentanglement is difficult, the polymer is strong. If not, the polymer is weak. In order for disentanglement to occur, the chains must slide past each other. Thus, polymers in which chain sliding is easy are weak, and those in which chain sliding is difficult are strong.

All polymers have some strength because of the large number of intermolecular forces between the polymer chains. Recall from chemistry that the magnitude of intermolecular forces affects the melting and boiling point of water, alcohol, chlorine, and many other chemicals. These forces are also weak compared to the covalent, ionic, or metallic bonding that holds the molecules/atoms together. However, because polymers are such large molecules, the large number of these interactions inhibit chain sliding and give the polymer strength.

There are four features which inhibit chain sliding:

- Connecting the chains through cross-linking is the most effective method of inhibiting chain sliding. Therefore, cross-linked polymers are among the strongest.
- Hydrogen bonding is the strongest of the intermolecular forces. If the polymer has a chemical structure in which hydrogen bonding can occur

between the polymer chains, chain sliding will be inhibited and the polymer will be strong. This is the case for nylons.

• If the polymer has benzene ring (aromatic) groups in the chain, chain sliding will be inhibited.

• If the polymer has large side groups, chain sliding will be inhibited, as the bulky side groups will interfere with chain sliding.

Hydrogen bonding can occur when there are F–H, O–H, or N–H bonds in the polymer. Note that C–H bonds do not participate in hydrogen bonding. This is shown in Figure 8.21. The hydrogen attached to the nitrogen becomes a bare proton, which is an extremely concentrated unit of positive charge. This can interact with the electron cloud from the adjacent molecule. Because there will be an extremely large number of hydrogen bonds between the polymer chains, the chain sliding is inhibited and the polymer is very strong. Hydrogen bonding cannot occur in molecules such as polyethylene, polyvinyl chloride, or polycarbonate. As shown in Figure 8.22, there are no F–H, O–H, or N–H bonds in any of the polymers.

Polycarbonate is a very strong polymer. As shown in Figure 8.22, polycarbonate has two benzene rings in the polymer chain. It also has two CH_3 side groups. The benzene ring in the polycarbonate chain, $-C_6H_4-$, although symbolized by a circle inside a hexagon, is quite large. Note also that it is not possible for polycarbonate as shown in Figure 8.22 or Kevlar as shown in Figure 8.21 to cross-link. There are no carbon–carbon double bonds in the polymer chain. Benzene rings do not contain carbon–carbon double bonds. The six electrons, symbolized by the circle, are shared equally among the six carbon atoms. This makes the ring itself very stable, and thus it will not break, enabling cross-linking.

Two of the polymers shown in Figure 8.22, polyvinyl chloride and polycarbonate, have side groups: the Cl in polyvinyl chloride and the two CH_3 groups in polycarbonate. These are moderate-size side groups and will inhibit chain sliding. Some polymers such as polymethylmethacrylate, which is shown in

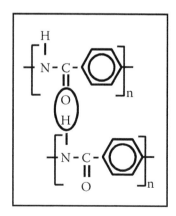

FIGURE 8.21 Hydrogen bonding between two Kevlar polymer chains.

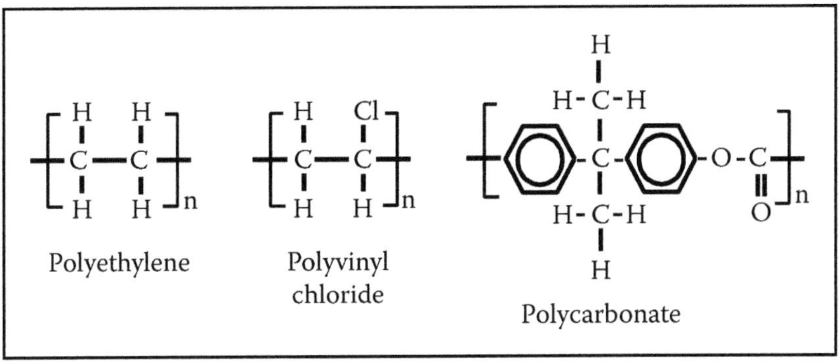

FIGURE 8.22 Polymers in which hydrogen bonding is not possible.

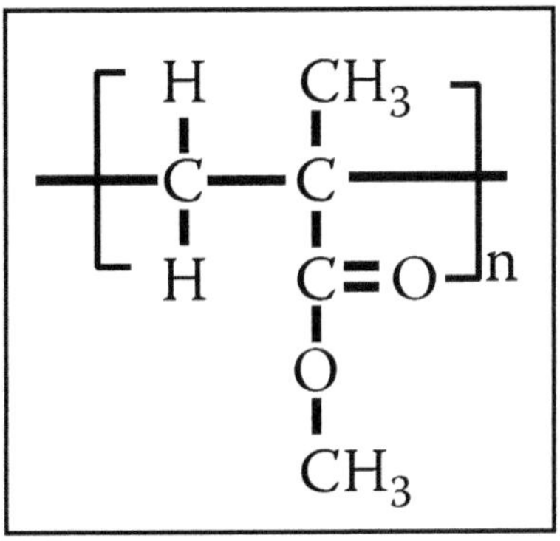

FIGURE 8.23 Polymethylmethacrylate.

Figure 8.23, have very large side groups. It would be very difficult to slide chains of this polymer past each other.

8.4.2 THERMOPLASTICS AND THERMOSETS

Polymers are often classified as thermoplastics or thermosets. The structure of the polymer determines whether it is a thermoplastic or a thermoset.

Thermosets cannot be reformed when heated. They do not melt — they sublime. Cross-linked polymers are thermosets. Because the chains are connected, it is not possible to use thermal energy to enable chain sliding.

Thermoplastics can be reformed on heating. Thermal energy enables chain sliding at moderate loads. Polymers that are not cross-linked are thermoplastics. Thermoplastic materials are recyclable.

8.4.3 CRYSTALLINITY

There are two types of polymers: amorphous and semicrystalline. No polymer is completely crystalline. Semicrystalline polymers are those in which there are crystalline and amorphous regions in the polymer chain. Crystalline regions are those in which the chain can neatly fold over on itself. Amorphous regions are those in which chain folding does not occur. Bulky side groups and benzene rings in the chain inhibit chain folding and increase the likelihood that the polymer will be completely amorphous.

8.5 TEMPERATURE-DEPENDENT MECHANICAL PROPERTIES

The mechanical strength of polymers changes dramatically between temperatures between −50 and 200°C. Polypropylene loses 30% of its strength when raised from 25 to 50°C. This means that normal weather conditions can cause marked decreases in the strength of polymers. This must be considered when using them in designs. Most metals and ceramics that would be considered for load-bearing applications do not show this behavior at temperatures we commonly experience.

As stated earlier, no polymer is completely crystalline. Thus, the glass transition temperature, T_g, is a critical design consideration. There are four cases to consider:

1. An amorphous polymer at temperatures below the glass transition temperature will behave as a rigid solid. It will be brittle and fairly strong. This is shown in Figure 8.24.
2. An amorphous polymer at temperatures above the glass transition temperature will behave as viscous liquid. Strength will be minimal, and the material will creep (flow under constant load) readily. This is shown in Figure 8.26.

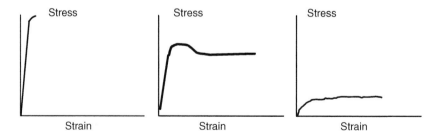

FIGURE 8.24 Stress–strain behavior of polymers at temperatures below T_g. FIGURE 8.25 Stress–strain behavior of semicrystalline polymers at temperatures above T_g. FIGURE 8.26 Stress–strain behavior of amorphous polymers at temperatures above T_g.

3. A semicrystalline polymer at temperatures below the glass transition temperature will behave as two rigid solids. It will be brittle and fairly strong. This is shown in Figure 8.24 and is similar to the first case in curve shape.
4. A semicrystalline polymer at temperatures above the glass transition temperature will behave as a rigid solid surrounded by a viscous liquid. The strength will be significantly reduced. This is shown in Figure 8.25.

The most dramatic change will occur at temperatures close to the glass transition temperature, T_g.

DEFINITIONS

Addition polymerization: Chemical reaction, usually requiring an initiator, where unsaturated (bifunctional) double-carbon-bonded monomer molecules are linked together.

Additives: Substances such as plasticizers, fillers, colorants, antioxidants, and flame retardants added to a polymer for a specific objective.

Alloy: In polymers, a blend of polymer or copolymer with other polymers.

Branched polymer: Main chain with attached side segments.

Catalyst: Substance that changes the rate of a reaction.

Condensation polymerization: Reaction of two monomers to produce a third, plus a by-product, usually water or alcohol.

Degree of polymerization (DP): Number of mers per molecule.

Engineering plastic: General term covering all plastics with or without fillers or reinforcements that have mechanical, chemical, and thermal properties suitable for use as a construction material.

Glass transition temperature T_g: Reversible change in an amorphous polymer at which the molecular mobility becomes sufficiently small to produce brittleness. Below T_g, polymers and glasses are brittle.

Kevlar: Organic polymer composed of aromatic polyamide (aramids) often used as fiber in composites (DuPont™).

Polyamides (nylons): Tough crystalline polymer with repeated nitrogen and hydrogen groupings, which have wide acceptance as a fiber and engineering thermoplastic.

Polycarbonate (PC): Thermoplastic polymer derived from a direct reaction between aromatic and aliphatic dihydroxy compounds with phosgene or by the ester reaction. A clear, tough engineering polymer.

Polyethylene terephthalate (PET): Saturated thermoplastic resin made by condensing ethylene glycol and terephthalic acid. It is used as fibers, films, and in injection molded parts.

Polypropylene (PP): Tough lightweight polymer made by polymerization of propylene gas.

Polyvinyl chloride (PVC): Thermoplastic material composed of polymers of vinyl chloride used for piping, containers, and outdoor applications.

QUESTIONS

8.1   What is the molecular weight of phenol? Of formaldehyde? Write the polymerization equation for the combination of these two to produce a giant molecule of phenol–formaldehyde. What is the mechanism of the reaction?

8.2   Synthetic fibers of the polyvinyl acetate copolymer have a molecular weight ranging from 16,000 to 23000 g. What is the range of the wt% of chloride mer to the total?

8.3   Using a reference, list the typical values of high-, medium-, and low-strength polymers and list one polymer for each of these strength ranges. What type of polymers are they?

8.4   Sketch the stress–strain curves for (a) a typical thermoplastic $T > T_g$ < Tm; and (b) a typical thermoset.

8.5   What occurs on a molecular scale during creep of a thermoplastic polymer?

8.6   Does crystallinity increase the strength of a thermoplastic?

8.7   Compare the modulus of common metals, ceramics, and thermoplastics and thermoset materials (use a reference).

8.8   In general, how do the strength and ductilities of the thermoset compare to that of the thermoplastics?

SUGGESTED READING

American Society of Metals (ASM), *Engineered Materials Handbook*, Vol. 2, Engineering Plastics, ASM International, Metals Park, OH, 1988.

Modern Plastics Encyclopedia, McGraw-Hill, New York, 2002.

Moor, G. R. and Kline, D. E., *Properties and Processing of Polymers*, Society of Plastics Engineers and Prentice Hall, Englewood Cliffs, NJ, 1984.

Society of Plastics Engineers, www.4spe.org.

9 Composites

9.1 INTRODUCTION

A composite can be defined as a mixture of two or more materials that are distinct in composition and form, each being present in significant quantities (e.g., >5 vol %). Another source defines a composite as "the union of two or more diverse materials to attain synergistic or superior qualities to those exhibited by individual members." There are about as many definitions of composites as there are textbooks on materials. Volume 1 of the American Society of Metals (ASM) *Engineered Materials Handbook* has elected the following definition: "a combination of two or more materials differing in form or composition on a macroscale. The constituents retain their identities. . . .and can be physically identified." The words *macroscale* and *identities* or *distinct* are the key words in any definition of *composite*. This automatically rules out solid solutions as well as most metallic alloys that have their properties enhanced via the presence of microconstituent phases. We can group the composite materials as:

- Fiber-reinforced materials (fiber-reinforced polymers, metals, and ceramics)
- Particulate composites (cemented carbides, contact materials, etc.)
- Laminar composites (metal sheet and glass fiber composite layers in Arall and Glare)
- Sandwich structures

Glass fiber composite products for ships and carbon fiber composites used in aircraft and Formula 1 cars are common examples of composites. The fibers mainly carry the load, and the matrix, a polymer in those cases, holds the fibers together; that is, the matrix is responsible for the shape. Many more composites exist, such as cemented carbide and concrete, which belong to the group of particulate composites.

In composites consisting of identifiable constituents, each possessing its own individual properties, the strength and moduli can be expressed by the well-known rule of mixtures:

composite strength = (strength constituent 1 X fraction constituent 1)
+ (strength constituent 2 X fraction constituent 2)

or in symbol form,

$$\sigma_c = \sigma_1 V_1 + \sigma_2 V_2 \tag{9.1}$$

where V_1 and V_2 are the volume fractions of the two constituents, and σ is the strength. Although we have not yet said anything about the orientation and distribution of these phases, constituents, components, or whatever we like to call them, whenever they are present on a macroscale and retain their identifiable properties, they obey the rule of mixtures or some modification thereof. Such modifications account for their respective shapes, orientations, and bond-interface characteristics. In most structural or load-bearing composites, the strong phase (e.g., a covalently bonded carbon fiber) is designed to carry most of the load, while the weaker phases, such as a polymer or ductile metal, provide the toughness and ductility of the composite.

Many natural composites exist. Wood and many other plant tissues contain the cellulose chain polymer in a matrix of lignin, a phenol–propane polymer, and other organic compounds in a type of cellular structure (Figure 9.1). The load-bearing parts of the human body are in essence composites composed of the fibrous tissue, ligaments, which are cords as strong as rope and are attached to the bones, thereby binding the bone segments together.

The first synthetically manufactured composites were developed because no single homogeneous material could be found that had all of the properties sought for a given application. Fiberglass is considered by many to be our first engineered composite. After invention these glass fiber-reinforced polymers were immediately put to use in filament-wound rocket motor cases as well as small boat hulls and a myriad of smaller applications.

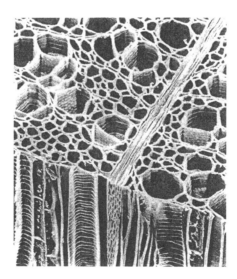

FIGURE 9.1 Cellular structure of basswood. (Courtesy of Tim O'Keefe, Cal. Poly.)

One can classify structural composites many ways. One way is according to the geometry of the load-bearing component, which can be in fiber, particulate, or laminar form, the latter often being made up of filaments or rows of fibers in matrixes that are shaped in a laminar arrangement. Particulate composites have no distinct arrangement of the phases or components, and the properties tend to be more isotropic than the fiber or laminar composites. Examples of particulate composites are cemented carbides, with tungsten carbide as hard particles embedded in a ductile metallic Co matrix; electric contact materials such as silver tin oxide, in which tin oxide particles (which reduce the wear caused by electric arcs during switching of high electric currents) are distributed in an Ag matrix (responsible for high electrical conductivity); and concrete. A more recently developed particulate composite is that of SiC particles in an aluminum alloy matrix.

Laminar composites are layered materials. They include the clad metals, often with a more corrosion-resistant metal bonded to a stronger one, fiber-reinforced laminates with different fiber orientations, and plywood and sandwich panels, in which two outer layers are separated by a less dense and lower-strength material. The latter includes the honeycomb core, which is placed between panels to form a sandwich-like structure. Figure 9.2 shows schematically a number of composite configurations, including a laminate honeycomb sandwich and a unidirectional fiber composite component contained in a single rotor blade. Fibrous composites are probably the largest composite category, so we will concentrate on this class of composites, particularly with respect to mathematical stress and modulus analysis. The advantage of composites is their high specific strength and ratio of strength to density. Figure 9.3 shows values for specific strength vs. specific modulus, which clearly reveals the advantage of composites over traditional

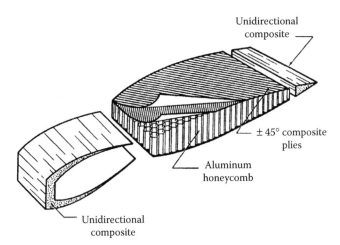

FIGURE 9.2 Composite rotor blade construction containing honeycomb, laminates, and unidirectional composites. (From R. L. McCullough, *Concepts of Fiber-Resin Composites*, Marcel Dekker, New York, 1971, p. 15. With permission.)

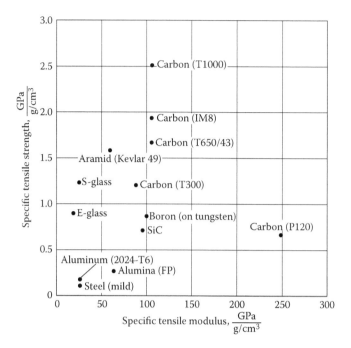

FIGURE 9.3 Specific tensile strength (strength-to-density ratio) vs. specific tensile modulus (modulus-to-density ratio) for various commercially available 65 vol % unidirectional epoxy-matrix composites and for steel and aluminum. Strength and modulus are based on fiber values. (From T. J. Reinhart, Introduction to composites, in *Engineered Materials Handbook*, Vol. 1, ASM International, Metals Park, OH, 1987, p. 28. With permission.)

alloys. In lightweight design fiber-reinforced composites are gaining increasing importance. In aircraft applications a number of components are made of fiber composites such as carbon fiber composite rudders and aileron, as shown for the Boeing 737-300 in Figure 9.4.

9.2 THE RULE OF MIXTURES

The rule of mixtures will be used throughout our discussions, so let us examine it in more detail. We use the fiber model for this purpose (that is, where fibers are used to reinforce a weaker matrix material). Several conditions must be met for our model to be of any analytical value. These are as follows:

- The fiber must be securely bonded to the matrix. We generally refer to this bond as a chemical bond, in the sense that atoms of each constituent react and bond together. However, the analysis will hold for a mechanically bonded interface as long as the interface is not the region of failure when the composite is subjected to a stress that exceeds the maximum value that the composite will sustain. Unfortunately, the interface is often the weakest region of the composite, and

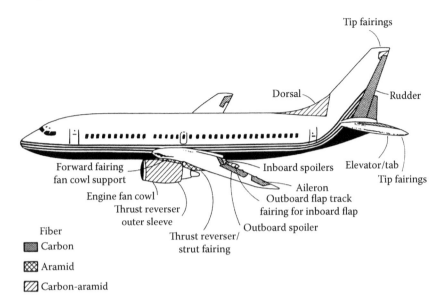

FIGURE 9.4 Composite applications on the Boeing 737-300. (From J. M. Anglin, Aircraft applications, in *Engineered Materials Handbook*, Vol. 1, ASM International, Metals Park, OH, 1987, p. 805. With permission.)

much research has been devoted to obtaining a strong chemical bond between the fiber and matrix. In general, a chemical bond is stronger than a mechanical bond, the latter being the point where the two components are interlocked by the penetration of the matrix material into the surface irregularities of the fibers.

- The fibers must be either continuous or overlap extensively along their respective lengths. Let us examine the extreme case of discontinuous fibers that do not overlap in one region of the composite, as depicted in Figure 9.5a by the region between the dashed lines. Since a chain is no stronger than its weakest length, the region containing no fibers will fail under a critical axial stress that equals the matrix fracture stress, and the fibers will not contribute to the strength of the composite. If, however, the fibers overlap, the load can be transferred to an adjacent fiber. This is the major role of the matrix: to transfer the load from the end of one fiber to an adjacent overlapping fiber (Figure 9.5b). Secondary roles of the matrix are to protect the fibers from surface damage and to blunt cracks that arise from fiber fractures.

- There must be a critical fiber volume, V_{fcrit}, for fiber strengthening of the composite to occur.

- There must be a critical fiber length for strengthening to occur. This critical length is also dependent on the fiber diameter d_f. The ratio of the critical length l_c to the critical diameter for that length, d_{fc}, is called the aspect ratio.

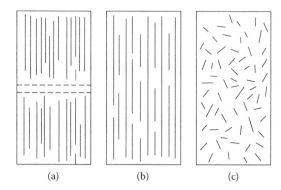

(a)   (b)   (c)

FIGURE 9.5 Schematics of a fiber composite (a) in which the parallel aligned fibers do not overlap in all regions of the composite; (b) in which the longitudinal aligned fibers overlap extensively (continuous fibers); and (c) of randomly oriented short fibers.

9.2.1 MATHEMATICAL ANALYSIS OF FIBER COMPOSITES

9.2.1.1 Continuous Longitudinal Fibers: Equal Strain and Elastic Conditions (Figure 9.6a)

When the fibers are continuous, and in this case we mean that the fibers are of sufficient length for extensive overlap, and when the composite is uniaxially stressed in the fiber alignment direction, the stress in the composite, σ_c, is simply that stated by the rule of mixtures:

$$\sigma_c = V_f \sigma_f + V_m \sigma_m \qquad (9.2)$$

where
  V_f = volume fraction of fiber
  σ_f = stress in fiber
  V_m = volume fraction of the matrix
  σ_m = stress in the matrix

Since $V_m = 1 - V_f$, Equation 9.2 becomes

$$\sigma_c = V_f \sigma_f + (1 - V_f)\sigma_m \qquad (9.3)$$

$$E_f \text{ (glass)} = 11 \times 10^6 \text{ psi} \quad E_m \text{ (epoxy)} = 0.5 \times 10^6 \text{ psi}$$

In most composites, the $V_f \sigma_f$ term is the predominate factor. The strain in the case above is made equal in both components by the conditions imposed. If

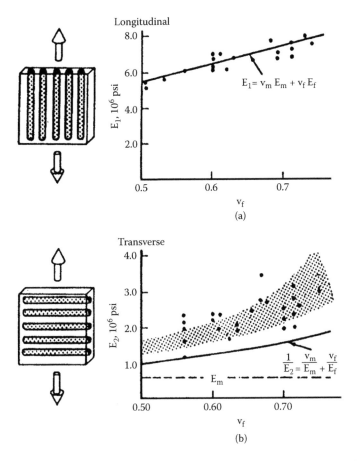

FIGURE 9.6 Comparison of experimental results with theory for the moduli of longitudinal and transverse loading of fibers. $E_1 = E_c$ for (a) and $E_2 = E_c$ for (b). (From R. L. McCullough, *Concept of Fiber-Resin Composites*, Marcel Dekker, New York, 1971, p. 31. Experimental data from L. C. Chamis and G. P. Sendeckyi, *J. Composite Mater.*, 2, 332, 1968. With permission.)

both components are stressed only elastically, $\sigma_f = E_f \varepsilon_\phi$, $\sigma_m = E_m \varepsilon_m$, and $\sigma_c = E_c \varepsilon_c$. Thus,

$$E_c \varepsilon_c = V_f E_f \varepsilon_f + V_m E_m \varepsilon_m$$

However, since all strains are equal, the composite modulus becomes

$$E_c = V_f E_f + V_m E_m \qquad (9.4)$$

The longitudinal arrangement of fibers in Figure 9.6a has been called the parallel reaction and applies just as well to thermal or electrical loads. Each fiber and the matrix will experience the same strain owing to the uniform load applied to the entire composite cross section. Again, the $E_f V_f$ is the major contributing term to the composite modulus E_c. If the fiber properties are two orders of magnitude greater than the matrix, the matrix contribution is negligible.

The equations above were derived for continuous fibers, but when the fibers are long and overlapping, as in Figure 9.5b, the equations are good approximations of the composite elastic strength and modulus. This is because the load (or stress) at the end of the fiber is transferred by the matrix to adjacent fibers. One can see this more clearly in Figure 9.7a, where the tensile stress and the shear stress at the fiber–matrix interface along the fiber length are depicted. At the fiber end points the normal stress is zero, and the high shear stresses cause the tensile stresses to be transferred to the adjacent overlapping fibers. Initial fiber fractures (Figure 9.7b) do not result in complete composite failure because of a coupling action between the fibers and the matrix. As the two broken fiber ends attempt to pull away from each other, the plastic flow of the matrix parallel to the stress (Figure 9.7c) resists this pulling action, thereby building up shear stresses in the matrix. However, the stress in the nonbroken fibers increases because they must carry a larger part of the load.

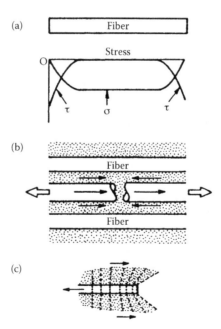

FIGURE 9.7 (a) Variation of the normal and shear stresses along the fiber when it is stressed along the fiber axis; (b) and (c) mechanism of load transfer from a broken fiber to adjacent fibers.

9.2.1.2 Continuous Fibers in a Transverse Arrangement (Figure 9.6b): Elastic Condition, Equal Stress

The equal stress state occurs when the stress is applied in a direction perpendicular to the fiber axes. In the case of fibers the cross-sectional area presented to the load in such a geometry is sufficiently small that the fibers may not make a significant contribution to the composite strength, except for composites of large volume percent fiber content (e.g., >30%). The stress is the same in both components, but the strain differs. The composite strain, ε_c, can be expressed as

$$\varepsilon_c = \varepsilon_f V_f + \varepsilon_m V_m \tag{9.5}$$

where

ε_f = strain in the fiber
V_f = volume fraction of the fiber
ε_m = strain in the matrix
V_m = volume fraction of the matrix

We can apply Hooke's law and express the composite modulus as

$$\frac{\sigma_c}{E_c} = \frac{\sigma_f}{E_f} \cdot V_f + \frac{\sigma_m}{E_m} \cdot V_m \tag{9.6}$$

However, since

$$\sigma_c = \sigma_f = \sigma_m$$

we can write

$$\frac{1}{E_c} = \frac{V_f}{E_f} + \frac{V_m}{E_m}$$

or

$$E_c = \frac{E_f \cdot E_m}{V_f \cdot E_m + V_m \cdot E'_f} \tag{9.7}$$

In the transverse arrangement the experimental results deviate considerably from the simple rule of mixtures. The numerical analysis shows that the behavior is dependent on:

- The shapes of the fibers
- Their packing geometry
- The distribution of spacings between fibers

Different mathematical analyses and modifications have been made of the simple rule of mixtures, but the comparison with experimental data shows good agreement with the simple rule of mixtures for the longitudinal situation but not for the transverse fiber geometry. However, the simple rule of mixtures gives a conservative estimate of the transverse moduli.

9.2.1.3 Continuous Fibers: Fiber Elastic and Matrix Plastic, Equal Strain Condition

In most cases of fiber composites, the strong fibers show little plastic deformation when the yield strength of the matrix has been exceeded. This is particularly true in the case of ceramic fibers in a metal matrix. Then the ultimate strength of the composite, σ_{cu}, is given by

$$\sigma_{cu} = \sigma_{fu}V_f + \sigma`_m(1 - V_f)$$ (9.8)

where σ_{fu} is the ultimate tensile strength of the fiber and $\sigma`_m$ is the flow stress of the strain-hardened matrix.

To obtain any strengthening via the fiber content, the ultimate strength of the composite, σ_{cu}, must be greater than the ultimate strength of the matrix, mu (i.e., cu > mu). Thus,

$$\sigma_{fu}V_f + \sigma`_m(1 - V_f) \geq \sigma_{mu}$$ (9.9)

This leads to a critical fiber volume, V_{fcrit}, for strengthening to occur. V_{fcrit} can be obtained by setting $V_f = V_{fcrit}$ in Equation 9.9 and solving as follows:

$$V_{f.crit} = \frac{\sigma_{mu} - \sigma`_m}{\sigma_{fu} - \sigma`_m}$$ (9.10)

A situation may arise from small fiber volumes, where all fibers are stressed to fracture and the strain-hardened matrix carries the load. Assuming such a condition, we can compute a minimum volume of fibers as follows:

$$\sigma_{cu} \geq \sigma_{mu}(1 - V_f)$$ (9.11)

which is arrived at by setting $\sigma_{fu} = 0$ in Equation 9.8. This expression defines the minimum volume fraction that must be exceeded for reinforcement to occur; thus,

$$V_{f,\min} = \frac{\sigma_{mu} - \sigma`_m}{\sigma_{fu} + \sigma_{mu} - \sigma`_m} \qquad (9.12)$$

However, $V_{f\min} < V_{f\text{crit}}$; therefore, $V_{f\text{crit}}$ is the fiber content that must be exceeded for fiber strengthening to occur. This analysis was intended for metal matrix composites; however, although polymer matrices do not strain-harden much and the strength of polymer fibers may be much less than that of ceramic fibers, the equations above are still applicable to such composites.

9.2.2 LOAD TRANSFER

For the load to be transferred to the adjacent fibers as illustrated in Figure 9.7, the matrix must be plastically deformed at the fiber end points. The transfer mechanism is the same whether the fiber breaks or whether it is merely the end point of the original fiber. The high shear stresses provide for this deformation. The fiber strength can be utilized properly only if the plastic zone in the matrix does not extend to the fiber mid-length before the stress in the fiber reaches the ultimate strength of the fiber. Equilibrating the forces that cause the shear stresses along the fiber over a length of half (the fiber is being loaded from both ends) to the force causing the tensile stress in the fiber, $_f$, we obtain

$$\sigma_f \frac{\pi d^2}{4} = \tau \cdot \pi d \frac{l}{2}$$

and solving for l yields

$$l_c = \frac{\sigma_f d}{2\tau_c} \qquad (9.13)$$

This length becomes the critical fiber length l_c, and becomes $_c$, which is the critical fiber–matrix bond strength, or the matrix yield strength, whichever is smaller. When $l > l_c$ the fiber can be loaded to its fracture stress, which is the objective of strength gain through fiber reinforcement by means of the load transfer via the plastic deformation of the matrix around the fiber end point. The fiber diameter, d, in Equation 9.13 is also the critical diameter d_{fc} for this critical length, and the ratio of l_c/d_{fc} is the critical aspect ratio, which can now be expressed in equation form as

$$\frac{l_c}{d_{fc}} = \frac{\sigma_{fu}}{2\tau_c} \qquad (9.14)$$

When $l \gg l_c$, somewhere in the vicinity of $l = 15 \, l_c$, and if l_c is also on the order of 100 fiber diameters (i.e., $l_c > 100 \, d_f$), the fibers behave essentially as continuous fibers. At the point in the composite where a fiber ends, the load is transferred to an adjacent fiber. Clearly, not all fibers break at the same stress or length. The average fiber length must be used in this analysis, but as long as $l > l_c$ and the bond holds, the load will be transferred as illustrated in Figure 9.7. Fiber breakage will continue, however, and probably at a faster rate as the load increases, since for every fiber fracture, additional load is transferred to adjacent fibers. Eventually, a sufficient number of fibers will, by virtue of their fracture, have lengths of $l < l_c$, which leads to incomplete load transfer and thus less contribution of the fibers to the composite strength. Continuation of fiber breakage will eventually lead to failure of the composite.

Discontinuous fibers are much shorter than the continuous ones and are used as both aligned and randomly oriented geometries, more often in the latter arrangement (Figure 9.5c). These randomly oriented short fibers include chopped glass fibers, aramid fibers, whiskers, and most carbon fibers. A correction factor based on the l/l_c ratio is needed for the case of discontinuous aligned fibers. When $l < l_c$ the average stress is simply $\sigma_{max}/2$. When $l > l_c$ a fraction of the fibers carry the same load as a continuous fiber, and the remaining fibers carry, on average, half of this stress. The various average stresses that the fibers carry, based on their lengths compared to l_c, are depicted in Figure 9.8. For discontinuous aligned fibers, when $l > l_c$ and $\sigma_f = \sigma_{fu}$, the strength of the composite can be expressed as

$$\sigma_{cu} = \sigma_{fu} \cdot V_f [1 - \frac{1}{2l}] + \sigma'_m V_m \qquad (9.15)$$

Composites of this type are mostly those of the short-chopped glass fibers; however, carbon and aramid aligned discontinuous fibers are also used to some extent. In the metal matrix composites, σ_m should be replaced by σ_m', the stress in the strain-hardened matrix, and σ_f by σ_{fu} the tensile strength of the whisker. For most aligned fiber composites the lengths are often hundreds of times l_c.

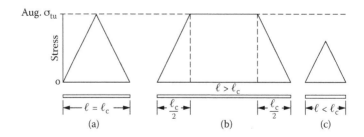

FIGURE 9.8 Relationship between the stress in the fiber and the fiber length. In (a) and (b) the fibers are being fully utilized in that they carry at some point along their length stresses equal to their ultimate tensile strengths using average fiber strengths, while in (c) the load-carrying capacity of the fibers is not fully utilized because of their short length.

For example, a fiber having an l/l_c ratio of 10 would in a composite exhibit an average fracture strength of only 5% less than if the fiber were continuous. Thus, appreciable reinforcement can be provided by discontinuous fibers when their lengths are much greater than the usual small critical lengths. For the case of discontinuous randomly oriented fibers, a different correction factor must be applied, and Equation 9.4 for the composite modulus becomes

$$E_c = \alpha E_f V_f + E_m V_m \qquad (9.16)$$

where α is a fiber efficiency parameter that is dependent on V_f, the E_f/E_m ratio, and fiber orientation. The efficiency of fiber reinforcement is often taken to be unity for an oriented fiber composite in the alignment direction and zero perpendicular to it. Experimentally, it is found to fall between about 0.2 and 0.6. The use of discontinuous randomly oriented fibers is quite common in glass fiber-reinforced polymers. They achieve isotropic properties in the composite and often ease of processing. Intricate shapes can be formed by a variety of processes.

9.3 FIBERS AND THEIR PROPERTIES

Many types of fibers are used in commercially available composites. The common ones include:

- Glass
- Carbon
- Boron
- Aramid polymer fibers

In the more advanced state of composite development we can also include silicon carbide, alumina (Al_2O_3), alumina silicates, mullite, and tungsten. There are also small fine fibers called whiskers that have been used in metal matrix whisker-reinforced composites, but because of their small length and diameter, they can be inhaled during manufacturing and maybe even during their service life. This may cause cancer, so whisker-reinforced materials have lost their importance.

The stress–strain curves for some common fibers are shown in Figure 9.9. The carbon (HM = high modulus and HT = high tensile strength) and boron fibers have the higher moduli and are lightweight and high-strength materials. The aramid fiber Kevlar 49 has a good combination of strength modulus and low density and accounts in part for its popularity. The aramid fibers are less expensive than the carbon and boron fibers. The specific moduli and specific strength of several fiber types, which take into account the fiber densities, are depicted in Figure 9.3. Some property data on the more common fibers are summarized in Table 9.1.

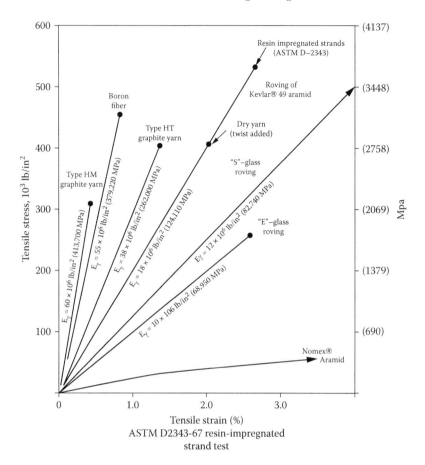

FIGURE 9.9 Stress–strain behaviors of various types of fibers. (From DuPont *Bulletin* K-5. September 1981. With permission.)

9.3.1 FIBER PROCESSING

9.3.1.1 Glass Fibers

Glass fibers are amorphous silica-based materials, chemically and physically existing of a three-dimensional network of an SiO_4–tetrahedric crystal structure. Depending on various additions to SiO_2, three glass fiber types are available: E-glass, S-glass, and C-glass. The approximate compositions of those glass fibers are given in Table 9.2.

As depicted in Tables 9.1 and 9.2, high strength and modulus values are achieved owing to strong covalent bonds between silicon and oxygen. However, the strength is dependent on the individual microstructural network and on addition of metal oxides. The amorphous structure is the reason for isotropic behavior,

TABLE 9.1
Properties of Selected Fibers from a Variety of Sources

Material	UTS [MPa (ksi)]		Modulus [GPa (msi)]		Density (g/cm³)
Al₂O₃	1380	(200)	2	(0.3)	3.95
Al₂O₃ silica coated, continuous	1900	(275)	2	(0.3)	3.95
SiC, continuous (Textron)	3450	(500)	400	(58)	3.0
Boron continuous (Textron)	3600	(520)	400	(58)	2.57
Kevlar 49 (DuPont)	2930	(420)	117	(17)	1.87
Nylon728	985	(143)	5.5	(0.8)	1.1
Stainless steel	1720	(250)	200	(29)	7.9
Tungsten	3900	(565)	350	(50.7)	19.3
E-glass	2410	(350)	69	(10)	2.55
Fused silica	3450	(500)	69	(10)	2.2
SiC whiskers	6900	(1000)	690	(100)	3.2
Discontinuous oxides (average of several types)	1500	(217)	175	(25)	2.75
Graphite	3200	(464)	490	(71)	1.9

TABLE 9.2
Approximate Composition in Percent of Various Glass Fibers

Constituents	E-glass	S-glass	C-glass
SiO₂	51–55	60	65
Al₂O₃	13–15	25	4
CaO	20–24	3	6
MgO			
B₂O₃	6–9	—	5
K₂O	<1	—	8
Na₂O			

Source: From D.M. Miller, Glass fibers, in *Engineered Materials Handbook*, Vol. 1, ASM International, Metals Park, OH, 1987, p. 45. With permission.)

whereas the above-described carbon and aramid fibers are anisotropic. The fiber diameters lie in the range of about 10 to 20 µm depending on the application.

The fiber production starts with melting quartz and additives at about 1400°C. After refining, which may take a couple of days, it will be forwarded to platinum bushings. The bushings are heated to a specific temperature so that the glass flows slowly through nozzle-like dies to filaments with a diameter of 2 mm. On high-speed coiling machines the fibers are stretched to their final diameters of, for instance, 14 µm.

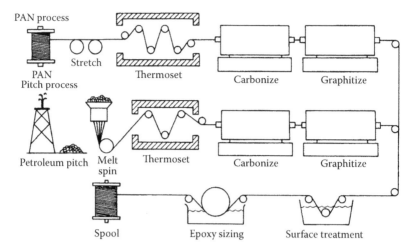

FIGURE 9.10 The processing sequence for PAN and mesophase pitch-based precursor fibers shows the similarities in the two processes. Highly oriented polymer chains are obtained in PAN by hot stretching, while high orientation in pitch is a natural consequence of the mesophase (liquid crystalline) order. (From R. J. Diefendorf, Carbon/graphite fibers, in *Engineered Materials Handbook*, Vol. 1, ASM International, Metals Park, OH, 1987, p. 50. With permission.)

9.3.1.2 Carbon and Graphite Fibers

Carbon and graphite fibers were originally developed for the aerospace industry but are now used in other applications such as sports equipment. Carbon fibers produced to date start with organic precursors, rayon, polyacrylonitrile (PAN), and isotropic and liquid crystalline pitches. The high-modulus fibers are made from liquid crystalline pitch precursors or from PAN, with the latter being the most common source. Figure 9.10 shows schematically the processing sequence for PAN and mesophase pitch-based precursor fibers.

The PAN fibers are first oxidized at 200 to 250°C to thermoset the fibers and then are carbonized at about 800°C to drive off the hydrogen and nitrogen. The resulting carbon fiber is passed through a higher-temperature (to 3000°C) furnace to graphitize the fiber. The modulus increases with the graphitization temperature, but the strength tends to maximize around 1500°C. Carbon fibers are available in tow (untwisted) bundles or rovings (yams, strands, or tows collected in a parallel bundle with no twist) containing many thousands of fibers. The PAN fibers vary in modulus from about 203 to 320 GPa, and the strength values range from 2400 to 5500 MPa. This is due to the orientation of the carbon chains in the fiber as depicted in Figures 9.11 and 9.12. The modulus of a carbon fiber is determined by the preferred orientations, microstructure, and elastic constants.

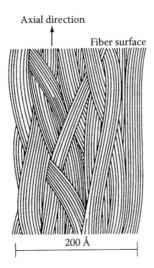

FIGURE 9.11 The undulating ribbon structure of the graphene layers for a PAN-based carbon fiber with a 400-GPa (60 × 106 psi) modulus. The ribbons at the surface have lower amplitude than in the core. There are about 20 graphene layers in the ribbons in the core and about 30 near the surface. (From R. J. Diefendorf, Carbon/graphite fibers, in *Engineered Materials Handbook*, Vol. 1, ASM International, Metals Park, OH, 1987, p. 50. With permission.)

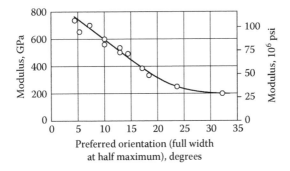

FIGURE 9.12 The modulus of a carbon fiber is determined by the preferred orientation, microstructure, and elastic constants. The relationship between modulus and preferred orientation for a pitch-based carbon fiber is shown. (From R. J. Diefendorf, Carbon/graphite fibers, in *Engineered Materials Handbook*, Vol. 1, ASM International, Metals Park, OH, 1987, p. 50. With permission.)

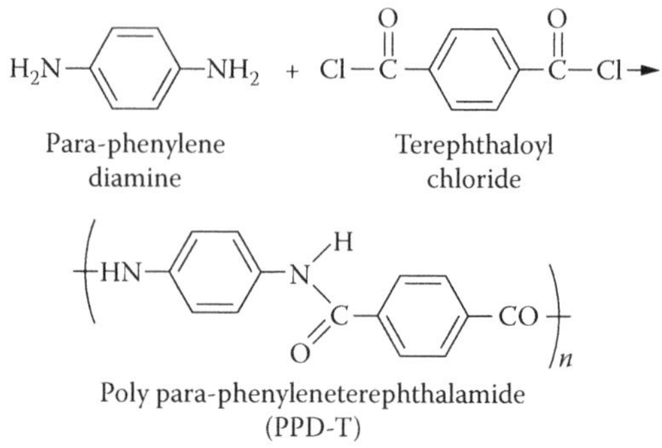

Para-phenylene diamine + Terephthaloyl chloride →

Poly para-phenyleneterephthalamide
(PPD-T)

FIGURE 9.13 Chemical structure of para-aramid fibers. (From J. J. Pigliacampi, Organic fibers, in *Engineered Materials Handbook*, Vol. 1, p. 54, ASM International, Metals Park, OH, 1987. With permission.)

9.3.1.3 Aramid Fibers

Kevlar aramid fibers are made from the condensation reaction of paraphenylene diamine and terepthaloyl chloride (Figure 9.13). They contain the aromatic rings, which lead to stiffness and high thermal stability. These fibers also belong to the rigid rod-like liquid-crystalline polymer class, because in solution they can form ordered domains in parallel arrays. When the solutions are extruded through a spinneret and drawn, the liquid-crystalline domains can align, giving an anisotropic high strength and high modulus in the fiber direction. There are different grades of Kevlar, with Kelvar 49 being the most popular. They are available in a range of yarn and tow counts as well as short forms. Tensile strengths range from 3400 to 4100 MPa, and the moduli range from 83 to 186 GPa.

9.3.1.4 Boron Fibers

Boron fibers are made by a chemical vapor deposition process whereby a tungsten wire is continuously drawn through a glass tube containing a mixture of boron trichloride and hydrogen. The BCl_3 vapors deposit boron on the hot wire, with the fiber diameter increasing as it passes through the reactor tube, usually attaining 100 to 400 μm diameter. The tungsten core is small (about 2 μm), thus yielding predominantly the lightweight boron of around 2.3 g/cm^3 density. Tensile strengths of 3600 MPa and moduli of 400 GPa are typical mechanical property values obtained on this wire. Most boron fibers are available in the form of a boron–epoxy *preimpreg*nated tape (called a prepreg) that the composite manufacturer uses to build ply layups and part fabrication. Boron–epoxy composites have been used in several military aircraft, the space shuttles, and helicopters. These fibers have also been used to reinforce metal matrices such as titanium–boron and aluminum–boron composites.

9.3.1.5 Other Fibers (SiC and Al$_2$O$_3$)

Silicon carbide fibers show a density of 3 g/cm^3, a tensile strength in the range 2800 to 4600 MPa, and a modulus around 400 GPa. Several variations of SiC fibers have been produced with carbon-containing silicon coatings to prevent reaction with aluminum matrices and others to be compatible with titanium matrices. These fibers have been used on a development basis for reinforcing aluminum and titanium and to some extent for organic and ceramic matrices.

Various ceramic fibers have also been developed. For example, oxide fibers with high alumina content have strengths in the range of 1700 to 2200 MPa and moduli in the range of 155 to 380 GPa.

9.4 POLYMER MATRIX COMPOSITES

Polymers are inexpensive, corrosion resistant, and easy to form into complex shapes, and they have low densities, just to mention a few advantages. They are increasingly used in automotive and many other applications. For high-load-bearing components, their strength is too low, but in combination with the high strength and moduli of fibers they reveal properties that can be achieved by no other material, especially high strength in combination with low density.

Therefore, polymers reinforced by glass, carbon, and aramid fibers are the materials of choice for many auto body panels, aircraft structural components, appliances, and a host of other industrial and consumer products. Almost all of the *neat* polymers discussed in Chapter 8 are also available with glass reinforcement, and many come with carbon and aramid fiber support. Figure 9.14 shows the increase in strength and modulus of several polymers as a function of glass content. The data were adapted from the 1990 issue of *Modern Plastics Encyclopedia*. The data in the encyclopedia showed a range of values, but for simplicity, average values are for Figure 9.14. These graphs were constructed to show the general trend of how the strength and moduli increase with glass fiber content. These data should not be used for design purposes. Some data are listed for both long and short fibers. The long fiber composites tend to have higher strengths for the same fiber volume content, but not by a large amount.

In the case of fibers in polymers, one should be aware that the equations for the critical fiber length must be modified to account for the fact that in polymer matrix composites the matrix is not capable of extensive plastic deformation. Usually the interfacial stress will cause delamination at the fiber end. Subsequent to delamination the matrix flows by the fiber and the tensile stress is transferred by a frictional force associated with this displacement. Thus, τ_c in Equations 9.13 and 9.14 should be replaced by μP, where μ is the frictional coefficient between the fiber and matrix and P is an internal pressure resulting from the shrinkage of the polymer matrix around the fiber during curing or incompatible lateral deformation. The product μP in polymer composites is on the order of one-tenth that of τ_c for metal matrix composites. In addition to increasing the strength and moduli of polymers, the glass fibers improve their high-temperature capabilities,

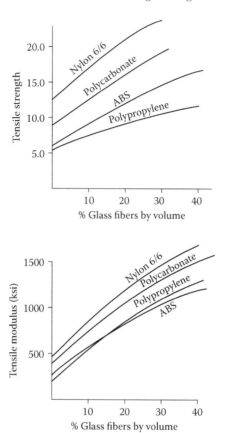

FIGURE 9.14 Tensile strength and moduli vs. volume percent of glass fibers for some common polymer–glass composites. (Adapted from *Modern Plastics Encyclopedia*, McGraw-Hill, New York, 1990.)

significantly increasing the heat deflection temperature and the continuous-use temperature, as depicted in Figure 9.15. There has been a tendency in the literature to classify polymer matrix composites into low-performance and high-performance categories based primarily on their strength and use temperatures.

Polyester and vinyl ester resins are the most widely used resin matrices. They are usually glass filled since they do not adhere well to carbon and aramid fibers (which are also more expensive). These composites fall into the low-performance category and are used in a multitude of applications, including automotive parts, bathtubs, showers, and piping. They do not require high-temperature strength. Most matrices are thermoset polymers, which have historically been the major matrix material for composites, although thermoplastics are making inroads.

In the high-performance category are three major matrices, all thermosets: the epoxies, phenolics, and polyimides. The epoxies are the most common resin matrix for the high-performance composites and have been used with just about

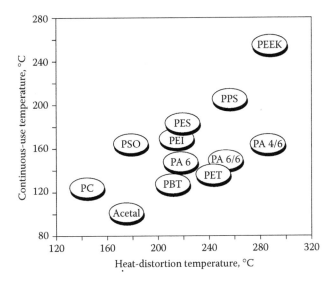

FIGURE 9.15 Continuous-use temperatures (10,000 h) and heat-distortion temperatures of glass-reinforced (30 vol %) thermoplastic polymers. (From D. F. Baxter, Jr., Green light to plastic engine parts, *Adv. Mater. Process.*, May, 28, 1991.)

all fibers, but the development of boron, carbon, and aramid fibers made the fabrication of advanced polymer matrix composites possible. Their strengths, reinforced with boron, carbon, and glass, are compared with the alloys of steel, titanium, and aluminum in the stress–strain curves of Figure 9.16. The alloys are not given, but the strengths are approximately those of a 2024-T6 aluminum alloy, a Ti-6% Al-4% V alloy, and a midrange value for steels. These curves show that boron composites are about equally stiff, but the boron composites are stronger and the weights of the composite materials are much less than those of the metals, even for the lightweight aluminum alloys. The epoxy matrix composites have been used in military aircraft structural components since 1972 and today they are common materials in commercial aircrafts. They are also utilized in sporting goods and printed circuit boards.

The reinforced phenolic (phenol-formaldehyde) resins are not used as much as the epoxy composites, one reason being their processing difficulties involving the water by-product that is released as a result of the condensation reaction, resulting in voids and loss of strength. However, phenolics have high heat resistance (200 at 500°F) and good char strengths. They also produce less smoke and toxic by-products on combustion and are often used as aircraft interior panels. Their good char strengths make them desirable materials for ablative shields for reentry vehicles, rocket nozzles, nose cones, and rocket motor chambers. They have been used with glass, high silica, quartz, nylon, and graphite fibers.

The polyimide thermoset matrices with boron, glass, and carbon fibers have been used to 150°C (302°F). They are candidate materials with carbon fibers for

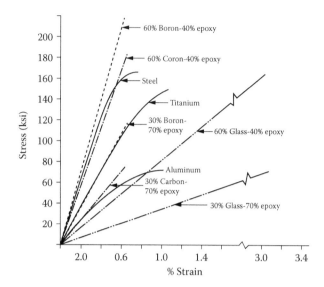

FIGURE 9.16 Strengths of epoxy composites compared to steel, titanium, and aluminum alloys. (From R. M. McCollough, *Concepts of Fiber-Resin Composites*, Marcel Dekker, New York, 1971, p. 9. With permission.)

use to 316°C (600°F). They cost more than epoxies, primarily because of processing problems.

Although the thermosetting polymer matrices have been the favorite for high-performance polymer composites, inroads are being made by certain thermoplasts. These consist primarily of the high-temperature thermoplasts such as Polyetheretherketone (PEEK), Polyphenylene Sulfide (PPS), Polyetherketone (PEK), and polyamideimide (PAI). The polyamideimide composites are molded as thermoplasts but then postcured in the final composite to produce partial thermosetting characteristics and high-temperature resistance to 190°C (375°F). They have better temperature resistance than the polyvinyl and polyvinyldene polymers and reportedly can be used continuously to temperatures as high as those of most epoxy composites. PEEK thermoplastic composites have shown continuous-use temperatures to 250°C (480°F). Another drawback to the thermoset composites is their brittleness, just as in nonreinforced thermosets. Some gains have been made to improve their toughness by the addition of about 20 wt % thermoplasts to the composite.

9.4.1 Fabrication of Composites

With the large variety of composites available today, together with the almost unique way in which each composite is manufactured, a complete coverage of all the fabrication methods would be a volume in itself. Generally, we can simplify a discussion of the fabrication methods by grouping them not according to materials but according to the geometry of the reinforcing material.

The small randomly oriented fibers, whiskers, and particulate composites can be grouped together because the fabrication methods, exclusive of fiber preparation, which was discussed earlier, are very similar, if not identical to the very common metal- and polymer-forming processes. The polymer matrix composites of these types can be formed in the same manner as the nonreinforced polymers. These methods include injection and pressed molding, extrusion, and sheet molding. The metals and some ceramics matrix composites are formed by similar methods whether or not small reinforcement particulates are present. Thus, we will not repeat all of the various pressing, casting, forging, extrusion, rolling, die casting, and molding processes. The flexibility provided by the large variety of forming methods is one of the more attractive features of these small-particle composites. Most of these forming methods, with a possible exception in some cases of extrusion, yield a composite material with isotropic properties. Some preforms and prepregs of these composites are also used in laminates, permitting additional design versatility.

The continuous and nearly continuous long-fiber composites as well as the laminar type are fabricated in a large variety of methods to give shapes with intended directional properties. These methods include filament winding, braiding, ply layup methods, pultrusion, impregnation and coating of fibers, and honeycomb structures.

9.4.1.1 Fiber Weaving

To obtain more isotropic properties in many fiber composites, multidimensional fiber weaving methods were developed. Figure 9.17 shows various fiber weaving construction forms. It is obvious that the plain fiber weavings lead to anisotropic composite behavior. Three-dimensional configurations, manufactured by automated computer-controlled three-dimensional weavers, are also extensively used. The objective of three-dimensional weaving is to produce orthogonal preforms as pictured in Figure 9.17g. This weaving machinery is described in the *ASM Handbook*. Fibers that have been woven include carbon, glass, silica, alumina, silicon carbide, and aramid.

9.4.1.2 Fiber Braiding

Braiding is a textile process similar to weaving, in which two or more yarns are interwoven in a variety of patterns. Braiding has many similarities to filament winding in that dry or prepreg yarns, tapes, or tow can be braided over a rotating removable mandrel.

9.4.1.3 Fiber Preforms

Fiber preforms for resin impregnation consist of fabric made by weaving, braiding, rovings, laid-up parts, and almost any configuration that lends itself to being impregnated by polymer resins. Fiber forms that are preimpregnated with matrix resin in the uncured state are known as prepregs. Prepregs consist of both

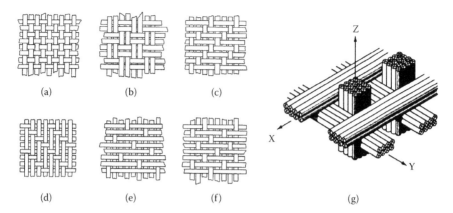

FIGURE 9.17 Fabric construction forms: (a) plain weave, (b) basket weave, (c) twill, (d) Crafoat satin, (e) eight-end satin, (f) five-end satin, and (g) geometry of a three-dimensional orthogonal weave preform. (From F. S. Dominguez, Woven fabric fibers, in *Engineered Materials Handbook*, Vol. 1, ASM International, Metals Park, OH, 1987, p. 148; F. P. Magin, Multidirectionally reinforced fabrics and preforms, in *Engineered Materials Handbook*, Vol. 1, ASM International, Metals Park, OH, 1987, p. 130. With permission.)

continuous and chopped fibers that the manufacturer uses to fabricate parts by molding, laminating, or rearranging to form a finished product. An example of a typical prepreg machine and spool of a graphite epoxy tape is shown in Figure 9.18. The manufacturer subjects these parts to heat and pressure to cure them to their final desired properties. The manufacturers of parts usually have nothing to do with the pregs, which are supplied by the larger composite fabricators. Often, the prepregs are divided into high-performance composites and composites with less stringent demands.

High-performance composites are used for the aerospace industry. They often include the carbon composites, of strengths around 1600 MPa (230 ksi) and moduli of 140 GPa (20×106 psi). For high-strength properties, a prepreg laminate layup is compacted to the final shape in an autoclave.

Composites that have less stringent demands are often made from *sheet molding compounds* (SMCs), suitable for consumer products as appliance housings and automotive parts. Sheet molding compounds contain polymer or glass fibers in polyester or vinyl resins, whose properties are about 50% of the high-performance prepreg composites. The principles of manufacturing an SMC are shown in Figure 9.19.

For all of these products, it must be realized that properties vary markedly depending on whether the resulting composite is formed in a unidirectional manner or as a combination of multidirectional layers, the latter having lower but isotropic properties compared to the higher unidirectional properties of a continuous-tape prepreg.

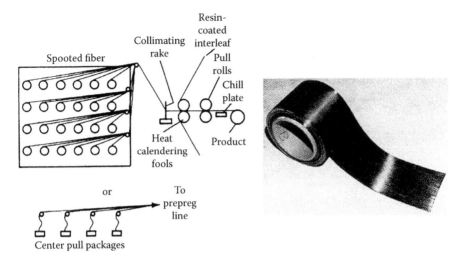

FIGURE 9.18 Typical prepreg machine and a spool of a graphite epoxy tape. (From F. S. Dominguez, Unidirectional tape prepregs, in *Engineered Materials Handbook*, Vol. 1, ASM International, Metals Park, OH, 1987, p. 143. With permission.)

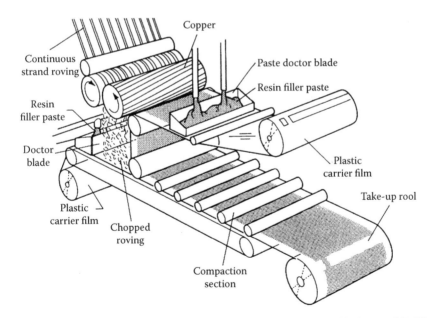

FIGURE 9.19 Sheet molding compound machine. (From J. J. McCluskey and F. W. Doherty, Sheet molding compounds, in *Engineered Materials Handbook*, Vol. 1, ASM International, Metals Park, OH, 1987, p. 157. With permission.)

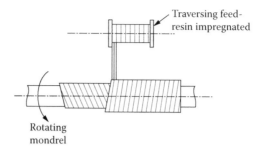

FIGURE 9.20 Filament winding fabrication process for continuous filaments.

9.4.1.4 Filament Winding

Filament winding is one of the older methods of fabricating Polymer Matrix Composites (PMCs) and was used to fabricate large rocket motor cases in the 1960s. Resin-impregnated rovings or tows are fed to a rotating mandrel, around which they are continuously wound (Figure 9.20). The mandrels are usually made of water-soluble sand, or the mandrels are collapsible for easy removal. In some cases, such as in pressure vessels, a metal mandrel remains as part of the load-bearing structure. The fibers can be laid down in a variety of patterns, each layer often being oriented a different direction. This is a rather economical process because the fibers and resin are essentially combined at the same time without the need for a prepreg.

9.4.1.5 Pultrusion

Pultrusion consists of pulling continuous fibers through a resin bath and then through a die into a curing chamber. Generally, the resin is a thermosetting polymer that reacts during the fabrication process to produce a shape that cannot be further altered or reshaped. This is a fairly complex process. The machinery must be capable of pulling, heating, infiltrating, and cutting the profiles to the desired length. In general, pultrusions can be produced in nearly any constant cross-sectional shape that can be extruded. It is a very versatile process that is expected to penetrate an increasingly large number of markets as it becomes better developed and recognized for its versatility and economy. For example, aluminum extrusions account for about 15% (and is growing) of all aluminum products; pultrusion now accounts for only about 5% of the reinforced polymer market. Pultrusions have been fabricated into I-beams, channels, and tubular shapes, much as in the metal industry. The number of materials that are amenable to this process are numerous. A typical pultrusion setup is shown schematically in Figure 9.21.

9.4.1.6 Layup Methods

The layup method was probably the first used to make a modern composite structure. It has a number of names, such as layup molding, wet layup, or contact

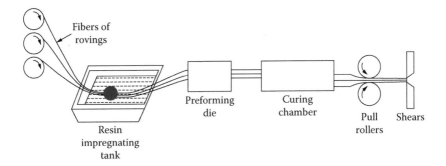

Fibers of rovings

Preforming die

Curing chamber

Pull rollers

Shears

Resin impregnating tank

FIGURE 9.21 Pultrusion process for fabrication of continuous-filament PMCs.

laminating. Basically, a mat of woven or braided fabric is saturated with resin, and then mats are laid on top of one another to obtain the desired thickness. The impregnation is done as the layers are placed in the mold. The wet layup process involves laying the dry fiber mat into the mold and then applying the resin. The easiest process is the wet layup method done by hand. It is suitable for prototypes and large components, for series of just a few pieces.

Another modification is the spray gun method, in which a mixture of chopped resin and fibers is sprayed into a mold. The prepreg method is a modification of the wet layup process. Usually, the prepreg is purchased in an impregnated tape or woven fabric that is cut and fitted into a mold. The prepreg method usually involves vacuum bagging to assist in compressing the plies while simultaneously withdrawing excess volatiles. Figure 9.22 shows schematically a vacuum bag installation procedure. Owing to the applied vacuum, the residual porosity will be decreased, and hence the properties of the composite will be improved. Vacuum processes are typically used for aerospace applications.

A further improvement in properties can be achieved by the autoclave method. An autoclave is a pressure vessel working at elevated temperatures. The prepregs can be applied to vacuum and external pressure. The temperature, pressure, and vacuum cycles can be controlled precisely in order to control the curing of the resin systems. An autoclave laminate stack layup is shown in Figure 9.23. The thermosets are manufactured at temperatures around 180°C. Typically pressures of 2 to 25 bar and vacuum of ca. 2×10^{-3} bar are applied. High-quality products can be produced this way. This technique is used for the aerospace industry to make rudders and many other parts as well as Formula 1 monocoques. However, this technique is expensive because it requires high investments, and because of the batchwise production, it is only applicable for small series.

It is beyond the scope of this book to give a comprehensive overview of the process routes of composites. The given layup methods are nonautomated or partly automated techniques. A number of automated processes exist that use various cold, hot, and wet pressing techniques. We emphasize once again the SMC process.

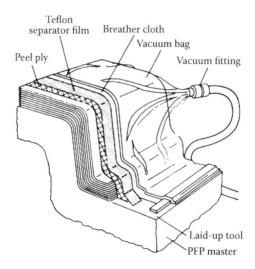

FIGURE 9.22 Vacuum bag installation over tool layup. (From B. D. Harmon, Graphite-epoxy tooling, in *Engineered Materials Handbook*, Vol. 1, ASM International, Metals Park, OH, 1987, p. 588. With permission.)

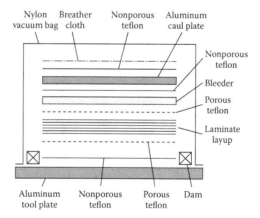

FIGURE 9.23 Autoclave laminate stack layup. (From R. A. Servais, Process modeling and optimization, in *Engineered Materials Handbook*, Vol. 1, ASM International, Metals Park, OH, 1987, p. 501. With permission.)

The SMC is composed of a filled, thermosetting resin and a chopped or continuous strand of glass fiber. The processing is fairly uncomplicated, as depicted in Figure 9.19, and a molding compound in sheet form is produced that is similar to rolled steel. After several processing steps the SMC material is cut into charges and the charges are placed in matched metal die molds made of machined steel. Under the combined application of heat and pressure in a hydraulic press, the SMC flows to all areas of the mold. Heat from the mold (149°C or 300°F) also activates the catalyst in the materials, and cure or cross-linking takes place. The part is then removed from the mold. The SMC process has many advantages:

- High volume production
- Excellent part reproducibility
- Low labor requirement per unit produced
- Minimum material scrap
- Excellent design flexibility (from simple to very complex shapes)
- Parts consolidation
- Weight reduction

All these advantages make this material useful for automotive applications, for example, for closures. Table 9.3 gives some typical mechanical properties.

9.5 METAL MATRIX COMPOSITES (MMCs)

MMCs can be divided into two subgroups:

1. Those reinforced by discontinuous particulates
2. Those reinforced by continuous fibers such as carbon, alumina, silicon carbide, and boron

9.5.1 PARTICULATE COMPOSITES

Particulate composites are metal matrix-based materials containing particles of a certain amount. Standard commercial products of this group are cemented

---

TABLE 9.3
Typical Mechanical Properties for Sheet Molding Compounds with 15 to 30 wt % Glass Fiber

Tensile modulus GPa (10^6 psi)	11–17 (1.6–2.5)
Tensile strength MPa (ksi)	55–138 (8–20)
Elongation to failure (%)	0.3–1.5
Compressive strength, MPa (ksi)	103–206 (15–30)
Flexural modulus GPa (10^6 psi)	124–207 (1.4–2.0)
Izod impact strength, notched J/m (ft lbf/in)	430–1176 (18–30)
Density in g/cm^3	1.7–2.1
Coefficient of thermal expansion 10^{-6}/K	15–22

Source: From J.J. McCluskey and F.W. Doherty, Sheet molding compounds, in *Engineered Materials Handbook*, Vol. 1, ASM International, Metals Park, OH, 1987, p. 158.)

carbides, electrical contacts (Ag–Ni, Ag–SnO$_2$–Ag/W, Ag–C), and a more recently developed AlSi25 material used for cylinder liners.

Cemented carbides (WC with Co binder) or cermets (TiC, TaC with metallic binder) contain hard carbides dispersed in a metallic matrix. They are widely used for cutting tools in machining operation. Tungsten carbide cannot be used as a monolithic tool because it is extremely brittle. To improve the toughness, the tungsten carbide particles are embedded in a cobalt matrix with about 3 to 15% Co. The ductility increases with increasing Co content. The tool tips must be manufactured by powder metallurgical processes. The basic steps are:

- Intensive mixing of WC and Co powders
- Pressing of tool tips of desired geometry
- Liquid phase sintering

The sintering procedure today first applies a vacuum until a closed porosity is achieved, and then the furnace switches to an external pressure, and a very dense high-quality tool tip is produced. Figure 9.24 shows microstructures of WC containing 10% Co. The particle size of the tungsten carbide varies from very fine to coarse. Hardness increases with decreasing particle size.

Materials for *electric contacts* are used in switches and relays. They must have a good combination of electrical conductivity and wear resistance. For achieving a good electrical conductivity, silver is used as a matrix material. In addition to having the highest electrical conductivity of all metals, it does not oxidize at high temperatures and hence does not form an insulating oxide layer on the contact surface. However, silver is too soft and it would show high wear. This problem can be solved by reinforcing the silver matrix with particulates such as nickel, tin oxide, or tungsten. For low currents Ag–Ni is used, whereas for higher currents Ag–SnO$_2$ shows better wear resistance. Most of the electric

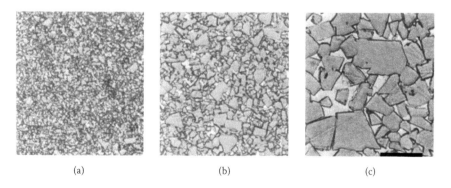

(a)   (b)   (c)

FIGURE 9.24 Microstructures of 90% WC—10% Co cemented carbide (straight grade). The light constituent is the cobalt binder; (a) fine grain size, (b) medium grain size, (c) coarse grain size. (From Cemented carbides and cermets, in *Metals Handbook*, Desk ed. 2nd ed., ASM International, Materials Park, OH, 1998, p 635. With permission.)

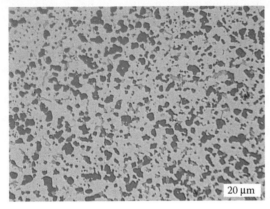

FIGURE 9.25 Cylinder housing of a DaimlerChrysler six-cylinder motor with inserted cylinder liners of $AlSi_{25}$ (top) and microstructure of $AlSi_{25}$-particulate composite material manufactured by a spray compacting process (bottom). (Courtesy of PEAK Werkstoff GmbH, Velbert, Germany.)

contact materials are produced by powder metallurgical processes. Silver and particulates are mixed and formed to an extruding billet by cold isostatic pressing, and the billet is extruded and rolled or drawn to profiles or wires. Contact plates are cut off from profiles and rivets are formed from wires. Pressing and sintering of individual parts and infiltrating methods (for Ag–W) are also common processes.

A recently developed aluminum alloy with high Si content ($AlSi_{24}CuMg_1$) is currently being used as a cylinder liner material in a DaimlerChrysler six-cylinder motor. As depicted in Figure 9.25, the Si particles are spherical and homogenously distributed in the Al matrix. The material has several advantages:

- High strength up to 450°C
- High Young's modulus (100 GPa)
- High wear resistance

- Good thermal conductivity
- Good formability

For high-silicon-containing aluminum alloys, such a homogeneous micro-structure cannot be achieved by standard casting processes. Powder/Metallurgical (P/M) routes have to be chosen. In this case a new technology, a spray compacting, or Osprey, process, was developed. In this process the AlSi material is atomized, and the atomized particles are immediately compacted on a rotating substrate. Atomizing and compacting occur in a chamber with an inert gas atmosphere. By slowly moving down the substrate in the atomizing chamber, a dense billet is produced. This billet can be extruded and rolled to achieve the desired geometry.

Many attempts have been made to reinforce high-strength aluminum alloys with ceramic particles. In 1991 Duralcan USA, a subsidiary of Alcan Aluminum, produced a molten aluminum composite with fine ceramic particles. However, the risks associated with using most MMCs is still very high, owing to an immature technology.

Aluminum reinforced by silicon carbide particulates up to approximately 70% by volume of SiC has been formed to near net shape by a pressureless metal infiltration process. The molten metal matrix alloy wets the filler (reinforcement) so well that the matrix infiltrates the filler spontaneously to form the composite. Compared to conventional aluminum alloys, these composites have low thermal expansion coefficients and high moduli (270 GPa) with little or no loss in thermal conduction and density. As a result, they are attractive candidates for packages, substrates, and support structures for electronic components.

9.5.2 FIBER-REINFORCED MMCs

Obviously, it is much more difficult to infiltrate a woven fiber with a liquid metal than with a resin. An Al alloy with 10 vol % SiC fiber composites formed by squeeze casting as reported strengths and moduli of 200 MPa (20 ksi) and 82 GPa (11.9 Msi), respectively. Squeeze casting involves the use of pressure via forging presses during the infiltration of the porous fiber compact. Considerable development work is still needed for this process because careful control of the many variables, such as preheat temperature, alloying elements, tooling temperature, and pressure levels and duration, is required to obtain reproducible properties. However, this process holds considerable promise for the future manufacturing of MMCs.

Continuous-fiber MMCs are expensive and have been, until recently, limited primarily to aerospace applications. Aluminum, titanium, and magnesium matrices have been studied extensively because of their light weight and potential aerospace applications.

SiC, SiC-coated boron, and boron have been the reinforcement continuous fibers of greatest importance. A 40% boron fiber content in a 2024-T6 aluminum matrix composite has strength values as high as 1459 MPa (211.5 ksi) and a modulus of 220 GPa. These composites have been used for structural tubular struts as the forms and truss members in the shuttle midsection fuselage. This composite is also being used as a heat dissipator for multilayer-board microchip carriers using its high thermal conductivity in conjunction with its lower thermal expansion coefficient.

SiC fibers manufactured by the chemical vapor deposition process and used at a 47 vol % level in 6061-T6 aluminum have achieved strengths of 1378 MPa (200 ksi) and a modulus of 207 GPa (30×10^6 psi). The boron and SiC composites' properties were both measured in the fiber direction. They are very anisotropic materials.

9.5.2.1 Boron and SiC Fiber Composites

As stated earlier, both boron and SiC fibers are made by depositing boron and SiC onto a hot wire by a chemical vapor deposition process. Once the fibers are formed, they are available to be turned into whatever composite is of interest. The metal foil is hot pressed around a fiber layer until it completely surrounds the fiber, forming a composite sandwich layer. These layers, or preforms, can then be stacked as layup plies to form a more extensive composite (Figure 9.26). Most often the fibers run in the same direction, but if desired they can be placed at right angles for high strength in two directions. These types of preforms have also been fabricated by spraying the fibers with plasma.

9.6 CERAMIC MATRIX COMPOSITES (CMCs)

Most of the current CMC work is directed toward improving the ductility and toughness of ceramics by the addition of modest amounts of short fibers. You can find a case study for a fiber-reinforced SiC ceramic in Chapter 7, Section 7.6.1.

One other direction that is being taken is to utilize the high moduli and strengths of carbon and ceramic continuous fibers in high-volume fractions in ceramic

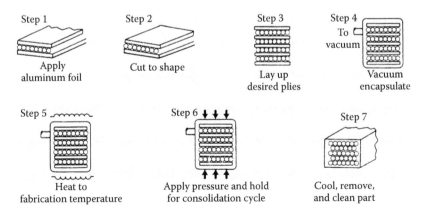

FIGURE 9.26 Typical fabrication process for boron and SiC fibers in metal matrices. (Courtesy of Textron Corp.)

matrices, the same as for polymers and metals, in an attempt to provide higher-temperature capability. Some success has been achieved in producing composites with improved properties for use in the low to intermediate temperature range, but a good high-temperature CMC that can withstand oxidizing conditions and have some significant degree of toughness remains to be developed.

The toughening mechanism in CMCs depends on the fiber type. For brittle fibers it comes about by the fiber–matrix debonding energy when the composite failure occurs via fiber pullout. When the reinforcement fibers are ductile metal wires, the plastic deformation of the wires during fracture absorbs energy and thereby increases the toughness. In laminate composites delamination between plies can absorb energy during fracture and can add to the toughness. In CMCs, contrary to the case for MMCs, it is more advantageous to have a weak fiber–matrix bond; otherwise, a matrix crack once initiated will propagate right through the fibers. The weak interfacial bond allows the fiber to slip so that the crack is deflected along the length of the fiber. Crack deflection increases toughness, since more energy is absorbed. This is similar to the crack deflection mechanism for improved toughness of zirconia.

Much research has been carried out on fiber-reinforced ceramics. The most common reinforcement material is SiC. Attempts have been made with Al_2O_3–SiC, SiC–SiC, ZrO_2–SiC, Si_3N_4–SiC, glass–SiC, and others. For example, SiC has been vapor deposited onto carbon and alumina fibers and has improved fracture toughness of CMCs by a factor of 2 to 4 times that of the uncoated fibers. Hence, coating fibers with certain materials weakens the fiber matrix interfacial bond strength and improves the overall strength properties as explained above.

This discussion of successes in the CMC arena might lead one to believe we are just a step away from having the high-temperature materials, so badly needed and sought after for many years, for utilization in the hot regions of gas turbine engines, but such is not the case. Ceramic products that can be substituted for superalloys remain elusive, despite widespread claims of new ceramic components that will solve all of our needs. Although reinforcement appears to be the most desirable and effective means of toughening ceramics, even considerably better than partially stabilized zirconia, problems still exist. Cost is one factor, but for aerospace applications, the lack of reliability and consistency is far more important.

9.7 GRAPHITE MATRIX COMPOSITES

9.7.1 CARBON–CARBON COMPOSITES

Carbon–carbon composites consist of an amorphous carbon matrix reinforced with carbon fibers. The fibers are based on rayon, pitch, or PAN and increase in cost in the order listed. The chief reason for using carbon–carbon composites is that their strength is still at temperatures to 2500°C (4532°F). They are biocompatible and can be made with directional properties. They oxidize readily and must have some type of protective coating, but their high-temperature properties, coupled with the fact that they do not degrade or outgas the way organic resins do in the vacuum of outer space, outweigh the necessity of protective coatings for reentry vehicles.

One of the most notable applications of carbon–carbon composites is in the nose cone and leading edge components of the space shuttle. Reentry temperatures are around 1650°C (3000°F). Not only can carbon–carbon composite materials withstand the heat, but they remain unscathed mission after mission and even increase in strength. The nose cone is made of a two-dimensional layup. Graphite cloth is preimpregnated with phenolic resin and then pyrolyzed, driving off the gases and moisture as the phenolic resin converts to a graphite matrix. This relatively soft composite is then impregnated with furfural alcohol and pyrolyzed three more times, with the density, strength, and modulus increasing each time. A ceramic coating of silica and alumina is applied, which after firing results in a SiC surface layer. The surface is further protected by impregnation with tetraethylothosilicate, which when cured leaves a SiO_2 residue throughout the coating.

Carbon–carbon composites are also used as aircraft brakes, where their high thermal conductivity and light weight are advantageous; as prosthetic devices, where their compatibility with body tissue and directional properties are desirable traits; and in other spacecraft components. Advanced carbon–carbon is used for turbine wheels on hypersonic aircraft. Continuous PAN fibers are used for this product. The manufacturing method is often tailored to fit the application. Carbon–carbon composites have also been used as lightweight housings for satellite solar cells and as lightweight materials for aerospace planes.

Carbon–carbon composites have a density of about 1.3 g/cm^3, compared to 2.7 and 4.5 for aluminum and titanium, respectively, but the densities and strengths of carbon–carbon composites vary considerably with manufacturing methods. Room-temperature strengths of around 450 MPa (65 ksi) are typical (although 2100 MPa [304 ksi] has been reported for advanced carbon–carbon material), with 600 MPa (87 ksi) for continuous fiber composites. Moduli in the range of 125 to 175 GPa (18 to 25×1016 psi) are common. Carbon–carbon composites are the strongest materials known at temperatures of 2204°C (4000°F). Figure 9.27 shows the specific strength vs. temperature characteristics for carbon–carbon composites compared to those of superalloys.

In the physical properties category the coefficient of expansion is of little concern since both the fiber and matrix form have essentially the same values. Thermal conductivities are on the order of 240 W/m K (0.142 Btu/ft-h-°F), which is a quite high value, comparable to metallic aluminum.

9.7.2 GRAPHITE–BORON COMPOSITE

Graphite matrices have been used with other fibers, notably boron. This composite is currently a very popular material for sports equipment, including golf club shafts, skis, and bicycles. Textron claims to have proof that golf clubs with

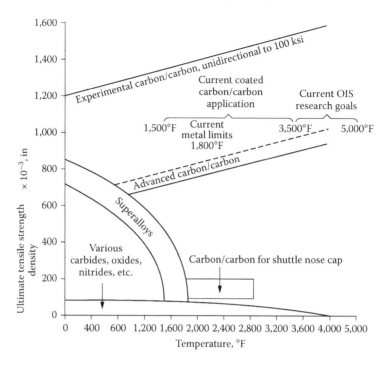

FIGURE 9.27 Specific strength vs. temperature for carbon–carbon composites compared to superalloys. (From A. J. Klein, Carbon/carbon composites, *Adv. Mater. Process.*, November, 66, 1986; data supplied by LTV Aerospace and Defense.)

boron–graphite shafts hit balls up to 20 m farther than clubs with titanium, graphite, or steel shafts.

9.8 LAMINAR COMPOSITES (ARALL AND GLARE)

Bimetallics, which are typical laminar composite materials, are used as temperature sensors. There are numerous possibilities for manufacturing laminar composites by cladding or coating. We cannot give a comprehensive overview of all laminar composites and their properties, so we will give an example of a new composite material that will be used in the new Airbus 380: Glare.

Laminates are layers of materials joined by an organic adhesive. Arall (aramid aluminum laminate) and Glare (glass aluminum laminate) have been developed as possible skin materials for aircrafts. The upper fuselage shell of the A380 will be fashioned from Glare, leading to a weight saving of about 800 kg.

In Arall an aramid fiber is prepared as a fabric or unidirectional tape that is impregnated with adhesive. In Glare, glass fiber tapes are used instead of aramid. The fiber tapes are laminated between layers of aluminum alloy, for example, 7075-T6 and 2024-T3. The density of Glare is about 10% less than aluminum. The composite has a high strength and stiffness combined with light weight. The laminate structure leads to improved fatigue properties. Furthermore, Glare is corrosion resistant and easy to repair.

9.9 SANDWICH STRUCTURES

Let us briefly mention sandwich structures, which have a thin layer of a facing material joined to a light filler metal. A good example is corrugated cardboard, in which a corrugated core of paper is bonded at both sides to flat, thick paper. In combination they form a rigid body, but neither the core nor the flat paper alone is rigid.

Another example is the honeycomb structure shown in Figure 9.28. Most honeycomb used today is an adhesively bonded core of aluminum that is subsequently bonded to face sheets to form a sandwich panel. The construction is an extremely lightweight structure that exhibits a high stiffness- and strength-to-weight ratio and hence is used in aircraft applications. Other common types of honeycomb cores presently being produced include metallic cores of corrosion-resistant steel and titanium- and nickel-based alloys fabricated by resistance welding, primarily used for elevated-temperature applications. The facings are usually attached by brazing or diffusion bonding.

9.10 FAILURE OF COMPOSITES

In discussing composite failure it is convenient to divide them into two types: (a) the continuous filament or laminate type and (b) the discontinuous particulate, whisker, or small randomly oriented fiber-reinforcement composites.

Fracture of continuous-fiber composites can occur in a number of ways. When the fibers are oriented in the direction of the load, the fibers are placed in tension, and thus the best strength characteristics are realized. In such loading the fracture can occur by:

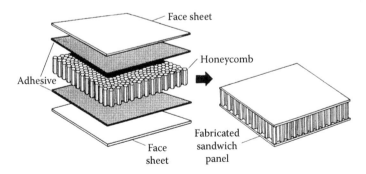

FIGURE 9.28 Example of a bonded sandwich assembly. (From J. Corden, Honeycomb structure, in *Engineered Materials Handbook*, Vol. 1, ASM International, Metals Park, OH, 1987, p. 721. With permission.)

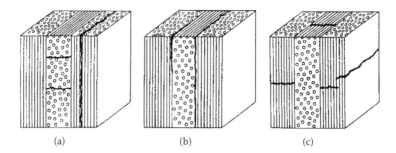

(a)   (b)   (c)

FIGURE 9.29 Different planes of separation in continuous-fiber-reinforced composites: (a) intralaminar, (b) interlaminar, and (c) translaminar. (From B. W. Smith, Fractography for continuous fiber composites, in *Engineered Materials Handbook*, Vol. 1, ASM International, Metals Park, OH, 1987, p. 787. With permission.)

- Fiber fracture
- Matrix failure
- Delamination at the fiber–matrix interface, commonly called the fiber pullout

Many continuous-fiber composites are made by alternating layers in which the fibers run in different directions, as illustrated in Figure 9.17. In such situations the failure takes place by what has been called intralaminar (Figure 9.29a), interlaminar (Figure 9.29b), and translaminar (Figure 9.29c) processes. Intralaminar failures are those occurring internally within a ply and are very close to, and may often be, failure along the fiber–matrix interface. Interlaminar failures are fractures that occur between the lamela used in the layups. This is more of a construction problem involving obtaining a good bond between the matrices of the respective laminates. Translaminar fractures involve fiber breakage, and although the matrix fractures at the same time, this type of fracture suggests that either the fiber was weak, it contained too many defects, or the lamela contained something less than the critical volume of fibers. Although the matrix also fractures in this case, in general one relies on the fibers to carry the load, and if this is the situation, the matrix cannot be blamed. A typical fracture surface of a carbon–epoxy composite is shown in Figure 9.30.

So far, we have been concerned with the failure of the composite in terms of its weaker components. In many cases it may be desirable, for example, to have a certain amount of fiber pullout, even though it may reduce the overall strength of the component. This is because fracture energy is absorbed during fracture via the pullout mechanism, and any increase in energy absorption increases the fracture toughness, or, in other words, it leaves less energy for crack propagation. Cottrell [1] suggested that the fiber length should be less than l_c, so that the fibers will pull out and prevent a fracture from propagating. Of course, this must be done with a sacrifice of strength since maximum strengthening occurs when $l > l_c$, but

20 μm

FIGURE 9.30 Fracture surface of a longitudinal (0°) carbon–epoxy specimen that failed in tension. The specimen was tested in the dry condition at 25°C (77°F). The fibers are broken at many different levels; the fiber breaks are perpendicular to the fiber (and load) direction. There is substantial epoxy adhering to the fiber surfaces, indicating that the interfacial bond strength is fairly high. (From *Metals Handbook*, 9th ed., Vol. 12, ASM International, Metals Park, OH, 1987; R. L. Stedfeld, Ed., Fractography, in *Atlas of Fraktography, Resin-Matrix Composites*, 1985, p. 476. With permission.)

in metals we often sacrifice strength for toughness, the tempering of martensite being the best example.

Owing to their high strength and low density, carbon fiber composites are often used in the aircraft industry. Hence, it is obvious that we must consider fatigue in composites, that deteriorating effect of cyclic stress with time (or number of cycles) that occurs in metals and polymers and is now known to occur in composites.

DEFINITIONS

Anisotropic laminate: One in which the properties vary with direction in the laminate.

Aramid: Highly oriented organic material derived from polyimide but containing aromatic rings. Kevlar and Nomex are examples of aramid fibers.

Aromatic: Unsaturated hydrocarbon with one or more benzene rings.

Borsic: Boron fiber that has an outer layer of SiC to prevent reaction with certain matrix materials.

Braiding: Process in which two or more systems of yarn are intertwined in the bias direction to form an integrated structure.

Carbonization: Process of pyrolyzing certain polymers in the range of 800 to 1600°C to obtain carbon fibers.

Cermet: Composite containing inorganic compounds (ceramics) with a metallic binder.

Char: To heat a composite in air until it is reduced to ash.

Chemical vapor deposition: Process of reacting metal compounds, usually halides, with a hot surface, whereby the compound decomposes and the metal deposits on the hot substrate.

Critical fiber length: Length necessary to cause fiber overlap and transfer of load to adjacent fibers. It is also a function of fiber diameter.

Cross-ply laminate: Laminate with plies usually oriented only at 0° and 90°.

Cure: To change the properties of a thermosetting resin irreversibly by chemical reaction.

Delamination: Separation of the materials in a laminate. Also applied to separation of fiber from the matrix.

Disbond: Area of a bonded interface between two adherents where adhesion has been lost.

E-glass: Family of glasses of calcium aluminoborosilicate with about 2% alkali content.

Felt: Fibrous material made of interlocked fibers.

Fiberglass: Composite of glass fibers in a polymer matrix.

Filament: Smallest unit of a fibrous material, usually less than 25 μm (1 mil) in diameter.

Filament winding: Process for fabricating a composite in which continuous filaments gathered into strands (or yarn or tape) are placed over a rotating removable mandrel to form a cylindrical composite when the filaments are impregnated with a resin.

Graphite fiber: Fiber made from a polymer precursor by a carbonization or graphitization process.

Graphitization: Process of pyrolyzation of certain carbon-containing compounds in an inert atmosphere at temperatures in excess of 2500°C, which converts carbon to its crystalline allotropic graphite form.

Hand layup: Process of placing successive plies of reinforcing material in resin-impregnated reinforcement in a mold by hand.

Helical winding: Winding in which a filament advances along a helical path not necessarily at a constant angle except for some cylinders. A shape frequently used in filament wound structures.

Honeycomb: Hexagonal cells made from numerous lightweight materials (paper, metal foil, glass fabric).

Hybrid laminate: Composite laminate consisting of laminae of two or more different composite materials.

Impregnate: To force resin, metal, or other matrix material into a fibrous body. Similar to infiltration.

Interface: Region between two different adhered materials.

Interlaminar: Event or process (e.g., fracture) occurring between two or more adjacent laminae.

Kaowool: Fiber derived from clay for insulation but also used in composites. Most often discontinuous. A Babcock and Wilcox trademark.

Kevlar: Organic polymer composed of aromatic polyimides with a parallel chain extending bonds from each aromatic nucleus (a DuPont trademark).

Knitted fabrics: Fabrics produced by interlooping chains of yarn.

Lamina: Single ply or layer in a laminate.

Layup: Process of placing reinforcing material into a mold, usually resin impregnated.

Macro: Scale of materials that pertain to their individual nature and of a size that can normally be seen without the aid of a microscope. Applies to macroscale.

Mat: Discontinuous fiber-reinforced material available in sheet, blanket, or ply form.

Matrix: Continuous phase of a composite that acts as the binder to hold together the reinforcing materials.

Particulate: Small metallic or inorganic reinforcing materials of a composite, often as chopped fibers or small particles.

Pitch: High-molecular-weight material left as a residue from the distillation of coal and petroleum products.

Ply: Fabrics or felts consisting of one or more layers. The layers that make a stack. A single layer of prepreg.

Precursor: Material preceding or used to make something else. For carbon or graphite fibers it is the rayon, PAN, or pitch that is pyrolyzed to form the fibers.

Preform: Preshaped fibrous reinforcement formed by distribution of chopped fibers or cloth over the surface of a perforated screen to the approximate thickness desired in the finished product. Also, a preshaped fibrous reinforcement of mat or cloth formed to the desired shape on a mandrel or mockups before being placed in a mold.

Prepreg: Either ready-to-mold materials in sheet form or ready-to-wind materials in roving form (e.g., mat or cloth) subjected to unidirectional fiber impregnation with resin and stored for use.

Pultrusion: Continuous process for manufacturing composites that consists of pulling a fiber-reinforced material through a resin impregnation bath and subsequently through a die and curing chamber.

Pyrolysis: Thermal process by which organic precursor fiber materials are chemically changed by heating into a fiber.

Reinforcement material: Strong material bonded into a matrix, usually to add strength, but could also be used for other purposes, such as improving thermal or electrical conductivity. Usually in the form of long or chopped fibers, whiskers, or articulates.

Resin: Solid or pseudoidal organic material, usually of high molecular weight, that tends to flow when subjected to stress. Most resins are polymers.

Roving: Number of yarns, strands, or tows or ends collected into a parallel bundle with little or no twist.

Tack: Stickiness of an adhesive- or filament-reinforced resin prepreg.

Tow: Untwisted bundle of continuous filaments.

Translaminar: Going through the laminae, such as in crack propagation.

Vacuum bag molding: Process in which a sheet of flexible materials is placed over a mold and sealed at the edges. A vacuum is applied to suck out entrapped air and volatiles and, simultaneously, to apply pressure.

Weaving: Manner in which a fabric is formed by interlacing yarns.

QUESTIONS AND PROBLEMS

9.1 For an epoxy resin with reinforced continuous fibers, what fiber would you select if:
(a) The tensile strength were the most important factor?
(b) The tensile modulus were the most important factor?
(c) The modulus and strength were of equal significance?

9.2 For a 2024 aluminum alloy reinforced with discontinuous fibers of SiC:
(a) Estimate the composite strength for 10, 20, and 30 vol % fibers.
(b) Plot a graph of strength vs. fiber content and extrapolate it to 65 vol % fiber content. Would you expect to obtain the strength that the 65% point shows on the graph? What problems would you encounter in fabricating such a composite?

9.3 Suppose that you used discontinuous fibers for Problem 9.2(a). Estimate the strength change compared to continuous fibers.

9.4 Describe how graphite (carbon) fibers are produced, including starting materials and process temperatures. How does the cost vary with the starting material?

9.5 Boron and SiC fibers are both made by the chemical vapor deposition method. Compare the advantages and disadvantages of the two fibers.

9.6 Compute the critical volume of E-glass for reinforcing pure aluminum. Assume that the composite will be strained until the aluminum has been plastically strained 10%. (You may need Chapter 6 for this one.)

9.7 What could you do to increase the strength of the aluminum composite of Problem 9.6 by a factor of 1.5 while holding the matrix strain to 10%? First list the variables that could be involved, including other fiber shapes and materials. Only the 10% strained aluminum matrix remains constant.

9.8   For a 10% strained aluminum matrix reinforced with continuous E-glass fibers, what is the critical fiber length for a 1-ml fiber diameter?

9.9   In Equation 9.15, what factors affect the efficiency coefficient α? Why is the range of α so large? Could it be the controlling factor for the moduli of discontinuous fiber composites?

9.10   Name two each of low- and high-performance PMCs.

9.11   Why do most commercial PMCs made to date have thermoset rather than thermoplastic matrices? What is one advantage of the latter?

9.12   Most composites have very low ductilities. Could you suggest a composite for which the matrix is considerably strengthened by fibers, yet the composite will have ductilities in excess of 20% elongation?

9.13   Ceramics are brittle. Why would CMCs be considered for structural purposes?

9.14   What is the primary advantage of carbon–carbon composites over other material combinations? What is a current application of carbon–carbon composites?

9.15   What is the difference between a preform and a prepreg?

9.16   Equation 9.10 expresses that a critical volume of fibers is required for reinforcement to occur. Would a critical volume be required even if the matrix did not strain-harden? Write a similar equation for a polymer matrix that does not strain-harden.

9.17   Many CMCs use fibers in a thermosetting epoxy that does not readily deform plastically. How is the load transferred for overlapping fibers in this situation?

REFERENCE

1.   Cottrell, A. H., Strong solids, *Proc. Roy. Soc. A*, 279, 282, 1964.

SUGGESTED READING

American Society for Metals, *Engineered Materials Handbook: Composites*, ASM International, Metals Park, OH, 1987.

DuPontBulletin K-5, September 1981.

McCullough, R. L., *Concepts of Fiber-Resin Composites*, Marcel Dekker, New York, 1971.

10 Electrical Properties of Materials

10.1 INTRODUCTION

The electron conductivity of solid materials gives an almost unambiguous way to classify them. Simply put, on the basis of electrical conductivity, materials are either insulators, semiconductors, conductors, or superconductors. Superconductors are a special class of materials that exhibit zero resistance below a certain temperature. They will not be considered here. The conductivity of all of the more common and widely used materials is shown in Figure 10.1. The range of conductivities is quite large. Where we draw the lines for these materials appears to be somewhat arbitrary, but we can define these three categories fairly precisely in terms of the number of electrons available for conduction. This number can be computed using the energy band structure for the valence electrons, a subject covered in the following section. Insulators and most polymers have a low conductivity because of their strong covalent bonds and the absence of free electrons, but in some polymers a conducting powder is mixed with the polymer to form a conducting composite. In a few others of the so-called conducting polymers, there exist some free electrons within the polymer structure, creating conductivity on the order of that found in crystalline semiconductors, and in some conducting polymers the conductivity approaches that of metals. There is a tremendously large variation in the conductivity of solids, being about a factor of 10^{25} from conductors to insulators.

Ohm's law can be used to express conductivity and its reciprocal, resistivity, which are not functions of specimen dimensions, and the conductance and resistance, which are functions of specimen dimensions. Resistance is related to resistivity by:

$$R = \rho \, l/A$$

where
R = resistance
ρ = resistivity (usually expressed in $\Omega \cdot m$)
l = specimen length
A = specimen area

and in terms of conductivity, σ,

$$\sigma = l/RA \quad \text{units are } (\Omega \cdot m)^{-1}$$

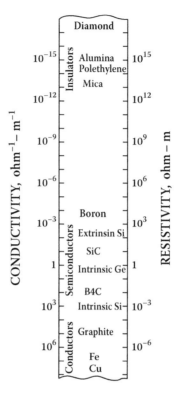

FIGURE 10.1 Range of electrical conductivities for various materials.

For semiconductors it is more desirable and convenient to express the conductivity in terms of the number of charge carriers and their mobility. To accomplish this feat, we state Ohm's law in a different fashion:

$$J = \sigma\varepsilon \tag{10.1}$$

where

J = current density or flux = Amp/m^2
σ = conductivity
ε = electric field strength = E/l = V/m

Now the current density can also be written as

$$J = nqV_d \tag{10.2}$$

where

n = number of charge carriers
q = charge on the carrier, usually that of the electron (1.6×10^{-19} C)
V_d = average drift velocity of charge carriers

Since the ampere is a rate of flow of 1 coulomb of electric charge per second, J is similar to a diffusion flux.

V_d is the only expression above that may be foreign to the reader and require an explanation. In any material there exists some resistance to the electron flow. In metals this resistance is primarily that caused by electron collisions with lattice ions. It is related to the acceleration of the electron as a result of the collision and the time between collisions. V_d is proportional to the electric field strength ε:

$$V_d = \mu\varepsilon \qquad (10.3)$$

where μ, the proportionality constant, is called the mobility of the carrier. It has the units

$$\frac{velocity}{fieldstrength} = \frac{m/s}{V/m} = \frac{m^2}{Vs}$$

Combining Equations 10.1, 10.2, and 10.3, we have

$$\sigma = nq\mu \qquad (10.4)$$

This is the equation we will be using in computations involving semiconductors. The mobility, μ, is somewhat different in metals than in semiconductors. In metals it is related to the mean free path and the average velocity between collisions. In semiconductors both negative and positive charge carriers may contribute to conduction. In such a situation, Equation 10.4 must be expanded for both contributions:

$$\sigma = n_n q_n \mu_n + n_p q_p \mu_p \qquad (10.5)$$

10.2 ENERGY LEVELS, BANDS, AND BONDS

10.2.1 ELECTRON CONDUCTION IN METALS

The most convenient way of explaining charge conductivity in crystalline solids is via the band theory of crystalline solids. We will begin with the energy levels for a single sodium atom. The levels for a single sodium atom of atomic number 11 are depicted in Figure 10.2a. The energy-level diagram indicates that three orbitals are associated with the $2p$ energy level. Two electrons occupy each of the $1s$, $2s$, and $2p$ orbitals. According to the Pauli exclusion principle, each level can be occupied by two electrons of opposite spin; that is, the two electrons are thus in separate energy states. No two electrons can occupy precisely the same energy state.

When many atoms of sodium, for example, a mole of atoms, are brought together, the energy levels are altered as depicted in Figure 10.2b. These atoms

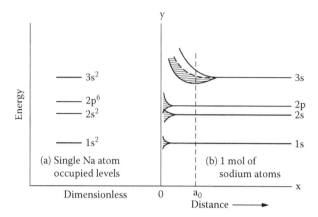

FIGURE 10.2 (a) Electron energy levels in a single sodium atom; (b) spread of the energy levels for the valence electrons when many atoms are brought together to form a solid crystal.

have now formed the metallic bond. In a metallic solid the atom core electrons are not directly involved in the bonding process, and their energy levels remain essentially unchanged, as those shown in Figure 10.2b at the equilibrium interatomic distance denoted by a_o. However, the large number of atoms produces an equally large number of energy-level splittings for the outer $3s$ orbital. This energy level splitting results in a pseudocontinuous energy band for the $3s$ electrons. Since these are the valence electrons that become involved in bonding, this energy band is also termed the valence band. This means that when we are considering the s valence band, we are considering the energy levels that each of the N atoms in the crystal contributes to the crystal as a whole, not to an individual atom. If we bring 1 mol of atoms together to form the crystal, there will be 6×10^{23} $3s$ levels within the crystal valence band. Since sodium is univalent there are 6×10^{23} electrons available to occupy these 6×10^{23} levels.

According to the Pauli exclusion principle, each state can hold two electrons of opposite spin. Accordingly, the $3s$ band will be only half full of electrons. Also, the lowest levels will be filled first, as depicted in Figure 10.2b by the shaded band. The empty portion, indicated by the absence of shading, is now the conduction band. If an electric potential is applied to the crystal, an electric field is set up within it. As a result, the valence electrons of the crystal, which are not associated with any single atom but are free to roam throughout the crystal, will be accelerated. By virtue of this acceleration, they gain energy. Fortunately, empty energy levels are available in the conduction band into which these electrons can move and occupy. This gives an unambiguous way to define a metal. *A material in crystalline form that contains empty electron energy levels immediately adjacent to the filled levels will conduct electricity when an electric field is applied. Such conductors of electricity will be called metals.*

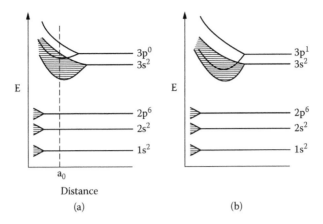

FIGURE 10.3 Valence electron energy levels for a large number (e.g., 1 mol) of (a) crystalline magnesium atoms; and (b) crystalline aluminum atoms.

Magnesium, a divalent metal, has the atomic number 12 and the electron configuration of $1s^2 2s^2 2p^6 3s^2$. This would suggest that the $3s$ band would be completely filled, and hence magnesium would not be a conductor. However, as depicted in Figure 10.3a, the $3s$ and $3p$ bands overlap to form a partially extended band, thus making magnesium a conductor. Aluminum has an atomic number of 13 and an electron configuration of $1s^2 2s^2 2p^6 3s^2 3p^1$. Since the $3s$ level is filled, one might expect aluminum to be univalent and have a half-filled $3p$ band and hence be a conductor on that basis alone, but again the $3s$ and $3p$ bands overlap, and both the $3s$ and $3p$ electrons are involved in bonding and in conduction (Figure 10.3b).

Many other elements are metals because of the overlapping of the valence bands. The transition metals involve the overlapping of the s, p, and d levels. Even some of the transition metal oxides have overlapping extended bands from the d and f levels and have been labeled as conductors. Table 10.1 compares the conductivity of the univalent, divalent, and transition metals. On average, the univalent metals have the highest conductivity, primarily owing to the larger number of carriers, since all of the electrons in the valence band are available for conduction. Aluminum has a high conductivity because it is trivalent and has an approximately half-filled $3p$ band.

10.2.2 Factors That Affect the Conductivity of Metals

Equation 10.4 expresses the conductivity of metals as a function of the number of carriers, their charge, and their mobility. Since the charge of an electron is a constant and the number of carriers for a given metal is fixed, conductivity changes are a function of situations that affect mobility. Since conductivity is the reciprocal of resistivity, we will use resistivity to explain factors that affect mobility. Resistivity is easier to measure, and hence more data exist for resistivity than for conductivity. Basically, anything that alters the perfect periodicity of the

TABLE 10.1
Relative Conductivity ($\Omega \cdot cm^{-1} \times 10^{-4}$)
of Monovalent, Divalent, and Transition
Metals, Corrected for Temperature

Monovalent		Divalent		Transition	
Li	12.9	Be	2.0	Ti	0.21
Na	24.0	Mg	8.1	Cr	0.51
K	15.3	Ca	11.1	Fe	1.14
Cu	9.1	Ba	1.0	Co	1.7
Ag	12.4	Zn	6.1	Ni	1.9
Au	8.1	Cd	4.5		
Avg.	13.6	Avg.	5.46	Avg.	1.09

atomic lattice structure can cause scattering of the electrons and thereby an increase in resistivity.

The electrical resistivity of metals can be divided into three components as follows:

$$\rho_t = \rho_{th} + \rho_d + \rho_i \tag{10.6}$$

where

ρ_t = total resistivity

ρ_{th} = contribution from thermal atom vibration

ρ_d = resistivity caused by lattice defects, primarily vacancies, dislocations, and grain boundaries

ρ_i = resistivity caused by impurity atoms, primarily those in solid solution (Some atoms may be added intentionally to form a solid solution alloy. Although they are technically not impurity atoms, they are included here as impurity atoms.)

For most metals at temperatures above about −200°C, the electrical resistivity varies almost linearly with temperature. The $\rho_d + \rho_i$ contribution is independent of temperature. Usually the defect contribution is small compared to the impurity or solute contribution.

The resistivity variation with temperature for a perfect metal and for a metal containing defects and impurity atoms is illustrated in Figure 10.4. The effect of temperature on the resistivity of a select group of metals is shown in Figure 10.5 and that of impurities on the resistivity of similar metals, in Figure 10.6.

The resistivity measurement at low temperatures (often done in liquid helium at 4.2 K) is a good indication of metal purity. In the semiconductor industry, where purity of conductors is of considerable importance, often the resistance ratio of resistivity at room temperature to that at 4.2 K is a specification. For ultrahigh-purity aluminum prepared by zone refinement the resistance ratio is about 40,000.

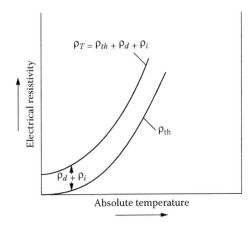

FIGURE 10.4 Resistivity vs. temperature for an idealized perfect metal compared to one containing defects such as vacancies, dislocations, and impurity atoms. (From ASM *Metals Handbook*, Desk Edition, Article by G. T. Murray, 1985, p. 14.2. With permission.)

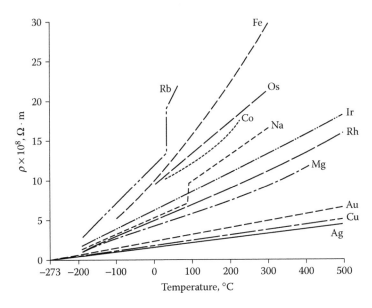

FIGURE 10.5 Effect of temperature on the electrical resistivity of selected metals. (Adapted from C. Zwikker, *Physical Properties of Solid Materials*, Pergamon Press, London, 1954, pp. 247, 249. With permission.)

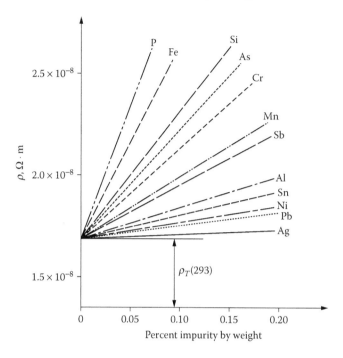

FIGURE 10.6 Effect of small amounts of impurity atoms on the electrical resistivity of selected metals. (From F. Pawlek and K. Reichel, *Z. Metallkd.*, 47, 347, 1956. With permission.)

The linear portion of the ρ vs. T curve is used to obtain the temperature coefficient of resistivity or the temperature coefficient of resistance. The temperature dependence of resistivity, $\rho(T)$, can be expressed as

$$\rho(T) = \rho_0 [1 + \alpha \Delta T] \tag{10.7}$$

where α is the temperature coefficient at the reference temperature, often that at 0°C. Likewise, the temperature coefficient of resistance is

$$TCR = \Delta R / \Delta T R_{0c} \tag{10.8}$$

10.2.3 The Fermi Level

Around 1900 it was thought that the electronic velocity distribution of the valence electrons in crystals was like that of an ordinary classical gas, that is, according to the Maxwell–Boltzmann distribution. However, with the advent of quantum theory and the necessity of applying the Pauli exclusion principle, the Maxwell–Boltzmann distribution was replaced by the Fermi–Dirac distribution. This distribution function governing the occupation of the quantum states may be shown to be

$$f(E) = \frac{1}{1 + e^{-[(E-E_f)/kT]}} \qquad (10.9)$$

where k is the Boltzmann constant and T is the temperature in degrees Kelvin. The value of kT at room temperature is about $1/40$ eV. The quantity E_f, which has the dimensions of energy, is termed the Fermi energy, or the Fermi level, and its value is dependent on the number of particles in the system. We have seen that at absolute zero the Fermi level is that energy level which in a metal separates the filled from the unfilled states. At any temperature above absolute zero we find from the distribution function (10.9) that for $E = E_f$, the value of $f(E) = {}^1/_2$; that is, the Fermi level is that energy level which has a 50% probability of being occupied by electrons. When $T = 0$ K is substituted into Equation 10.9, the distribution function has the values $f(E) = 0$ for $E > E_f$ and $f(E) = 1$ for $E < E_f$. The variation of E_f with temperature is shown schematically in Figure 10.7. The discussion of the Fermi level up to this point is quite general and applies to metals, semiconductors, and insulators. We will find in the following that E_f values and positions in the bands are quite different for semiconductors.

10.3 ENERGY BANDS IN INSULATORS AND SEMICONDUCTORS

The most important group of elements for this subject is the four-valency Group IV elements, which include carbon (in diamond form), silicon, and germanium. All crystallize in the diamond structure and have a band gap that decreases progressively from carbon to tin. Lead, which is the remaining Group IV element, has a face-centered cubic structure and is metallic. Gray tin also transforms to a tetragonal form near room temperature and is metallic.

The diamond structure is a tetrahedral bonded structure consisting of four hybrid sp^3 orbitals that bond with four sp^3 orbitals from four other carbon atoms. Most semiconductors, both elemental and compound, crystallize in the diamond crystal structure. Silicon has the atomic number 14 and an electron configuration

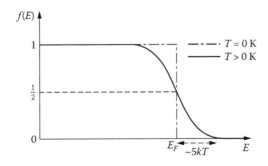

FIGURE 10.7 Variation with temperature of the Fermi energy level for metals.

of $1s^2 2s^2 2p^6\ 3s^2 sp^6$. The s and p levels of the isolated atoms combine in the solid to form the hybrid sp^3 orbital. This gives a band structure consisting of a lower, or valence, band and an upper conduction band separated by an energy gap. The total of $8N$ states of the isolated atoms, $2N$ from s and $6N$ from p levels, combine in the solid to give $4N$ states in each of the two bands. Since the silicon is quadrivalent, the electrons participating in the bonding of the solid fill all of the available states in the valence band, and the conduction band is unoccupied at $T = 0$ K. This band structure is shown in Figure 10.8. This band structure also applies to germanium, diamond, and many compound semiconductors.

Generally, when we show the band structure for a crystalline material, we take a small slice near a_0 from Figure 10.8 and sketch it in a magnified fashion as shown in Figure 10.9. The filled valence band is represented by the shaded area, while the conduction band is shown as being empty, similar to the procedure for metals. The difference between metals and semiconductors is that the latter have a forbidden region or energy gap, E_g. No electron within the crystal is permitted to possess energy that falls within this region. An electron wave of this energy destructively interferes with itself. Most semiconductors used in devices

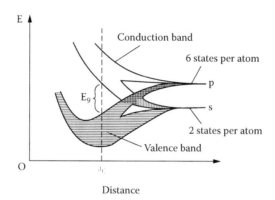

FIGURE 10.8 Energy-level band structure for semiconductors and insulators.

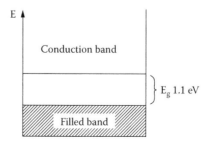

FIGURE 10.9 Energy band structure for silicon at the equilibrium interatomic spacing.

have energy gaps of the order of 1 eV. These materials have a slight conductivity at room temperature. The thermal energy is sufficient at room temperature to cause a few electrons to be excited to the conduction band (that is, they jump across the forbidden region). Silicon has an E_g of 1.1 eV and germanium of 0.67 eV at 20°C. Both of these elements are semiconductors. Carbon, in diamond form, has a large forbidden region with an E_g value of about 5.4 eV at 25°C. The thermal excitement of the electrons at room temperature is not sufficient to cause a measurable quantity of electrons to jump the gap. Therefore, diamond is an insulator. The only difference between a semiconductor and a crystalline insulator is the size of the energy gap. The E_g division between semiconductors and insulators is about 4 eV.

It should be pointed out here that the above description of insulators is the band gap model. Many insulating materials, both electrical and thermal, will be discussed elsewhere in this book, where the band gap does not necessarily apply. These include ceramics, glasses, and polymers. In many of these materials the insulating characteristics can be explained to a large extent based on their strong covalent bonds.

10.4 SEMICONDUCTORS

10.4.1 INTRINSIC SEMICONDUCTORS

The electron conductivity in a pure nondoped semiconducting element or compound is referred to as *intrinsic* conductivity; that is, it is a function of this substance and temperature only. It can be shown that the Fermi energy E_f for semiconductors lies in the middle of the forbidden region. Now the energy level at the bottom of the conduction band becomes

$$E = E_f + E_g/2 \tag{10.10}$$

Accordingly, Equation 10.9, which was developed for metals, can be revised to read

$$f(E) = \frac{1}{1 + e^{[E_f + E_g/2 - E_f/kT]}} = \frac{1}{1 + e^{E_g/2kT}} \tag{10.11}$$

Using this function, it can be shown that the number of electrons n that can jump the forbidden region gap is

$$n = n_o e^{-E_g/2kT} \tag{10.12}$$

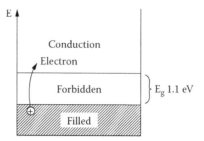

FIGURE 10.10 Electron energy band structure for an intrinsic semiconductor such as silicon. At room temperature there is sufficient thermal energy to excite a small fraction of the electrons from the filled band to the conduction band, leaving behind a positive charge (hole) in the valence band.

where n_o is a constant that is a function of the effective mass of the electron and of temperature. (In essence it is related to the density of states in the conduction band.)

When an electron jumps to the conduction band, a deficiency of negative charge is left behind in the valence band, as illustrated in Figure 10.10. This acts as a positive charge and is often referred to as a hole. It moves in an electric field in the opposite direction to that of the electron. (This is somewhat analogous to the substitution diffusion mechanism, where a vacancy moves in the opposite direction to that of the diffusing atoms.) Now the conductivity becomes

$$\sigma = n_e q_e m_e + n_h q_h m_h \qquad (10.13)$$

where the subscripts e and h represent electrons and holes, respectively. Since $n_e = n_h$ and $q_e = q_h$, we can write

$$\sigma = nq(\mu_e + \mu_h) = n_o q(\mu_e + \mu_h e^{-Eg/2kT}) \qquad (10.14)$$

Table 10.2 lists the values of E_g and the mobilities of electrons and holes for a number of materials. Note that the mobilities of the electrons are larger than those of the holes.

10.4.2 EXTRINSIC SEMICONDUCTORS

The intrinsic semiconductors have a very small conductivity, which is measurable but has found little practical application other than as temperature sensors. The conductivity is very sensitive to temperature changes, and with a sensitive meter (e.g., an electrometer or picoammeter) it is a very good method for making temperature measurements. To fully exploit the semiconducting properties of these materials, a higher conductivity is needed. This conductivity is brought about by the introduction of minute quantities of foreign elements called dopants. Depending on our choice of dopant, we can make the charge carrier become

TABLE 10.2
Values of the Band-Gap Energy and Carrier Mobility for Selected Semiconductors[a]

Material	E_g(eV)	μ_e(m²/V·s)	μ_h(m²/V·s)
C	5.3	0.18	0.16
Si	1.1	0.14	0.48
Ge	0.7	0.39	0.19
GaAs	1.4	0.85	0.40
GaP	2.3	0.11	0.007
InSb	0.2	8.0	0.075
CdTe	1.5	0.03	0.006
Zns	3.54	0.02	0.005
CdS	2.4	0.03	—
ZnTe	2.26	0.03	0.01

[a] Mobility values vary considerably from source to source and must obviously be very structure sensitive and thus a function of processing history.

either the negative electron or the positive hole. These extrinsic semiconductors are referred to as *N* and *P* type, respectively.

10.4.2.1 *N*-Type Semiconductors

If we add elements of a valency of 5 (e.g., P, As, Sb, and Bi) to silicon or germanium, we introduce an extra electron for each atom added. This electron can be viewed as lying outside the covalent bond (Figure 10.11) and is somewhat free to move under the influence of an electric field. Each 5-valency dopant atom added introduces an extra electron, so that the total number of charge carriers is almost temperature independent but is dependent on the concentration of the impurity element. Such a semiconductor is an extrinsic, or impurity, semiconductor. Note that the dopant is a substitutional impurity atom.

Instead of the physical picture depicted in Figure 10.11, we usually view extrinsic semiconductors in terms of the energy band structure sketched in Figure 10.12 for an *N*-type extrinsic semiconductor. The dopant charge is *screened* out by the silicon ions, which redistribute their electron states slightly in response to the local positive charge. As a result of this screening action, the extra electron is bound rather loosely to its ion. This binding energy is referred to as the ionization energy of the dopant (i.e., the energy required to remove the extra electron from the impurity atom). For phosphorus in germanium, this ionization energy is 0.012 eV. As a result, this electron energy state resides in the forbidden region of the semiconductor energy band structure, very close to the conduction band. This ionization energy is represented in Figure 10.12 as E_d, the donor

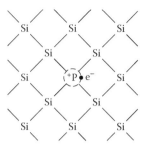

FIGURE 10.11 Extra electron bound to a 5-valence donor impurity that is added to the silicon crystal.

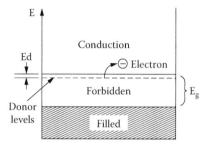

FIGURE 10.12 Energy band structure for an *N*-type extrinsic semiconductor showing the donor impurity energy levels.

energy-level gap, since it can, by thermal energy at room temperature, donate an electron to the conduction band. For arsenic in silicon, E_d is only 0.05 eV. Again, this electron energy state lies in the forbidden region of the semiconductor, very close to the conduction band. For all donor dopants, E_d is very small, 10% or less of the E_g value in an intrinsic semiconductor. Now many electrons can jump the gap from the donor levels to the conduction band, and we can write

$$\sigma \approx n_d q \mu_e \approx n_{0d} \exp\left(-\frac{E_d}{kT}\right) q \mu_e \qquad (10.15)$$

Note that we do not have an equivalent hole or positive charge for each electron that enters the conduction band as was the case for the intrinsic semi-conductors. Furthermore, the entire contribution of the intrinsic electrons and holes is insignificant at room temperature compared to the extrinsic effect, when $N_d \gg N$-intrinsic. Also, in the exponential term we have kT instead of $2kT$ since for the intrinsic case the factor 2 arose because the Fermi level E_F occurred in the middle of the forbidden gap. Furthermore, the ionization energy in the expo-nential term is the small E_d value rather than the larger E_g value of the intrinsic semiconductor.

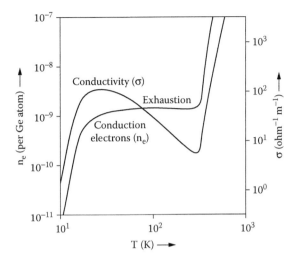

FIGURE 10.13 Variation of conductivity vs. temperature for an N-type semiconductor. (Adapted from E. M. Cromwell. *Proc. IRE*, 40, 1327, 1952. With permission.)

Now let us examine the plot of conductivity for the N-type extrinsic semiconductor as a function of temperature, as shown in Figure 10.13 for germanium. At the lower temperature just a slight increase in temperature will result in many donor electrons moving into the conduction band. At a certain temperature and a fixed concentration of dopant atoms, all of the impurity electrons will have entered the conductor band. This is portrayed in Figure 10.13 as the *exhaustion* region. There may be a slight contribution from the intrinsic electrons at this temperature, but to a large extent the conductivity will be constant until higher temperatures are reached, at which point the intrinsic contribution becomes significant and begins to increase measurably with temperature. Usually, one controls the dopant concentration so that the exhaustion region of the σ vs. T plot will be in the vicinity of room temperature, because this is the temperature at which most semiconducting devices operate, and a conductivity independent of temperature fluctuation is desirable. The conductivity curve does not follow exactly the number of electrons curve because of the mobility factor. The decrease in conductivity in the exhaustion region is associated with the decrease in mobility as the temperature increases.

10.4.2.2 *P*-Type Semiconductors

If instead of adding 5-valency dopant atoms we add 3-valency impurities (e.g., B, Al, Ga, and In); we obtain a positive hole bound to an acceptor impurity as illustrated in Figure 10.14. Since one of the bonds to the impurity atom is lacking, the hole acts as a positive charge. This corresponds in the band picture to a localized hole state. If the vacant state migrates away from the impurity atom, the latter finds itself associated with an extra electron density (i.e., it becomes

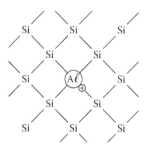

FIGURE 10.14 Positive hole bound to an acceptor impurity in a P-type semiconductor.

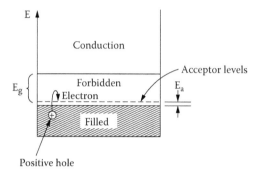

FIGURE 10.15 Energy band structure for a P-type semiconductor showing the empty acceptor levels near the filled valence band.

negatively ionized). In the band picture (Figure 10.15) the impurity or acceptor levels are very close in energy to the filled valence band and the electrons can jump from the filled band to the acceptor levels by the room-temperature thermal energy of 1/40 eV. When the electron jumps from the filled band to the acceptor level, a positive charge or hole is left behind. The conductivity occurs by the hole movement and becomes

$$\sigma = n_h q_h \mu_h \approx n_{0a} \exp\left(-\frac{E_a}{kT}\right) q_h \mu_h \qquad (10.16)$$

where

n_h = the number of acceptors or holes

n_{0a} = constant

E_a = the acceptor energy gap

Such extrinsic semiconductors are called acceptors or P-types. If we plot the conductivity vs. temperature (Figure 10.16), a curve similar to that depicted in

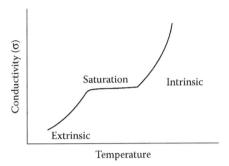

FIGURE 10.16 Variation of conductivity with temperature for a *P*-type semiconductor.

Figure 10.13 is obtained, except now the constant-temperature conducting region is called *saturation* and occurs when all the receptor levels are filled.

10.4.3 COMPOUND SEMICONDUCTORS

10.4.3.1 Group 3-5 Compounds

When certain elements of 3-valency combine to form a crystalline compound with 5-valency atoms, the resulting crystal structure, net valence, and energy band structures are identical to those of the group 4 elements carbon, germanium, and silicon. The only difference between the pure elements and compounds, shown in Figure 10.17, is that in the latter each atom has four nearest neighbors of unlike atoms, in contrast to like atoms in carbon, silicon, and germanium. Both structures are of the diamond cubic lattice type. GaAs, GaP, and InSb are such group 3-5 compounds that have been used as semiconductors, with GaAs being by far the most famous of the group. The band-gap energies and carrier mobilities of these compound semiconductors are given in Table 10.2.

GaAs and GaP were prepared in single crystal form and used as light-emitting diodes (LEDs) in the early 1970s in applications such as the first hand-held calculators and wrist watches. These materials lost out to the liquid crystals for most displays many years ago. There are some applications where LEDs are still in use. This light emission occurs due to electron transitions between the conduction and valence bands or between impurity levels in either band and is a result of the recombinations of electrons and holes. The frequency and hence the color of the light, are a function of the difference in the energy levels involved. Figure 10.18 shows how recombination could occur. This phenomenon is called P-N junction injection electroluminescence. By varying dopants and composition of the compounds, a number of different colors of light emission will result (Figure 10.18b). N-doped Ga (As,P) can emit green, yellow, amber, or red by varying the Ga-As-P ratios. In GaP, green light is emitted from a nitrogen donor to the valence band, and red light from a ZnO impurity level to a valence-band transition. Both GaAs and the ternary AlGaAs are used to fabricate diodes that emit visible in the visible light in the range of 500 to 700 nm wavelength. GaAs

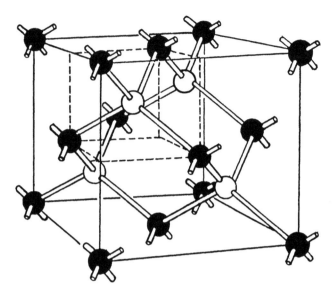

FIGURE 10.17 GaAs and GaP crystal structure. This structure is known as the zinc blende type (ZnS). Except for the unlike atoms, it is the same structure as that found in diamond, silicon, and germanium.

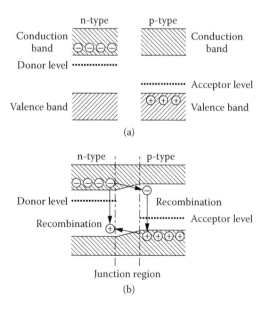

FIGURE 10.18 Energy band and dopant levels in a GaAs *p-n* junction injection light-emitting diode.

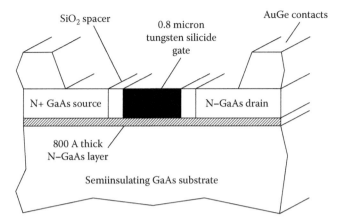

FIGURE 10.19 Cross-section of a GaAs *N*-channel field-effect transistor. (Courtesy of DM Data, Inc.)

has also been used in lasers and photovoltaic cells. But the principle use of GaAs is for integrated-circuit transistors. It will never be a threat to the silicon integrated circuits because of its high cost. But GaAs circuits switch at faster rates, pass signals more rapidly, consume less power at high speeds and operate over a wider temperature range than do similar Si circuits. GaAs is used in monolithic microwave integrated circuits as diodes or as bipolar or metal-semiconductor field-effect transistors. Since GaAs does not form an oxide, an MOS per se cannot be made, but metal-semiconductor field-effect transistors (MESFETs) can be and are made.

The base structure of a GaAs field-effect transistor is shown in Figure 10.19. It is an *N*-channel device formed from several layers of doped and undoped GaAs. A refractory metal gate of tungsten silicide modulates a thin channel of lightly doped *N*-type GaAs that connects the source and drain. GaAs has been used for discrete devices in microwave and satellite communication applications where high frequencies are required.

10.4.3.2 Group 2-6 Compounds

When atoms of 2-valency combine with atoms of 6-valency, we have a net 4-valency. Many of these compounds have the diamond cubic structure shown in Figure 10.17. ZnS (zinc blende) was one of the earliest used compounds of this type, and all of these diamond cubic compounds are referred to as having the zinc blende structures. ZnS has been used as a phosphor on television screens. Electrons bombard the coated screen and excite electrons across the 3.54-eV energy gap of the ZnS. Other group 2-6 semiconducting compounds include CdS, ZnTe, and CdTe.

10.5 DIELECTRIC MATERIALS

A dielectric material, either via induced or spontaneously existing dipoles, can increase the charge on capacitor plates by virtue of its presence. The dielectric constant K' can be expressed as follows:

$$C = \frac{K'\varepsilon_0 A}{d} \quad \text{or} \quad K' = \frac{Cd}{\varepsilon_0 A} \tag{10.17}$$

This is a material constant, but it varies with the frequency of the applied electric field. Dielectric constants are usually presented at 60 Hz, but these constants are also often stated at much higher frequencies. They can be found in many handbooks together with the dielectric strength (i.e., the volts/[thousandth of an inch] mil of thickness that will break down the dielectric material) and their resistivities. A number of such materials and their properties are listed in Table 10.3. The equation for the dielectric constant was defined in terms of the capacitance, permittivity, and capacitor plate dimension. It can also be expressed simply as

$$K' = \frac{\varepsilon}{\varepsilon_0} \tag{10.18}$$

which is the ratio for the permittivity of the material to the permittivity in vacuum. It is a dimensionless quantity. We used K' so as not to confuse it with the thermal conductivity K. The most useful dielectrics are those with high dielectric constants, such as the titanates. In these compounds (e.g., $BaTiO_3$) polarizations exist in the normal crystal since the Ti^{4+} and O^{2-} ions are displaced slightly from their face-centered positions, causing them to be permanently polarized (Figure 10.20a). When an ac alternating field is applied, the titanium ion moves back and

TABLE 10.3
Properties of Selected Dielectric Materials

Material	Dielectric Constant 60 Hz	Dielectric Constant 10^6 Hz	Dielectric Strength (10^6 V/m)	Loss Factor (10^6 Hz)	Resistivity (W · m)
Polyethylene	2.3	2.3	20	0.00023	10^{16}
PVC	3.5	3.2	40	0.15	10^{10}
Soda-lime glass	6.9	6.9	10	0.063	10^{13}
$BaTiO_3$	—	3000	12	—	—
Al_2O_3	9.0	9.9	9.8	0.0004	10^9–10^{12}
Nylon	8	4	12–24	0.14	10^{11}–10^{13}
Steadites	6	6	8–14	0.007–0.025	$>10^{12}$
Fosterites	—	6.2	8–12	0.001–0.002	$>10^{12}$

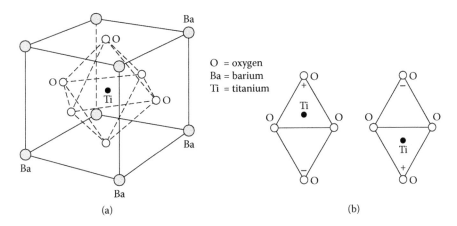

O = oxygen
Ba = barium
Ti = titanium

(a)

(b)

FIGURE 10.20 $BaTiO_3$ unit cell. The Ti^+ ion can be in either the (a) Perovskite structure; or (b) central structure position. This permanent dipole will switch from one position to another by application of an electric field. (From C. W. Richards, *Engineering Materials Science,* Brooks/Cole, Monterey, CA, 1961, p. 453. With permission.)

forth between its two allowed positions (Figure 10.20b). Some amorphous polymers have sufficient mobility to polarize, although their dielectric constants are low compared to the titanates. The polarization is greater with larger dielectric constant and can be expressed as $P = (K' - 1)\varepsilon_0\xi$, where ξ is the strength of the dielectric field in volts/mil.

Since dielectrics are used at high frequencies, their dipoles must switch often and thus give rise to a dipole friction energy loss. The more difficult the dipoles are to orient, the greater the loss. There are several types of polarization. Electronic polarization is that where electron rearrangement occurs in the presence of an applied field. The electrons move to the side of the atom nucleus nearest the positive end of the field. Ionic polarization involves an elastic shift of ions in an ionic solid. In molecular polarization molecules or atoms shift in the presence of a field to form strong dipoles. Some materials have permanent molecular dipoles even in the absence of a field. Molecular polarization is the type of interest in our present discussion. Electronic polarization, which involves no atom shift, occurs easily at high frequencies and hence incurs little energy loss. This is not so for molecular polarization. In a perfect dielectric the current will lead the voltage by 90°. However, owing to losses, the current leads the voltage by 90° − δ, where δ is the *dielectric loss angle*. The power lost in the form of heat, called the *dissipation factor*, is given by

$$\text{dissipation factor} = \tan \delta \qquad (10.19)$$

and the dielectric loss factor is

$$\text{dielectric loss factor} = K' \tan \delta \qquad (10.20)$$

The total power loss P_L is a function of the dielectric field, the frequency, and the dielectric constant. The dissipation factor and is given by

$$P_L = 5.556 \times 10^{-11} \; K' \tan \delta \xi^2 fV \qquad (10.21)$$

where the electric field strength ξ is given in volts/meter, the frequency f in hertz, the volume V in cubic meters, and the loss in watts.

10.6 FERROELECTRIC AND PIEZOELECTRIC BEHAVIOR

A few materials have atom arrangements that afford polarization of the ions without the assistance of an external electric field. This characteristic is called *spontaneous polarization,* and the materials are called *ferroelectrics.* They have nothing to do with ferrous materials. The name *ferroelectric* was selected to be analogous to *ferromagnetic* materials, which are materials that are spontaneously magnetized in the absence of an external magnetic field. Obviously, the field of magnetism was studied before the recognition that ferroelectric behavior existed.

In spontaneous polarization permanent dipoles exist within the crystal. The most widely studied and used material of this type is barium titanate ($BaTiO_3$). It has the *perovskite* structure and displaced charges as in Figure 10.20. The position of the titanium ion can be in either of the two positions shown in Figure 10.20b. In both cases the unit cell has a permanent dipole because of the unequal distribution of the electric charge within the cell. The titanium ion can also occupy similar positions in the other principal directions in the cell, for a total six possible polar axes. Barium titanate is used as a dielectric material and as a ferroelectric material.

Ferroelectric materials form domains, in which the dipoles of adjacent cells are aligned in the same direction, as depicted in Figure 10.21. In $BaTiO_3$ there are six possible directions, but in some ferroelectric materials there may be only two such directions. The domains tend to form because this is a lower-energy state. When an external field is applied, the domains rotate toward the same direction, thereby increasing the polarization as depicted in Figure 10.21. As the electric field is increased, the degree of alignment increases until it attains a maximum saturation polarization value, denoted as P_s in Figure 10.21. At this point all of the domains are aligned. If the field is reversed, the domains begin to fall back to their random arrangement but do not quite attain it when the field is reduced to zero. To obtain random orientation of dipoles, the field must be reversed. If we continue this field reversal, a cycle will be completed that results in the hysteresis loop depicted in Figure 10.21. Ferroelectric domains of various orientations can be observed in barium titanate by chemical etching and viewing of the resulting microstructure at a magnification of about 500×. Ferroelectric

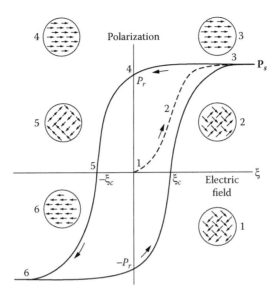

FIGURE 10.21 Ferroelectric hysteresis loop. P_s represents the point where all domains are polarized in the same direction. (From D. R. Askeland, *The Science and Engineering of Materials*, 2nd ed., PWS-Kent, Boston, 1989, p. 696. With permission.)

materials possess a temperature called the *Curie temperature*, again analogous to magnetic behavior, above which spontaneous polarization does not occur. In $BaTiO_3$ this temperature is 120°C; it is 233°C in $PbZrO_3$ and 490°C in $PbTiO_3$.

10.6.1 PIEZOELECTRIC EFFECT

Piezo is derived from the Greek word *piezen*, which means *to press*. The piezoelectric effect is related to the relationship between the pressure applied to a material and the electric field created thereby. Compressive stresses reduce the dipole distance and accordingly the dipole moment. This change in the dipole moment changes the charge and the voltage between the ends of the sample. The voltage is proportional to the strain. Conversely, the application of a voltage will cause a dimensional change, with the strain being proportional to applied voltage. Thus, electric energy can be changed to mechanical, and vice versa. These actions are all called the *piezoelectric effect*. All ferroelectric materials have piezoelectric properties, which also has been referred to as *electrostriction*, again because of a similar term and effect that exists in ferromagnetic materials. This pressure on ferroelectrics is illustrated schematically in Figure 10.22. Piezoelectrics are invaluable as electromechanical transducers because their large response makes them more efficient than magnetic transducers. Such devices are used for measuring pressure or strain, sonar, audio devices, ultrasonic vibrators, and radio tuners.

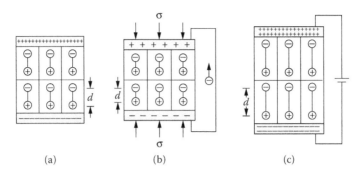

FIGURE 10.22 Piezoelectric effect. Application of pressure changes the net dipole moment and thus the voltage between the plates. Conversely, a voltage application produces dimensional changes. (From L. H. Van Vlack, *Elements of Materials Science and Engineering,* 3rd ed., Addison-Wesley, Reading, MA, 1975, p. 279. With permission.)

DEFINITIONS

Acceptor level: Permitted energy levels that are formed in the forbidden region but near the filled valence of an extrinsic semiconductor when it is doped with 3-valence impurity atoms.

Brillouin zone theory: Theory that determines energy levels in three-dimensional reciprocal lattice space at which electrons are allowed.

Conductivity: A measure of charge flow. The reciprocal of resistivity.

Donor level: Energy level in a semiconductor that is formed in the forbidden region of the semiconductor near the conduction band.

Dopants: Impurity atoms added to an intrinsic semiconductor to make it an extrinsic type.

Extrinsic semiconductor: One that conducts on the basis of small additions of other atoms (dopants).

Hole: An absence of charge related to the addition of an atom of lower valence than the matrix atoms of the semiconductor.

P-type semiconductors: One whose conduction is by hole movement.

Resistivity: An electrical property that measures the resistance to electron flow and is independent of specimen dimensions.

QUESTIONS AND PROBLEMS

10.1 Sodium has a lattice parameter of 4.289 Å at 20°C and a temperature coefficient of thermal expansion of 71×10^{-6} per degree Celcius. Assuming that this coefficient is maintained (i.e., constant to 0 K), what lattice parameter will sodium have at 0 K?

10.2 Compute the number of allowable energy levels in 1 cm³ of copper. The density of copper is 8.96 g/cm³.

10.3   Estimate the probability that an electron in Sn can gain sufficient energy at 20°C to enter the conduction band.

10.4   What fraction of the charge in intrinsic Si is carried by the electrons? Compute the same fraction for a GaAs intrinsic semiconductor.

10.5   What is the resistance of a 0.001-in-diameter gold wire of 10 ft in length at (a) 0°C and at (b) 100°C?

10.6   Compute the conductivity of copper at 120°C.

10.7   Sketch the energy band structure of semiconductors, metals, and insulators and explain their differences.

10.8   How many phosphorus atoms must one add to 1 mol of germanium atoms to give a resistivity of $10 \; \Omega \cdot cm$? What will be the concentration of phosphorus in parts per million (ppm)?

10.9   List all the factors that affect the conductivity of metals in order of the smallest to the largest effect.

10.10 At 300 K the electrical conductivity of intrinsic germanium is 22 $(W \cdot m)^{-1}$. What is its conductivity at 300°C?

10.11 For a parallel-plate capacitor of 1 cm^2 area, a 1-cm separation of the plates, and a 10-V applied potential, what will be the capacitance for (a) a vacuum between the plates and (b) a material with a dielectric constant of 10?

10.12 The intrinsic electrical conductivities for a semiconductor material at 20°C and 100°C are 0.5 and 5 $(W \cdot m)^{-1}$, respectively. Estimate the E_g gap width in electron volts for this material.

10.13 Explain the difference in the variation of conductivity with temperature between metals and semiconductors.

10.14 Several years ago copper was marketed for electrical conduction purposes as oxygen-free high-conductivity (OFHC) copper, yet chemical analysis showed oxygen contents of thousands of parts per million. This oxygen content did not appreciably affect the electrical conductivity. How is this possible?

10.15 A *P*-type semiconductor is made by adding aluminum to silicon. The resistivity of such a doped silicon is measured to be $10^{-3.}$ m. Using the hole mobility from Table 10.2, what is the aluminum content in ppm?

SUGGESTED READING

Metals Handbook, Vol. 2, 10th ed., ASM International, Metals Park, OH, 1990.

Nishi, Y. and Doering, R., *Handbook of Semiconductor Manufacturing Technology*, Marcel Dekker, New York, 2000.

Schaffer, J. P. et. al., *The Science and Design of Engineering Materials*, McGraw-Hill, New York, 1999.

11 Magnetic Materials

11.1 MAGNETIC FIELDS

A moving charge creates a magnetic field. If this moving charge takes place as current in a wire, the magnetic field flows around the wire. When the wire is formed into a coil the magnetic lines of flux reinforce each other and thereby increase the field within the coil (Figure 11.1a), which is given by

$$H = \frac{4\pi NI}{10l} \tag{11.1}$$

where
  H = magnetic field strength
  N = the number of turns within the coil
  I = current
  l = coil length

In what used to be the more common centimeter–gram–metric (cgs) units, H was expressed in oersteds when I was in amperes and l was in centimeters. The System International (SI) units for H is ampere turns per meter, more often expressed as amperes per meter (A/m).

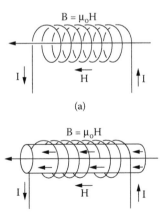

(a)

(b)

FIGURE 11.1 Current passing through a wire creates a magnetic field around the wire (a). When the wire is wound into a coil, the magnetic flux density is reinforced and increases. When a core material with $\mu > 0$ is inserted into the coil, the flux is further increased (b).

TABLE 11.1
Magnetic Property Conversion Units

M (magnetization): 1 A/m = $4\pi \times 10^{-3}$ Oe
H (applied field): same as magnetization
B (magnetic induction): 1 Wb/m^2 = 1 T = V-s/m^2 = 10^4 G
μ_0 (permeability in vacuum): $4\pi \times 10^{-7}$ H/m = Wb/m
μ_r (relative permeability): dimensionless = μ/μ_0
$\chi = M/H$ = susceptibility (dimensionless)

The units in magnetism are many in number and often confusing to the reader. A conversion chart and list of units used herein are presented in Table 11.1. When a material is placed within the coil (Figure 11.1b), the magnetic flux is increased if the material offers a lower-resistant path than does the air of Figure 11.1a. The magnetic field strength produces a magnetic flux density B, which is proportional to the field H, the proportional constant being μ, the permeability of the material. In equation form the B–H relationship becomes

$$B = \mu H \qquad (11.2)$$

When H is given in amperes per meter, B is expressed in webers per square meter (tesla) and μ as henrys per meter; when H is in oersteds, B is in gauss. Most readers will have heard of a gauss meter and perhaps a tesla meter or a tesla coil but probably not a webers-per-square-meter meter. The gauss was originally thought of as the number of lines of magnetic force penetrating unit area. Perhaps the observance of the lines of iron filings when such filings were placed in a magnetic field had something to do with the lines of force concept (Figure 11.2). In vacuum, μ_r becomes dimensionless because

$$\mu_r = \frac{\mu}{\mu_0} \qquad (11.3)$$

The permeability is somewhat like the dielectric constant, in that the magnetic field is increased by the presence of a material (e.g., as a core material in a coil over that in air or vacuum). (Diamagnetic materials would decrease the field strength.)

Equation 11.2 is similar to the equation often used to express the *magnetization* of a solid, which is written as

$$B = \mu_0(H + M) = \mu_0 H + \mu_0 M \qquad (11.4)$$

where M has the same units as H. We can also write

$$M = \chi H \qquad (11.5)$$

FIGURE 11.2 Iron filings reveal the magnetic field around a bar magnet. Note that the magnetic lines of force leave the bar at one pole and return at the other.

where χ is called the susceptibility. M can be viewed as being related to the magnetic flux increase when a material of magnetization M is placed in the field, because, except for diamagnetic materials, $\mu_0 H < \mu_0 M$, and for ferromagnetic materials, $\mu_0 H \ll \mu_0 M$. Equation 11.3 can also be written as

$$B = \mu_0 H(H + M) = \mu_0 \mu_r H \qquad (11.6)$$

Therefore,

$$\mu_0 = \frac{H + M}{H} = 1 + \frac{M}{H} = 1 + \chi \qquad (11.7)$$

Now that we have become sufficiently confused with units, let us return to the fundamentals of magnetism.

11.2 SOURCES OF MAGNETISM IN THE ATOM

An electron revolving in orbit around a nucleus is a moving charge and creates a magnetic field. In most atoms these orbits are randomly oriented and the fields nearly cancel. When an external field is applied to a material that derives its magnetic moment from orbital motion, if the orbits reorient to oppose the field, and if this is the only effect of the field, the material is said to be *diamagnetic*. Cu, Bi, Zn, and the inert gases are diamagnetic. Their permeability is negative and about the order of $\mu \approx 1 - 1.0005$ (diamagnetic). When the orbital moments do not cancel and when they align with an applied field, and if this effect outweighs any diamagnetism in the material, the material is weakly paramagnetic and has a permeability of about $\mu \approx 1.005$ (weak paramagnetism). Many materials are weakly paramagnetic.

Electron spins create a magnetic field, and when there are an unequal number of parallel and antiparallel spins, either a somewhat strong paramagnetic or a *ferromagnetic* material exists. For the elements, these two types of magnetism occur in the transition element groups and result from the noncancellation of electron spins. The somewhat strong (relative to the weak paramagnetism) paramagnetic transition elements have permeabilities of approximately $\mu \approx 1.01$. In the transition elements that are ferromagnetic, the permeabilities are very large and varied, but for now let us say that they are about $\mu \approx 10^6$ (ferromagnetic). In all of the following discussion of magnetic materials, we will be dealing with ferromagnetism only, because these are the materials of commercial interest. It should be mentioned that there exists a nuclear magnetic effect that is small but useful for magnetic resonance applications, most notably in equipment used in the medical field.

11.3 FERROMAGNETISM

What is the source of ferromagnetism, and why are only a few of the transition elements ferromagnetic? It has already been mentioned that a necessary but not sufficient condition for ferromagnetism to exist in an atom is that it must have an imbalance of electron spins in the unfilled d energy band. This imbalance also occurs in the f band of higher-atomic-number elements. These include many of the rare earths. Here we deal primarily with the first series of transition elements, which include Sc, of atomic number 21, through Ni, of atomic number 28 (that is, we will be concerned with the unbalanced spins in the $3d$ level of these elements). Ferromagnetism occurs in transition elements that have an offset between $m_s = +\,^1/_2$ and $m_s = \,^1/_2$ energy levels in their $3d$ band. Because of this effect, there arises an *exchange force* between adjacent atoms that causes their magnetic dipole moments to align in the same direction, and this occurs without the application of an external magnetic field (i.e., they possess the characteristic called *spontaneous magnetization*). The theories that explain this exchange force are rather complex and beyond our present discussion, but it is related to the ratio of the atomic diameter (or atom spacing) to that of the $3d$ shell diameter and we will simply express it as

$$\text{exchange force} = \text{function of} \left(\frac{\text{atom diameter}}{3d \text{ cell diameter}} \right)$$

When this ratio is greater than 3, the exchange force is positive and causes the alignment of the magnetic moments of adjacent atoms. For the elements of interest here the number of the 3d and 4s electrons and their respective spins are shown in Figure 11.3. As the atomic number increases, the number of unbalanced spins increases from Sc through Mn. The 3d level can hold 10 electrons and becomes filled with Cu, of atomic number 29. As the 3d levels are filled in these transition elements, their spins are in the same direction until we reach Fe, of atomic number 26, which has six electrons in the 3d level. The maximum number of unbalanced spins is five and is found in the elements chromium and manganese. (Chromium is an anomaly since it contains only one electron in the 4s level.) One would thus expect that chromium and manganese would possess the largest ferromagnetic character, but when we examine the atom diameter to 3d shell diameter ratio, we find that only Fe, Co, and Ni possess a ratio greater than 3 (Figure 11.4). Thus, they are the only elements of this series that possess a positive exchange force and as such are the only ferromagnetic elements of this series.

		3d electrons	4s electrons
	No.	Spin directions	
Sc	1	↑	2
Ti	2	↑ ↑	2
V	3	↑ ↑ ↑	2
Cr	5	↑ ↑ ↑ ↑ ↑	1
Mn	5	↑ ↑ ↑ ↑ ↑	2
Fe	6	↑ ↑ ↑ ↑ ↑ ↓	2
Co	7	↑ ↑ ↑ ↑ ↑ ↓ ↓	2
Ni	8	↑ ↑ ↑ ↑ ↑ ↓ ↓ ↓	2

FIGURE 11.3 Spin directions of the 3d electrons in the first series of transition metals. Note the opposing spins in Fe, Co, and Ni.

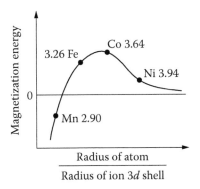

FIGURE 11.4 Magnetization energy as a function of the ratio atom diameter/3d shell diameter.

The spin magnetic dipole moment of a single electron is given by

$$\mu_b = \frac{eh}{4\pi m_e} \qquad (11.8)$$

where
  e = charge on electron
  h = Planck's constant
  m_e = mass of the electron

and is called a Bohr magneton. In SI units it has the value of 9.27×10^{-24} A · m². Again, one would expect from Figure 11.3 that Ni, which has three unbalanced spins, would have the largest ferromagnetic character of this group of three elements, but such is not the case. In Table 11.2 the net Bohr magnetons are listed as 2.2, 1.7, and 0.6 for Fe, Co, and Ni, respectively. This is simply because not all electrons take on the character with regard to unbalanced spins depicted in Figure 11.3. Instead, we must deal with an average number of unbalanced spins. Thus, Fe is the most ferromagnetic of the group. When its magnetization is saturated at 0 K, the net dipole moment is 2.2 Bohr magnetons per atom. This means that on average, 2.2 electron spins per atom are unbalanced. Not all atom moments align as shown in Figure 11.3. The ferromagnetic elements lose their ferromagnetic character as the temperature is increased. Thermal energy tends to break the spontaneous magnetic alignment of the moments of adjacent atoms until at a certain temperature, called the Curie temperature, θ_c, the exchange forces are no longer positive and the material becomes paramagnetic (Figure 11.5). The Curie temperatures of some ferromagnetic elements are listed in Table 11.3. Gadolinium is ferromagnetic at temperatures below what we usually consider to be room temperature, 20°C (68°F).

TABLE 11.2
Net Magnetic Moments per Atom for the Ferromagnetic Metals, Fe, Co, and Ni[a]

	Element[a]		
	Fe	Co	Ni
Total number (s and d) electrons	8	9	10
Average number in s band	0.2	0.7	0.6
Average number in d band	7.8	8.3	9.4
Average number with parallel spins	5.0	5.0	5.0
Average number with antiparallel spins	2.8	3.3	4.4
Net moment per atom (number of Bohr magnetons)	2.2	1.7	0.6

[a] These are not whole numbers as predicted by Figure 11.3.

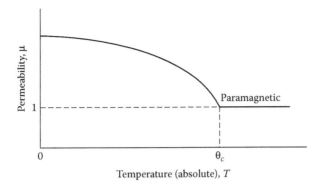

FIGURE 11.5 Decrease of permeability of iron with increasing temperature. θ_c is the Curie temperature, where the iron ceases to be ferromagnetic.

TABLE 11.3
Curie Temperature of Some Ferromagnetic Metals

Element	Curie Temperature (°C)
Fe	780
Co	1125
Ni	365
Gd	16

11.3.1 FERROMAGNETIC DOMAINS

One more characteristic of ferromagnetism must be explained, and that is why certain ferromagnetic materials appear to be magnetic only in the presence of an external field. They are still ferromagnetic, but the internal fields of ferromagnetic materials can be arranged so that they cancel out, making it appear that the material is not magnetic. This phenomenon occurs because the aligned moments of individual atoms break up into regions called *domains*, where in one region the moments of the atoms are opposed to those of the moments in another region of the material.

Domain formation occurs because the presence of domains lowers the energy of the material. Domains in general are small in size (e.g., 10^{-2} to 10^{-5} cm across, smaller than most grains). American Society for Testing and Materials (ASTM) grain size 7 represents a grain diameter of about 1/8 in, or approximately 0.3 cm. This lowering of the energy, called the *magnetostatic energy*, can be explained with the aid of Figure 11.6. The magnetostatic energy is the potential magnetic energy of a ferromagnetic material that can be produced by an external field. In Figure 11.6, the moments of all the atoms are aligned in a single-crystal bar.

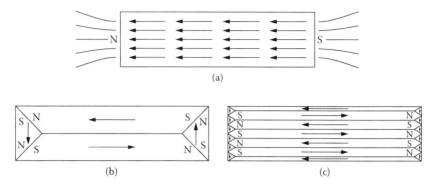

(a)

(b)   (c)

FIGURE 11.6 Single crystal of a ferromagnetic material subdividing into domains to reduce its magnetostatic energy. (From C. W. Richards, *Engineering Materials Science*, Brooks/Cole, Monterey, CA, 1961, p. 431. With permission.)

However, this is not a desirable situation because the return magnetic flux path from the N back to the S pole is outside the bar. This path in air has a permeability on the order of one millionth (10^{-6}) of that in the ferromagnetic material. In this single-crystal bar of iron, all the magnetic dipoles prefer to align in a <100>-type direction. This is the direction of easy magnetization in iron, and when this occurs, the energy is also reduced.

This reduction in energy by virtue of the crystallography of the iron crystal is known as the *anisotropy energy*. Without it, domains could not form and the *magnetostatic energy* could not be reduced. In body-centered cubic iron there are six such directions all along the cube edges. Thus, the bar magnet breaks up into little domains in which the moments align in more than one of the six directions (e.g., a [100] in one region, while in another region they are aligned in a [010] direction at 90° to those in the other region). By so doing, all of the magnetic flux can find an easy path through the ferromagnetic bar of a high permeability of about 10^6 without passing through the air or vacuum outside the bar. Thus, a big reduction in the magnetostatic energy occurs when domains are formed. The crystal breaks up into small domains that are arranged such that the magnetic flux can stay completely within the bar. This situation is depicted in Figure 11.6b for this single crystal. In essence, small regions act as individual magnets possessing N–S poles. We know from basic physics principles that magnetic poles of opposite signs attract. In Figure 11.6b the magnetic flux flows around the crystal in an internal circuit that consists of a series of N–S poles, all of which are in a <100>-type direction to reduce both the anisotropy and magnetostatic energies.

Figure 11.6b does not show the lowest-energy state. If the crystal forms more domains, the magnetostatic energy is further reduced. This situation is depicted in Figure 11.6c. A new energy, however, must be considered, and that is the domain wall energy. The magnetic moments in the wall or boundary separating the domains are not aligned with those in either domain. Thus, the domain wall is a region of higher energy. As the crystal subdivides into smaller domains, its volume magnetostatic energy is reduced, but a gain in energy is experienced by the creation of more domain walls. When the point is reached where the decrease in volume energy is just balanced by the increase in energy owing to the increase in domain wall volume, no further subdivision takes place.

In a polycrystalline metal each grain or crystal subdivides into domains just as described above for the single crystal, resulting in a large number of small flux paths all going around in small circuits in <100> crystallographic directions. The magnetic fields of all of these randomly oriented crystals with their regions of small domains effectively cancel each other, so it appears as if the bar is not magnetic at all. When a magnetic field is applied, however, the domains shift their dipole moments to align with the field. This can occur in two ways. One method is by domain wall motion, as one domain becomes large at the expense of its neighbor. This occurs at low magnetic fields. First there is a reversible wall motion where the domain wall returns to its original position when the field is removed. The reversible displacement is followed by an irreversible displacement as the field is increased in strength (Figure 11.7). At higher fields the moments of individual atoms rotate. As a consequence, the variation of B with H is nonlinear. Equation 11.2 is still valid.

Eventually, all domains will be aligned at a certain field strength. This is called the saturation magnetization, B_s, shown in Figure 11.8. Each domain by itself is always in a saturated condition, which is the spontaneous magnetization characteristic of ferromagnetism. When the field strength is reduced from the overall saturation, the domains begin to return to their random orientations. Even though the field is reduced to a zero value, not all of the domains return to the original random orientation, and the remaining magnetic field is called B_r, a residual flux called the *remanence,* or remanent, induction. A field, H_c, known as the *coercive force*, is required to reduce the internal field to zero. During a

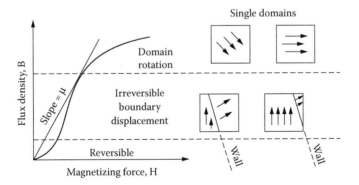

FIGURE 11.7 Effect of a magnetic field on domain alignment and resulting flux density B. Alignment occurs first by domain wall movement, but as the magnetizing force is increased, eventually the magnetic moments of adjacent atoms switch instantaneously to align in the same direction, thereby removing a domain wall.

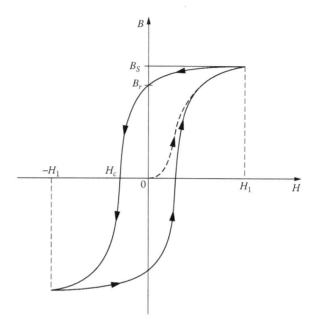

FIGURE 11.8 Hysteresis loop produced by reducing field from H_{max} (H_1) to the same magnitude in the opposite direction, $-H_{max}$ ($-H_1$). The dashed line shows the original magnetization curve.

complete cycle of the field H, the domains are forced to go through irreversible changes in direction, and as a result, energy is expended that is in proportion to the area of the hysteresis loop.

11.4 HARD MAGNETIC MATERIALS

It has been found that the area of this hysteresis loop (that is the energy dissipated per cycle of H) is very structure sensitive. The remanence, coercive force, and area of the hysteresis loop are higher for the materials containing solute atoms, second-phase particles, and dislocations. These types of defects tend to interfere with domain wall motion. Such materials are said to be magnetically hard, so named because many of these same factors that make a material mechanically hard also increase the loop size and make the material magnetically hard. Such materials are used as permanent magnets because they retain their overall alignment of domains to a large degree after the field is removed. The magnetically hard materials include the ferromagnetic aluminum, nickel, and cobalt (AlNiCo) alloys, ferromagnetic rare-earth alloys, and the ferromagnetic rare-earth alloys. The most important commercial permanent magnetic materials are the alnicos. They have very high coercive forces, Hc and BH$_{max}$ products, the terms often used for specifying permanent magnets. The addition of a few %Ti increases the coercive force. Fe-Cr-Co alloys have been developed which are more easily deformed than the alnicos. Applications for the alnicos include relays, motors, loud speakers, and magnetos. The cold deformable Fe-Cr-Co magnets are often found in telephone receivers. The best, but most expensive, permanent-magnet materials are the rare-earth alloys such as SmCo5 and Sm-Pr-Co compounds. These materials are ferromagnetic because of unbalanced spins in the 4f electrons. SmCo5 is made by powder metallurgy methods. A precipitation-hardenable composition $Sm(CoCu)_{7.5}$ contains fine precipitates when aged around 450°C. These precipitates pin the domain walls thereby inhibiting wall movement. These magnets are used in small motors, in medical devices, electronic watches, and direct current synchronous motors and generators. Their chief disadvantage is the high cost of the rare-earth materials. Neodymium-iron-boron permanent magnets were introduced in the mid 1980s. They are produced by both powder metallurgy techniques and by rapid-solidification methods. The latter process produces ferromagnetic grains surrounded by nonferromagnetic Nd-rich phase, which interferes with domain reversal within the ferromagnetic grains.

Cobalt is a relatively expensive and strategically important material. If possible one would prefer to select a hard magnetic material that does not contain cobalt. The ferromagnetic hexagonal ferrites of compounds such as barium and samarium iron oxides have relatively high Hc values (~240) and are a logical choice. These less powerful permanent magnets are used in motors, relays, speakers, latches, toys, and so forth.

Since magnetic storage devices require a material that can be easily magnetized and that retains most of the magnetization after the field is removed, these materials should be classified under the heading of hard magnetic materials. However, a square hysteresis loop is desirable so that it can be easily erased by application of a magnetic field. The BH product is not as large as that for most permanent magnets. Most magnetic storage tapes consist of polymers coated with a monolayer of iron oxide or chromium dioxide particles.

The spontaneous magnetization in a ferromagnetic crystal is accompanied by an elongation or contraction of the material called magnetostriction. It is associated with electrical anisotropy of the crystal structure and the easy directions of magnetization. The saturation magnetostriction strain is very small. Nickel has the highest value, about $- 40 \times 10^{-6}$. The property of magnetostriction is used in transducers, but ferrostriction (discussed under ferroelectrics) is larger and more often used for this purpose.

Some typical hard magnetic materials and their properties are listed in Table 11.4.

11.5 SOFT MAGNETIC MATERIALS

Magnetically soft materials are those with a small hysteresis loop. The loop sizes for soft and hard magnetic materials are compared in Figure 11.9. Soft magnetic materials are needed for applications where the material is easily magnetized and demagnetized. The materials in transformer cores which are continually subjected to alternating currents is the most important application for soft magnetic

TABLE 11.4
Magnetic Properties of Some Hard Magnetic Materials

Material	Hc	BHmax
AlNiCo 5 8AL, 14Ni, 25Co, 3Cu, bal.Fe	51	44
AlNiCo 8 7Al, 15Ni, 24 Co, 3 Cu, bal.Fe	150	40
SmCo 65 Co, 35 Sm	675–1200	160
Sm-Pr-Co	805	207
Fe-Cr-Ni 59 Fe, 30 Cr, 10 Co, 1 Si	46	34
MO.Fe2O3 (M = Ba, Sr,) Ferrite	238	28

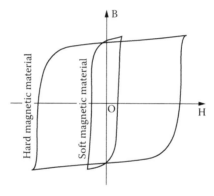

FIGURE 11.9 Comparison of the hystersis loops for soft and hard magnetic materials.

materials. Both hysteresis loss and eddy current losses must be minimized. A fluctuating magnetic field induces electric currents in the material called eddy currents. High electrical resistivity minimizes eddy current losses. The most common alloys for this application are the well annealed Fe-Si alloys. These large grain structures are strain free, single phase alloys, and are magnetically soft. The addition of silicon increases the resistivity. Further reduction of the BH loop can be attained by cold-working the material prior to the final anneal. This procedure tends to align the atoms in the grains in the easily magnetized crystallographic directions. For high frequency applications the cubic and rare-earth ferrites become the material of choice. They possess high resistivities in combination with relatively small hysteresis loops. The cubic ferrites have smaller hysteresis loops while the rare-earth ferrites have higher resistivities. The latter are preferred for microwave devices.

Metallic glasses, which are noncrystalline amorphous materials, have some important soft magnetic properties. They are alloys of Fe, Co, and Ni containing some B and Si, and are produced by the rapid solidification process, that is by using cooling rates from the molten state of the order of $10^{6\circ}$C per second. Since

TABLE 11.5
Magnetic Properties of Some Soft Magnetic Materials

Material	Maximum Permeability	Bs (kg)	Resistivity ($\mu\Omega \cdot cm$)	Applications
	DC Properties			
99.9% Fe	8,000	21.5	10	Flux carriers
Fe-1.1% Si-0.5% Al	7,700	21.0	25	Transformers
Fe-4.0% Si-0.5% Al oriented	18,500	19.5	25	Transformers
	AC Properties			
Fe- (Si + Al)-1.9% ASTM A677	1,300 (60 Hz and 15 kG)	20.7	—	Rotating machinery
Fe- (Si + Al)-3.2% ASTM 876 oriented	>1,800 at 10 Oe:	20	—	Power transformers
79% Ni-5% Mo-15% Fe	90,000 at 40 G	7.9	16	Shielding

there are no grain boundaries in these metallic glasses, domain walls move with ease when the fields are reversed. They also have high resistivities. Multilayered metallic glasses can be used as transformer cores.

Some typical soft magnetic materials and their properties are listed in Table 11.5.

DEFINITIONS

Anisotropy energy: Energy by which a magnetic material is reduced when the field is applied in the direction of the easiest magnetization.

Coercive force: Magnetic field required to switch domains in a magnetic material back to the random arrangement that they had before a field was applied.

Diamagnetic: Material in which the magnetic dipole moments oppose an applied field and the magnetic permeability is less than unity.

Domain: Region in a ferromagnetic material in which the magnetic dipole moments of all atoms are aligned in the same direction.

Exchange force: Force (or energy) that causes spontaneous magnetization in ferromagnetic materials.

Ferromagnetic: Characteristic of a magnetic material that causes magnetic electron moments to align even in the absence of a field.

Magnetostatic energy: Reduction of energy in a ferromagnetic material when domains are formed.

Magnetostriction: Characteristic of a magnetic material whereby a dimensional change affects the magnetic properties and vice versa.

Remanence: Magnetization remaining in a ferromagnetic material when a magnetic field has been applied to reverse the spin moment alignment and this field has been reduced to zero.

Susceptibility: Proportionality constant between the magnetization M of a material and the applied field strength H.

QUESTIONS AND PROBLEMS

11.1 The saturation magnetization of iron at 0 K is 2.2 Bohr magnetons. What is this value in ergs/G?

11.2 Discuss all of the similarities (i.e., analogous characteristics) of ferroelectric and ferromagnetic behavior.

11.3 What are the two requirements for a material to be ferromagnetic?

11.4 Why do domains form in a ferromagnetic material? What factor limits the minimum domain size?

11.5 The maximum relative permeability for 99.91% pure iron and 99.95% pure iron is 500 and 180,000, respectively, yet the saturation magnetization for each is the same (2500 G). Explain.

12 Optical Materials

12.1 INTRODUCTION

In the classification of materials the exact placement of optical materials is to some extent a matter of preference. The gallium arsenide light-emitting diodes are generally covered under semiconductors (see Chapter 10), but semiconducting gallium arsenide and other compound semiconductors have also been used for lasers. The optical properties of liquid crystals appear to fit better under polymers than with the optical materials. In the general broad field of optoelectronics, the optical and electronic properties of a given material often overlap. The optical properties of materials have become of more interest in recent years because of their application in the high-tech industries. The telecommunications industry is dependent on optical fibers. Lasers are used in many high-tech devices, and luminescence materials are widely used in many types of displays. In this chapter we will confine our presentation to refraction, reflection, absorption, luminescence, optical fibers, and lasers.

12.2 REFRACTION, REFLECTION, AND ABSORPTION OF LIGHT

The light spectrum from the infrared to visible wavelengths is shown in Figure 12.1. The visible spectrum includes the wavelengths from about 0.40 to 0.70 μm (micrometers). Light is a form of electromagnetic radiation consisting of waves, but light also can be considered as particles called *photons*. The energy E of a photon is related to its frequency by

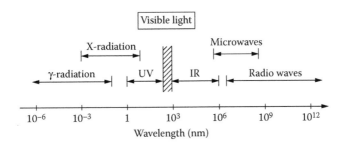

FIGURE 12.1 The electromagnetic spectrum of radiation. The visible light wavelengths range from 400 to 700 nm. (From J. F. Shackelford, *Introduction to Materials Science for Engineering*, 5th ed., Prentice Hall, Upper Saddle River, NJ, 1999, p. 591. With permission.)

$$E = h\nu \qquad (12.1)$$

where h is Planck's constant (6.62×10^{-34} Js [Joule–second]). The frequency ν is related to wavelength by

$$\nu = c/\lambda \qquad (12.2)$$

where c is the light velocity (about 3×10^8 m/s) and λ is the wavelength. Thus, E can be expressed as

$$E = hc/\lambda \qquad (12.3)$$

Light can be refracted, reflected, absorbed, or transmitted when it impinges on the surface of a given medium (Figure 12.2). The medium material in Figure 12.2 is assumed to be transparent; that is, there is no absorption.

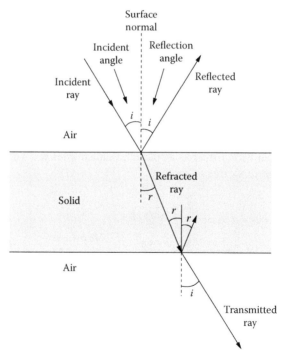

FIGURE 12.2 Refraction, reflection, and transmission of light as it passes from vacuum (or air) into a solid. (From J. P. Schaffer, A. Saxena, S. D. Antolovich, T. H. Sanders, Jr., and S. B. Warner, *The Science and Design of Engineering Materials*, 2nd ed., WCB McGraw-Hill, New York, 1999, p. 494. With permission.)

TABLE 12.1
**Index of Refraction for
Selected Materials**

Material	Index of Refraction
Silica glass	1.46
Borosilicate glass (Pyrex)	1.47
Quartz	1.55
Alumina	1.76
Diamond	2.41
Silicon	3.49
Polyethylene	1.50–1.54
Polystyrene	1.58–1.60
Polymethylmethacrylate	1.48–1.50
Polyisoprene	1.48–1.52
Polyamides	1.53

The *refractive index n* is defined as

$$n = \text{velocity of light in vacuum/velocity in a medium}$$

It is expressed as the ratio of the angle of incidence to the angle of refraction (Figure 12.2). These angles are a function of velocity, and n can be expressed as

$$n = \sin \theta_i / \theta_r \qquad (12.4)$$

where θ_i is the angle of incidence and θ_r is the angle of refraction. Typical values of n for some glasses and polymers are presented in Table 12.1. The values for glasses are in the range of 1.5 to 2.5, while for polymers they are about 1.4 to 1.6. The high value of 2.4 for diamond gives rise to the attractive sparkle from the multifaceted jewels. The refractive indexes from one medium of refractive index n through another of n' are related to the corresponding incident angle θ and refractive angle θ' by the relation

$$n/n' = \sin \theta / \sin \theta' \qquad (12.5)$$

The light reflected is expressed by the *reflectance, R*, which is defined as the fraction reflected or

$$R = (n - 1/n + 1)^2 \qquad (12.6)$$

where n is the refractive index for the refracting medium. The way R is defined here, it applies to light only of normal incidence, but it is a good approximation for angles up to about 20°. For a polished glass surface, n is usually small. For a common silicate glass with a refractive index of 1.5, R = 0.04, that is, a reflectivity of 4%. The light is usually reflected from the interface at an angle equal to that of the incident light. When light passes from a medium of high refractive index to one of low refractive index, at some critical angle of incidence, θ_c, all light will be reflected. This situation of *total internal reflection* is an important factor in the fiber optics industry, where it is desirable to have the light rays stay within the wave guide.

The absorption of light is represented by the ratio of light entering a medium, I_0, to the fraction of light, I, exiting by

$$I/I_0 = e^{-\alpha t} \tag{12.7}$$

where α is the linear absorption coefficient and has the units cm^{-1} if the thickness t is measured in centimeters. In glass plates the amount of light absorbed is generally very small, being about 1 or 2%. In polymers the absorption varies over a wide range depending on the degree of crystallinity and the presence of impurities or fillers.

12.3 LUMINESCENCE

This is another subject that overlaps the electronic behavior of materials to some extent. When photons are reemitted after being absorbed, we have the *luminescence* phenomenon. In the broad sense it is defined as the emission of light accompanying the absorption of other forms of energy. The input energy could be thermal, mechanical, chemical, or high-velocity electrons. If light readmission occurs rapidly, for example, in less than 10 ns, the phenomenon is termed *fluorescence*, and for longer periods it is known as *phosphorescence*.

Phosphors are materials that have the ability to spontaneously emit lower-energy, longer-wavelength radiation after absorbing high-energy, short-wavelength radiation. A wide variety of materials exhibit these phenomena, including sulphides and oxides. The color, or emission spectrum, is controlled by the addition of impurities (activators). These activators provide discrete energy levels in the energy gap between the valence and the conduction bands. Light emission takes place after excited electrons drop to lower energy levels. Television screens have red-, green-, and blue-emitting phosphors, often made of ZnS (or ZnCdS) with appropriate activators. The common fluorescent lamp is coated with a phosphor coating that is excited by ultraviolet radiation from a low-pressure mercury arc.

The luminescence process can also be classified according to the excitation source. A photon source would be termed *photoluminescence,* while an electron source would be referred to as *electroluminescence*. The ubiquitous cathode ray tubes, used in oscilloscopes, television screens, electron microscopes, and more, are examples of electroluminescence, but often the term *cathodeluminescence* is used for this process when an energized cathode generates the impinging electron beam.

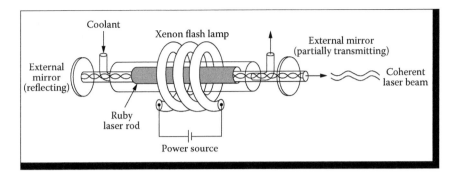

FIGURE 12.3 Schematic diagram of a pulsed ruby laser. (From W. F. Smith, *Principles of Materials Science and Engineering*, 3rd ed., McGraw-Hill, New York, 1996, p. 846. With permission.)

12.4 LASERS

The word *laser* is an acronym for light amplification by stimulated emission of radiation. Normal light sources, for example, a fluorescent light, emit light randomly in all directions, with the light wave trains being out of phase. In lasers the wavelengths are in phase-producing *coherent* radiation. The many types of lasers include solid state, gas, excimer, dye, and semiconductor lasers.

Solid-state ruby lasers (Figure 12.3) are made from Al_2O_3 crystals which have been doped with Cr^{+3} ions to about a 0.05% content. These ions act as fluorescent centers when excited and thereby permit electrons to drop to lower energy levels. Excitation is provided by a xenon flash lamp. Some electrons can decay directly back to the ground state, but others decay to a metastable state for about 3 ms before decaying to the ground state. This process sets off a stimulation–emission chain reaction producing an intense, coherent monochromatic light beam. The waves in the beam move in phase in a parallel direction. One end of the ruby rod is silver coated to reflect waves back and forth from the other partially coated silver end. The intensity increases as more emissions are stimulated. After about 6 ms the beam reaches sufficient intensity to cause a pulse (about 694 nm wavelength) to be transmitted through the partially coated end. This is a pulsed-type laser, in contrast to a continuous-wave laser. The latter are produced in inert gas and compound semiconductor lasers.

Neodymium–yttrium–aluminum–garnet crystals, which are somewhat easier to grow than the alumina ones, have also been used instead of ruby for similar applications. They are frequently referred to as YAG garnet crystals. For lasing properties neodymium is added as the dopant. Nd:YAG and ruby lasers are used for welding, cutting, drilling, and other such types of machining operations.

Gaseous lasers include the continuous-wave He–Ne lasers, which emit over a range of about 0.5 to 3.4 wavelength and 1 to 50 mW of power. Carbon dioxide gaseous lasers emit in the middle infrared region with a few milliwatts of continuous

TABLE 12.2
Some Commercial Lasers

Laser	Wavelength (nm)	Power (Watts)	Applications
Ruby	694	Pulsed — ~5	
InGaAsP	1200–1600	Continuous 0.1	Compact disks, printers
He–Ne	543–3400	Continuous 0.001–0.050	Optical scanners
Nd:YAG	1060	Continuous — to 250	Welding, cutting
Carbon dioxide	5000–10000	Continuous	Cutting, surface treatment

power and pulses of much higher power. These lasers are also used for cutting and welding operations.

Diode lasers, usually made from semiconducting GaAs, are the smallest lasers produced. They consist of a *P–N* junction, which has a large-enough band gap to produce visible light in the 400- to 700-nm range. Visible light photons are emitted by virtue of the recombination of electron–hole pairs after a voltage excitation is applied that excites electrons from the valence to the conduction band. Efficiency has been improved by sandwiching a thin layer of p-GaAs between the *P*- and *N*-AlGaAs. This causes the holes to be confined within the *P*-GaAs laser. These heterojunction lasers are used for compact disks. The GaAs lasers, when built in larger arrays, are used in laser printers and CD players. Some commercial lasers and their characteristics are given in Table 12.2.

12.5 OPTICAL FIBERS

The area of telecommunications has seen a transition from metals to optical glass fibers for the transmission of information. Any type of digital data can be transmitted as photonic light pulses over optical fibers rather than as electronic signals over metal cables. The optical fibers must have extremely low light loss for long distance transmissions. Around 1970 Corning glass works introduced an optical fiber with a loss of 20 dB/km (decibels per kilometer) at a visible range wavelength of 630 nm. The light loss (attenuation) is related to the intensity of the light by

$$- \text{loss (dB/km)} = 10 \log_{10} (I/I_0) \tag{12.8}$$

The Bell Systems in the mid-1970s commercially used glass fiber bundles. Silica fibers have now been developed with losses as low as 0.2 dB/km at 1.6 m wavelength. The fiber material must be of high purity, particularly with respect to Fe^{2+} ions. Other defects that can contribute to losses include bubbles, grain boundaries, impurities in general, and compositional fluctuations.

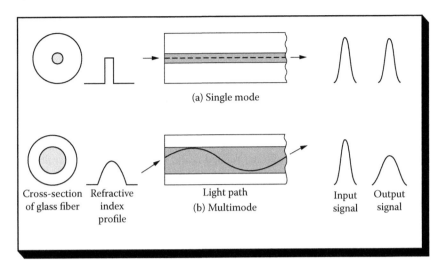

(a) Single mode

Cross-section   Refractive   Light path   Input   Output
of glass fiber   index   (b) Multimode   signal   signal
                   profile

FIGURE 12.4 Single-mode and multimode optical fibers. (From W. F. Smith, *Principles of Materials Science and Engineering*, 3rd ed., McGraw-Hill, New York, 1996, p. 852. With permission.)

The terms *single-mode* and *multimode* are often encountered in the optical fiber industry. In the single-mode fiber the glass core is surrounded by another glass cladding of a lower index of refraction (Figure 12.4). This results in almost total internal reflection and is the type used in most communication systems. In the multimode construction there exists a step-index fiber arrangement in which a large core and a sharp step-down index of refraction exist at the core–cladding interface. This results in a zigzag pattern of light travel along the fiber. These fibers are most often used for short transmission distances, for example, medical endoscopes. There is also a multimode type with a graded index of refraction at the interface that causes the light to travel in a helical rather than a zigzag path of the step-index fibers. These fibers may be used in local area networks.

Optical glass fibers for communication are often manufactured by a modified chemical vapor deposition process. In the ordinary chemical vapor deposition process, for example, in purifying chromium, chromous iodide is first manufactured in a separate chamber and then made to decompose by heating in a vacuum chamber. Here the pure chromium vapor is deposited onto a cool substrate, where it condenses as solid pure chromium. In the modified process for making high-purity optical fibers, $SiCl_4$ containing some $GeCl_4$ and some fluorinated hydrocarbons is passed through a rotating pure-silica tube along with pure oxygen. An external heat source allows the contents to react to form high-purity silica glass. GeO_2 is added to the silica in this way and increases the refractive index of SiO_2. The glass particles condense onto the tube wall and are later heated to the glass softening point, where the glass particles gather to form a solid rod. The glass fibers are later drawn from this rod.

Most optical-fiber communications use single-mode fibers in combination with a InGaAS–InP heterojunction laser diode transmitter at a wavelength of 1.3 μm. Another advanced type uses an erbium-doped optical-fiber amplifier. The fiber doped with erbium, when used with light from an outside semiconductor laser, amplifies the power of the light passing through it.

DEFINITIONS

Absorption coefficient: A number related to the fraction of light transmitted through a material.

Electroluminescence: Session from an electron source of excitation.

Fluorescence: The process whereby light is reemitted within 10 ns of excitation.

Lasers: Acronymn for light emission by stimulated emission of radiation. Materials or devices that produce lasing action.

Optical fibers: Glass fibers over which digital photonic light pulses are transmitted.

Optoelectronics: The title given to the subject or field of discussions and devices involving both optical and electronic characteristics.

Phosphorescence: Readmission of light after a short (nanoseconds) delay when excited.

Photoluminescence: Related to light emission from a photon source of excitation.

Reflectance: Fraction of light reflected from the surface of a material.

Refractive index: Ratio of velocity in vacuum to velocity of light in medium of interest.

YAG: Yttrium–aluminum–garnet.

QUESTIONS AND PROBLEMS

12.1 A photon in a ZnO seminconductor drops from an impurity energy level of 1.2 eV below its conduction. ZnO has a band gap E_g of 3.2 eV. What is the wavelength of the radiation emitted?

12.2 Ordinary incident light impinges on a pyrex plate of 1.0 cm thickness, a refractive index of 1.47, and an absorption coefficient of 0.03 cm^{-1}. What fraction of the light is absorbed by the pyrex plate?

12.3 The attenuation in an optical fiber is 5 dB/km. What fraction of light is lost over a distance of 50 m?

12.4 θ_c is defined as the critical angle of incidence beyond which no light passes through an interface. For light going from silica glass to air, what is θ_c?

12.5 What is the wavelength of the light emitted from
(a) A GaAs laser?

    (b) A GaP laser?

    (c) In what region of the electromagnetic spectrum are these two radiations located?

12.6   Why are metals opaque to visible light, whereas most glasses are transparent?

12.7   List three defects that contribute to light attenuation in optical fibers?

SUGGESTED READING

Becker, P. C., *Erbium-Doped Fiber Amplifiers*, Academic Press, New York, 1999.

Shaffer, J. P. et al., *The Science and Design of Engineering Materials*, McGraw-Hill, New York, 1999.

Smith, W. F., *Principles of Materials Science and Engineering*, McGraw-Hill, New York, 1996.

13 Semiconductor Processing

13.1 INTRODUCTION

Modern alchemists are doing a far better job than the legendary ones who tried to turn a variety of metals into gold. Today silica, in the form of sand or quartz, is turned into a product worth thousands of dollars per ounce. This process of conversion of sand to electronic devices takes place in what is known as the microelectronics industry. Microelectronic device fabrication encompasses materials purification, crystal growth, circuit design, wafer preparation, and chip fabrication. Equipment suppliers for these processes are also a vital part of the industry. Many of us own millions of transistors in the form of integrated circuits in computers, cars, communication devices, and a multitude of other products.

Silicon wafers, thin slices (approximately 0.2 mm thick) of a silicon single crystal, have grown in size from 50 mm in 1970 to 300-mm-diameter wafers around the year 2000. The surface of this thin slice is flattened, highly polished, and then scribed to produce lines that allow chips to be separated from the wafer after the microcircuits, using a variety of complex procedures, have been formed. The chips are about 5 to 25 mm on a side depending on the type of circuit. Concurrently with the increase in wafer diameter there has been an increase in circuit density, that is, the number of components in a circuit or on a chip. In 1965, Gordon Moore, a founder of Intel, noted that the number of transistors per chip was doubling about every 18 months. In 2001 the transistor count per chip was 55 million. This increase in circuit density has been brought about by a large decrease in the feature size of the individual components. The feature size is the minimum width of pattern openings or spaces in a device. The line width is an indicator of the feature size, and in 2001 it was 150 nm for a memory chip. Dramatic improvements in the imaging process, known as lithography, and the use of multiple layers of conductors have largely been responsible for this decrease in feature size. The entire fabrication process involving steps from sand or quartz to finished device is summarized in Figure 13.1.

With all of the different procedures and the multitude of steps involved in microchip fabrication, it is impossible in one chapter of an introductory materials engineering text to do more than skim the surface of this subject. Reference [1] is an excellent source for more detailed information.

Most integrated circuits are designed and processed using either a metal–oxide semiconductor (MOS) or a bipolar configuration. The faster switching speed of bipolar circuits has made them favored for logic circuits, that is,

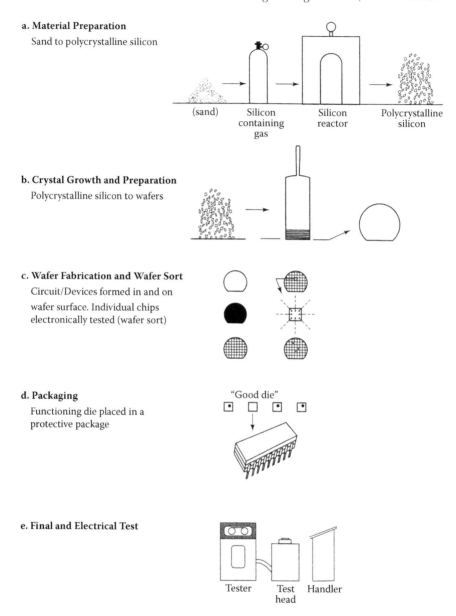

a. Material Preparation
Sand to polycrystalline silicon

(sand) Silicon Silicon Polycrystalline
containing reactor silicon
gas

b. Crystal Growth and Preparation
Polycrystalline silicon to wafers

c. Wafer Fabrication and Wafer Sort
Circuit/Devices formed in and on
wafer surface. Individual chips
electronically tested (wafer sort)

d. Packaging
Functioning die placed in a
protective package

"Good die"

e. Final and Electrical Test

Tester Test Handler
head

FIGURE 13.1 Stages of semiconductor microchip production. (From P. Van Zant, *Microchip Fabrication*, 5th ed., McGraw-Hill, New York, 2004, p. 15. With permission.)

circuits that perform a specified logical operation on the incoming data. MOS circuits, with their smaller component dimensions, have been incorporated into memory circuits. Some MOS circuits use a *field effect transistor* (FET), sometimes called a MOSFET. Another version later developed is known as a

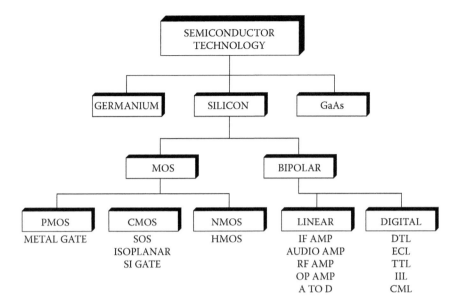

FIGURE 13.2 Semiconductor technology tree.

complementary MOS (CMOS) circuit. As shown in Figure 13.2, either Ge, Si, or GaAS wafers can be, and most often are, used in integrated circuits, with Si being used in all but a few devices.

13.2 SILICON PURIFICATION AND CRYSTAL GROWTH

The ideal silicon crystal wafer would be free of impurities, contaminants, and crystalline defects such as dislocations and vacancies. The purity of the silicon used for crystal growth is of the utmost importance. The first step in the conversion of silica to silicon is the production of metallurgical-grade silicon. This commercially pure silicon is produced by the carbothermic reaction between carbon electrodes and quartz in an electric arc furnace. The reaction is

$$SiO_2 + 2C \rightarrow Si + 2CO$$

Most of this silicon is used in the metallurgical steel and aluminum industries. Only a small fraction is converted to semiconductor-grade silicon, the second step in the overall process. It commences with the formation of a high-purity gas, usually trichlorosilane. Anhydrous HCl is reacted with finely crushed metallurgical-grade silicon in a fluidized bed at a temperature of about 300°C according to the chemical equation

$$3HCl + Si \rightarrow SiHCl_3 + H_2$$

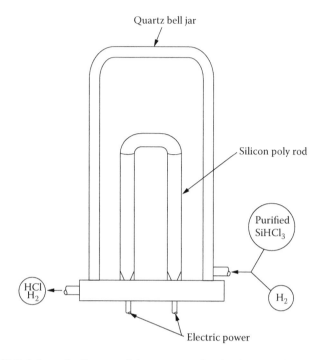

FIGURE 13.3 Schematic diagram of the apparatus for the decomposition of trichlorosilane.

All but the purest of this trichlorosilane is used in the silicone industry [2]. The purest part is separated from the other chlorosilanes and further purified by distillation, where the boron and phosphorous impurities are removed as BCl_3 and PCl_3. This trichlorosilane is sufficiently pure to make semiconductor-grade silicon by hydrogen reduction according to the reaction

$$SiHCl_3 + H_2 \rightarrow Si + 3HCl$$

The silicon is deposited on a thin high-purity silicon rod in a reactor that has been referred to as a Siemens reactor, and the process is referred to as the Siemens process (Figure 13.3). The resulting polycrystalline solid silicon rods, of 99.9999999% purity, are crushed to irregular chunks to make the charge material for the single-crystal growth process. These crystals must be nearly free of defects such as grain boundaries, dislocations, and vacancies.

The Czochralski process grows large silicon single crystals, up to diameters of 8 in, where a small silicon seed crystal is made to come into contact with the molten silicon. As the seed is slowly withdrawn while it is simultaneously being rotated, the molten silicon adheres to the seed and solidifies onto the seed, resulting in a giant crystal of the same crystalline orientation as the seed crystal (Figure 13.4). The interface temperature must be slightly lower than the melting

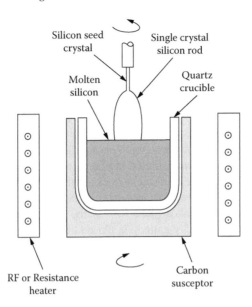

FIGURE 13.4 Silicon crystal growth by the Czochralski process.

point of the solid seed. The seed acts as a heterogeneous nucleation site permitting the liquid–solid transformation to take place at a temperature slightly above the temperature for homogeneous nucleation. The surface tension between the seed and the melt promotes adhesion of the molten film to the seed crystal. The growth rate must be very slow in order to maintain the correct temperature and surface tension at the interface. Slow growth rates also favor high crystalline perfection. The seed crystal, and likewise the finished crystal, are oriented such that the planes of the wafers are of either the {100} or {111} crystal family of planes, the {100} orientations being used for MOS devices, while the wafers whose planar surfaces are of the {111} type are used for bipolar devices.

Prior to the commencement of the growth of silicon crystals, the dopant atoms, either *N*- or *P*-type, are added to the molten silicon and are incorporated into the crystal as it solidifies. Since the dopant controls the conductivity, uniformity of dopant concentration is important. Uniformity of dopant concentration, as well as crystalline perfection, is aided by rotating the crucible and seed in opposite directions. The rotation and withdrawal rates, as well as the melt temperature, are integrated and carefully controlled.

13.3 OTHER MATERIALS AND THEIR CRYSTAL GROWTH METHODS

Other crystal growth methods have been used for semiconductors, including the horizontal and vertical crucible Bridgeman techniques, and the float-zone method.

In the horizontal boat method a long boat-type crucible is enclosed within a glass tube, often quartz glass. The chunk or rod-shaped solid charge material is placed in the boat, and a heat source, either an induction heating coil or a resistance wound furnace, is passed along the charge to solidify it into a continuous shape conforming to the boat. A seed crystal is then placed in one end of the boat, and the heat source is passed slowly along the boat commencing at the seed–charge interface. This narrow molten zone allows crystal growth to occur at the liquid–solid interface, resulting in a crystal of the same orientation as that of the seed. This method has been employed for the production of GaAs crystals.

A variation of this method, called the horizontal gradient freeze method, is a static method in which the heat source does not move, but instead a temperature gradient profile is built into a resistance-type furnace such that as the temperature is reduced, solidification takes place first at the seed end. Instead of using a preformed charge, high-purity gallium is placed at one end of the boat and high-purity arsenic at the other, with a porous quartz plug placed between the two materials. The arsenic end is maintained at about 620°C to yield approximately 1 atm of arsenic pressure. The gallium is at the higher temperature end, with the gallium metal above the melting point of GaAs. The arsenic vapor reacts with the gallium metal to form molten GaAs. Then the hot zone temperature is slowly decreased, and single crystal growth is achieved. This method has also been used for growth of InP crystals. In the vertical Bridgeman method the furnace and crucible are mounted in a vertical arrangement, and the furnace is slowly moved from the seed end. The static gradient freeze technique can also be used where the temperature is slowly reduced. GaAs, GaP, and InP crystals of low dislocation density have been grown by this technique. In some horizontal and vertical Bridgeman methods, the glass tube is pressurized to prevent the evaporation of the high vapor pressure arsenic.

In some special situations the float zone technique may be employed. In this method the charge material in rod form is held with grips at each end inside of a vacuum or inert gas chamber. A heat source is slowly passed along the rod to which a seed crystal has been "welded" to the rod end by the same heat source. The heat source forms a narrow molten zone that is passed along the rod as crystal growth takes place. The molten zone is held in place by the surface tension of the molten material. The distinct advantage here is that the material does not come into contact with a crucible, thereby eliminating any contamination that could arise from crucible contact. Such an arrangement is shown in Figure 13.5. The disadvantage of this technique is that the diameter of the crystal is limited by the surface tension of the material. Some research crystals have been grown by the float zone method. The heat sources are usually either an induction coil or an electron beam source. In the latter case the chamber must be evacuated. Many of the metals used for microcircuits in the thin film metallization process are purified by the electron-beam float-zone method (Figure 13.6).

Currently it appears that the preferred method of crystal growth for GaAs and InP is the Czochralski method, where the material is enclosed in a quartz ampoule to prevent evaporation and decomposition of the compounds. A variation

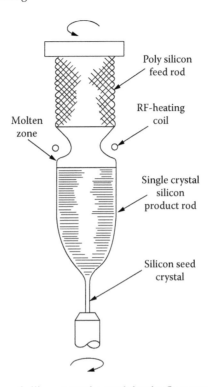

Poly silicon
feed rod

RF-heating
coil

Molten
zone

Single crystal
silicon
product rod

Silicon seed
crystal

FIGURE 13.5 Diagram of silicon crystal growth by the float-zone method.

of this technique is the liquid encapsulation Czocharlski method, where the melt is covered by a thin layer of molten of B_2O_3. A counter-pressure of an inert gas, which is higher than the partial pressure of the group V element, is maintained on top of the B_2O_3. Pyrolytic boron nitride crucibles are used instead of quartz to avoid silicon doping from the quartz. This method is used for the larger-diameter crystals. The Bridgeman-grown wafers have lower dislocation densities and are used for optoelectronic devices such as lasers. The details of these crystal growth methods can be found in References [3] and [4].

Although not usually considered when discussing crystal growth techniques, the epitaxial layer method should be mentioned here. The term *epitaxial* means the growth of a thin single-crystal semiconductor film on a single-crystal substrate. The epitaxial layer has the same crystal atom arrangement as that of the substrate. Epitaxial layers of silicon have found use in advanced bipolar devices, CMOS circuits, and as silicon on sapphire. GaAs layers have also been formed in this fashion. One advantage of the epitaxial crystal layer is that precise doping by gaseous diffusion can be obtained. In some cases polycrystalline silicon may be deposited on an amorphous substrate. Even though this is not single-crystal growth, it is often, perhaps erroneously, referred to as an epitaxial layer. Vapor phase and molecular beam deposition methods are frequently employed to create epitaxial films.

FIGURE 13.6 Purification of titanium by the electron-beam float-zone method. (From W. H. Class and G. T. Murray, *Solid State Technol.*, May, 1975. With permission.)

13.4 WAFER PREPARATION

Wafer preparation includes all steps involved in getting the wafer ready for circuit deposition. The first steps consist of cropping the ends of the crystal ingot, mechanically grinding the surface to achieve diameter control, and checking the crystal for orientation and conductivity, that is, *N*- or *P*-type. The amount of dopant along the crystal length is determined more precisely using a four-point probe method. There will be some variation along the length, resulting in wafers that fall into several resistivity specification ranges.

Once the crystal is oriented on the cutting block, a flat is ground along the length to indicate the major crystalline plane, that is, the {100} or the {111} type surfaces. It is used to place the first pattern mask on the wafer so that the orientation of the chip surface is always a major crystal plane. On larger-diameter wafers a notch may be ground along the length to indicate the orientation. Often a smaller secondary flat may be ground as a code to indicate the conductivity type as well as which of the major orientations are involved. Specially designed diamond-coated inside diameter cutting saws are used to obtain thinly sliced crystalline wafers. Laser marking, using bar codes or laser dots for 300-mm-diameter wafers, identifies the wafers for further processing.

A rough polish is used to provide a flat surface, which is followed by a chemical–mechanical polish that removes several microns of surface to further ensure flatness and a surface that is ready for the detailed masking and deposition processes. Many surfaces are oxidized before shipment to protect the surface from scratches and to save a later manufacturing step. Protective packaging under clean-room procedures further protects the surface now rather expensive wafer during shipment to a wafer fabrication company or to transfer it to the manufacture's in-house wafer fabrication facility. Increasingly, wafer fabrication companies are requesting wafers with top-side layers, such as epitaxial silicon or silicon deposited on insulators such as sapphire or diamond.

13.5 WAFER FABRICATION

13.5.1 INTRODUCTION

While *wafer preparation* applies to all of the above processes, the term *wafer fabrication* is used to define all of the processes required to manufacture circuits on the prepared wafers. In most cases these are integrated circuits. These processes include layering, patterning, etching, doping, and heat-treating. The fabrication of semiconductor devices involves the creation of a series of *P–N* junctions located and patterned to form both active transistors and the passive elements such as resistors, capacitors, conductors, and insulators. *Chip, die, device, circuit,* and *microchip* are all terms used to describe the microchip circuit patterns that cover the wafer surface. Scribe lines are lines that are introduced as spacers between the chips that permit separation of the chip from the wafer. A finished processed silicon wafer is shown in Figure 13.7.

Silicon Wafer

Circuit Die "Chips"

Wafer Flat

Packaged circuit

FIGURE 13.7 Processed silicon wafer.

13.5.2 LAYERING, PATTERNING, DOPING, AND HEAT TREATING

Layering is the operation involving creation of thin layers of the desired circuit elements on the wafer surface. The deposited layers may be formed by physical vapor deposition (PVD), chemical vapor deposition (CVD), and electrodeposition methods. Silicon dioxide layers are grown by exposing the heated wafer surface to oxygen, if this process has not already been done during the wafer preparation operations. The general term for this oxidation step is *passivation*. It plays an important role in protecting semiconductor devices from contamination. Electroplating is used to deposit gold metallization on high-density integrated circuits. PVD includes all techniques based on evaporation deposition such as electron beam or hot-boat deposition, reactive evaporation, ion plating, and sputtering, either by plasma or some type of ion beam. Today sputtering is used in all but a few special cases for which electron beam or ion beam evaporation may be used.

Today most sputtering equipment uses a diode source known as a magnetron. In the sputtering operation a target material of the desired layer is connected to a high negative potential in a vacuum chamber. A small quantity of argon gas is then introduced into the chamber, where the atoms become ionized, forming a plasma. Via the high negative potential on the cathode (target), the ions bombard the target, thereby dislodging atoms, which deposit onto the wafer substrates.

The vapor deposition operation also takes place in a vacuum chamber, but here the material can simply be evaporated by electron beam heating of the metal. The electron beam impinges on the charge material, which is contained in a water-cooled copper crucible. Film thickness is controlled by shutters accompanied by feedback information to the electron-gun power supply. The charge vapor deposits on the substrate. Evaporation is a less expensive operation, but sputtering usually gives more adherent layers. The wafer is rotated to create a more uniform thickness, particularly on steps in the circuit pattern. Quartz heaters in the chamber also aid step coverage.

Chemical vapor deposition is used in certain special operations. Contact holes, called via holes, are etched into layers down to the first metal layer. Tungsten is used in a variety of structures to fill the vias. Here the CVD process is employed using the reaction

$$2WF_6 + 3Si \rightarrow 2W + 3SiF_4$$

Tungsten can also be deposited selectively as a barrier metal over aluminum. These depositions are performed at temperatures of about 300°C, which makes the process compatible with aluminum deposition. Other examples of the use of the CVD process include the deposition of titanium, silicon, and silicon dioxide.

When the sputtered or evaporated films consist of conductors or metallic resistors, these processes are frequently referred to by the collective term *metallization,* but it should be kept in mind that dielectrics, capacitors, and insulators can also be formed by sputtering.

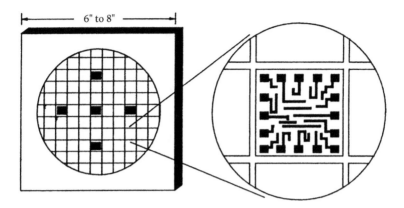

FIGURE 13.8 Example of a pattern formed by photomasking.

The patterning processes include masking, photomasking, lithography, photolithography, and microlithography. Patterning is used in conjunction with photoresist materials. The layers previously discussed, that is, the circuit components, are created one layer at a time. Portions are removed with the patterning process to leave a desired shape. A typical pattern formed by photomasking is illustrated in Figure 13.8. For the photo process an intermediate step employs a rectile. A rectile is a hard copy of the individual circuit drawing that is recreated in a thin layer of chromium that has been deposited on a glass plate. The rectile may be used directly or may be used to produce a photomask. Some companies specialize in making rectiles and masks for each circuit type and supply them to wafer fabricators. Individual masks are aligned and placed in contact with the photoresist for the exposure cycle. To obtain high resolution, short-wavelength light or radiation may be used. Electron beams and x-ray equipment have been developed for special applications. In electron beam lithography an electron beam source that produces a small diameter spot is used in the production of high-quality masks and rectiles. A set of masks, on the order of 10, are required to form interacting patterns for processing each layer of a circuit structure. Typical photo masking schemes are depicted in Figure 13.9. The number of material layers involved in the construction of a typical CMOS integrated circuit is illustrated in Figure 13.10. All of the masking steps have been omitted here.

Doping is the process of injection of the P- or N-type elements to yield the desired conductivity. These injections are made either by thermal diffusion or by ion implantation. In thermal solid-state diffusion the dopant materials are introduced in the vapor phase at temperatures of the order of 1000°C. The dopant atoms diffuse into the wafer surface to form a thin layer of doped material. In ion implantation the dopant atoms are shot into the semiconductor material. Additional movement of the atoms into the surface layer takes place by solid-state thermal diffusion.

In addition to heat treatments for dopant diffusion, other desirable effects are obtained from the application of heat. Heat treatments provide for the bonding

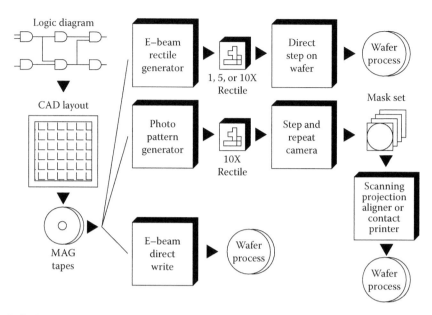

FIGURE 13.9 Typical photomasking approaches.

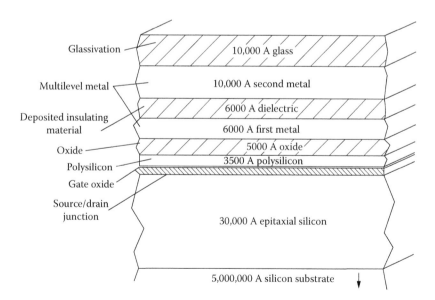

FIGURE 13.10 Material layers involved in the fabrication of a typical CMOS device.

of the conductor stripes onto the wafer surface. When photoresist layers are present, heat is required to drive off solvents that interfere with accurate patterning. The ion implantation process can also introduce crystalline defects in the wafer surface that can be removed by annealing. The basic steps of layering,

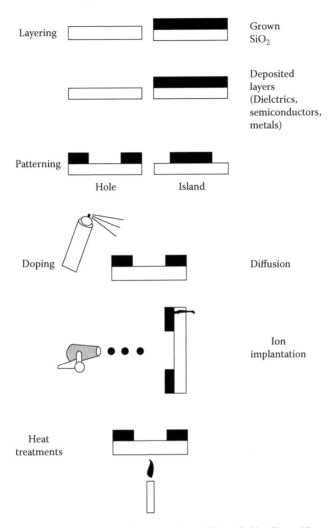

FIGURE 13.11 Basic wafer fabrication operations. (From P. Van Zant, *Microchip Fabrication*, 5th ed., McGraw-Hill, New York, 2004, p. 75. With permission.)

patterning, doping, and heat treatments are schematically summarized in Figure 13.11.

As circuits have advanced to very large-scale integration levels, advanced photolithography processes have played a major role. The increased surface density and more surface layers involve improvements in all of the elements of the basic patterning processes including mask materials, alignment, exposure sources such as x-ray and electron beams, and planorization of deposited surface layers by chemical mechanical polishing. Although we now speak of feature sizes on about 100 nm, it is predicted that by 2016 these sizes will be about 20 nm.

13.6 PURIFICATION OF METALS FOR METALLIZATION

In the later steps of circuit fabrication, often referred to as the *back end of the line,* the metals necessary to connect the devices and different layers are added to the chip. Since only metals are involved here, fabricators use the term *metallization* to describe the deposition of metals. The electronic industry places stringent demands on the purity of the metals used for metallization. It is the semiconductor industry that has stimulated the development and manufacture of high-purity metals.

In the purification of the metals discussed here, one commences with the highest commercial purity metals available. These commercial purity metals are usually about 99 to 99.9% pure, this purity being achieved by a series of mechanical and chemical processes that commence with the natural ore.

13.6.1 FRACTIONAL CRYSTALLIZATION

Fractional crystallization, a liquid phase purification method that has been used for many years, relies on the difference in solubility of the impurity element between the liquid and solid phases. The material to be purified is dissolved, often in organic solvents, whereby the impurity element is much more soluble in the solvent at elevated temperatures. On cooling, the product precipitates out of solution while the impurities concentrate in the solution. This process may be repeated many times, using a new batch of solvent each time.

Probably the metal that uses this method and is of most interest to the semiconductor industry is gallium. Most gallium is found in bauxite, an aluminum ore, although small quantities are also found in zinc deposits. A relatively impure gallium metal is produced by electrolysis of alkaline solutions of the hydroxide or sodium aluminate liquors. Repeated fractional crystallization steps result in gallium on the order of 99.9999% pure.

13.6.2 ZONE REFINEMENT

Zone refining is also a liquid phase technique and is probably the most widely used of all metal purification processes to obtain semiconductor grade purity. The classic experiments by W. G. Pfann of Bell Telephone Laboratories in the 1950s led to the achievement of germanium of sufficient purity for semiconductor use, and it was from this material the first transistor was made. In essence, zone refining relies on the difference between the solubility of the impurity element in the liquid and solid phases. A molten zone is made to move very slowly along a metal, which can be placed in a boat crucible (Figure 13.12) or suspended vertically in the same type of equipment described earlier for crystal growth. The solute is more soluble in the liquid than the solid phase at equilibrium. At equilibrium at the liquid–solid interface, the partition coefficient K_0 is defined as

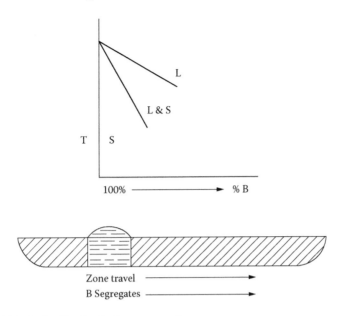

FIGURE 13.12 Purification by boat zone refinement.

$$K_0 = C_s/C_l$$

where

C_s = concentration in solid phase
C_l = concentration in liquid phase

In practically all conditions of freezing, equilibrium is not obtained, and the effective distribution coefficient K_e is a function of the freezing velocity, solute diffusion, and thickness of the diffusion layer. If solidification is permitted to take place by the slow movement of a molten zone from one end of the bar to another, as illustrated in Figure 13.12, the concentration C_s in the solid phase at distance x from the starting end after one zone pass of length l is given as follows:

$$C_s/C_l = 1 - (1 - K_e) \exp (- K_e\, X/l)$$

where $K_e < 1$, as in most cases additional zone passes result in more concentration of solute at one end of the bar. After many zone passes this end is removed and discarded. Many lower melting point metals are zone-refined in a long boat. When the metal is suspended in a vertical position, the same as in crystal growth, the molten metal is held in place by its own surface tension. Although contamination by the boat material is avoided, the diameter of the bar is somewhat limited. The float zone method is particularly advantageous for the reactive metals Ti, Zr, Nb, Ta, V, W, and Mo, and they are usually refined in this manner, while Au, Ag, Cu, Al, Zn, Pb, Sn, and Bi are zone-refined in a boat. The dividing line in terms of melting point is about 1200°C. As in crystal growth, the heat sources used include induction heating, resistance wire wound furnaces, and electron bombardment. Electron bombardment is normally employed for the higher melting point metals. It takes less power, and the electron beam–heated region can be confined to a small region by use of electrostatic focusing shields. The high vacuum employed in electron bombardment offers additional purification by boiling off the higher-vapor-pressure elements and gases.

13.6.3 Vacuum Melting

Zone refinement of metals is often conducted in dynamic vacuum, but many metals, particularly those with high melting points, can be purified to a significant degree by the vacuum melting process alone. Although this process may not achieve the degree of purity that one obtains by zone refinement, it is less expensive and for some applications yields a metal of the required purity. In vacuum melting purification occurs by:

- Degassing, that is, the removal of oxygen, nitrogen, and hydrogen, as well as CO or CO_2 formed by side reactions of oxygen with carbon
- Vacuum distillation of high-vapor-pressure elements

The removal of gaseous elements takes place by virtue of the change in solubility with the partial pressure of the same elements in the surrounding media. The relationship is

$$S = \sqrt{P}$$

where
S = the solubility of the gas in the liquid phase
P = the partial pressure of the same gas in the surrounding media

This purification process is dependent on the ability of the vacuum to maintain a sufficiently low gas partial pressure near the molten surface, on the diffusion of the gas atom through the liquid to the surface, on the presence of any stirring action that might enhance the gas atom transport in the liquid phase, and on the composition of the starting material.

Vacuum distillation during the melting operation is based on the preferential evaporation of solute. To a first approximation, the degree of purification is dependent on the ratio of the vapor pressure of solute to that of the solvent. In addition to this ratio, it is essential that the vapor pressure of the solute be high relative to its partial pressure in the immediate vicinity of the molten surface. After the first few minutes of the vacuum melting operation, transport in the liquid phase becomes the rate-controlling process. A high pumping speed of the vacuum system is required. The power sources for the melting process can be either an induction type or an electron beam, the latter being accomplished via drip melting into a watered-cooled copper crucible.

13.6.4 CHEMICAL VAPOR DEPOSITION

This method was briefly discussed under layering of circuit deposits (Section 13.5.2), but it is also a method of purification of certain metals. This is a type of vapor purification process whereby the starting material is first reacted to form a compound that is subsequently decomposed in the vapor state. The metal vapor is allowed to condense to a solid, leaving the impurities behind in the residue.

One of the more popular and widely used processes of this type is the iodide process, which has been used extensively to purify titanium, zirconium, and chromium. For all three metals the starting metal of commercial purity of about 99.9% is reacted to form a volatile iodide compound that is decomposed to separate the iodine from the metal and deposit purified metal on some type of substrate. If proper temperatures are maintained, oxygen, nitrogen, hydrogen, and carbon as well as many metallic impurities are removed. Typical purity of the deposits is about 99.96% for titanium and 99.996% for chromium. In the case of titanium further purification can be obtained by electron-beam float-zone refinement (Figure 13.6) or by vacuum melting using an electron-beam drip-melting process.

13.6.5 FABRICATION

Despite all of the effort one may expend in purification, it will be of little significance unless this purity is maintained when fabricating the metal to form some type of sputtering target or a charge material for circuit deposition. All mechanical means of fabrication must be done at room temperature to avoid oxygen and nitrogen pick-up from the atmosphere and hydrocarbons from the lubricants in the fabrication processes. All dies used in rod and wire forming and rolls for sheet formation must be free of contamination. The metal must also be periodically cleaned by degreasing and chemical etching to remove lubricants. These cleaning procedures are quite involved and vary from metal to metal. For some brittle metals such as chromium and certain alloys, sputtering targets are cast in either a vacuum or inert atmosphere furnace. The final cleaning and packaging of the metal must be done in a clean-room environment.

13.7 CIRCUIT ELEMENTS

13.7.1 P–N JUNCTION

As the name implies, such a junction involves bringing a P-type semiconductor into contact with an N-type material. In the intrinsic silicon the Fermi level is at the center of the forbidden region. In extrinsic semiconductors this is not so. At temperatures near and below the exhaustion and saturation regions, conduction comes from the impurity atom effects. The Fermi energy level must now lie near the impurity energy levels, since these are the levels from which the electrons are excited. When the P and N types are joined, the Fermi levels are at different positions; thus, the electrons will move to the region of lowest energy to equalize the Fermi level.

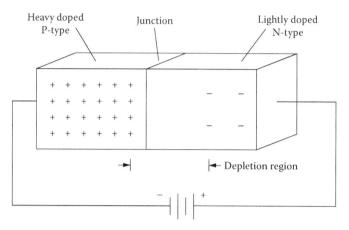

Heavy doped P-type   Junction   Lightly doped N-type

Depletion region

FIGURE 13.13 Reverse biased *P–N* junction diode. (Courtesy of DM Data, Inc.)

A similar flow of holes occurs by the same analysis. In essence we have electrons from the *N*-type and holes from the *P*-type flowing toward the junction. As soon as equilibrium is attained with a uniform Fermi potential, the flow stops. If we make the junction part of a circuit by introducing a potential source (Figure 13.13), the electrons on the *N* side are continually removed, and the Fermi level is decreased. This application of a potential creates a reverse-bias condition (i.e., the *N*-type material is connected to the positive terminal and the *P*-type to the negative side). The potential source causes a movement of majority carriers away from the junction, forming a depleted region that causes the current to cease flowing. If the current is altered so that the positive potential is applied to the *P* side, we have a forward bias where both holes and electrons flow toward the junction and recombine. Electrons are continually drawn from the battery to replace those that are lost by recombination. A new hole is formed when an electron leaves the *P*-type material. Thus, we have a rectifier in which current flows in one direction and not the other. This is called a rectifying diode, the same term that was used in the vacuum-tube days. A rectifying diode converts as current to a half-wave dc current. The current–voltage characteristics are shown in Figure 13.14.

13.7.2 TRANSISTORS

Transistors are devices that amplify electronic waves, or signals as they are sometimes called. Although most students today are not familiar with how the triode vacuum tube accomplished this feat, we will remind them of it here. It was a triode tube (i.e., three electrodes) in which the central electrode, called a grid (or base), accelerated electrons from the cathode (or emitter) to a plate (or collector). A small signal on the grid produced a large signal at the plate. The transistor does this via *N–P–N* junctions as depicted in Figure 13.15. A *P–N–P* junction configuration will accomplish this same feat. The *P* region, called the

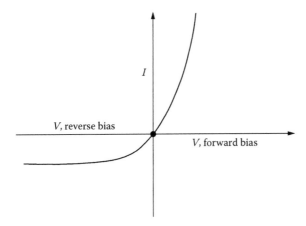

FIGURE 13.14 Current-voltage characteristics of a *P–N* junction diode.

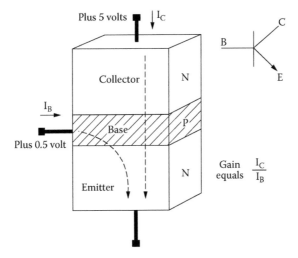

FIGURE 13.15 *N–P–N* bipolar transistor schematic showing how amplification occurs. (Courtesy of DM Data, Inc.)

base in an *N–P–N* transistor, is a very thin section of a semiconductor (just as is the *N*-type layer in a *P–N–P* transistor) that is sandwiched between two larger *N* sections, one being the emitter and the other being the collector. The positive voltage on the collector collects the electrons emitted by the emitter. There must be some current flow into the base, in the form of a small signal, to turn on the major flow between the collector and emitter. The *P* base is thin, so that although some electron–hole recombination occurs, most of the electrons go right through the base to the collector. The ratio of the flow out of the collector to the input current is the gain of the device, and values typically range from 30 to several hundred. A discrete transistor can obtain higher performance (e.g., power, gain, or frequency) than that obtained in integrated-circuit form. This junction transistor is known as a bipolar transistor. The basic bipolar integrated-circuit element is the *N–P–N* transistor. A typical bipolar transistor in a plan-view picture and a plan-view schematic layout, together with the cross-section construction, is shown in Figure 13.16. The basic bipolar manufacturing process involves 33 steps requiring a seven-mask processing sequence.

The bipolar transistor was the first transistor developed by W. Shockley and was a discrete device. As pointed out earlier, only 2% of the components on a typical printed circuit board are discrete transistors. The metal–oxide semiconductor (MOS) emerged later and has replaced a lot of the bipolar transistor applications. The basic *N*-channel MOS transistor cross section is shown in Figure 13.17. One can also have a PMOS device (i.e., a *P*-channel MOS).

When a chip contains both types, it is referred to as a complementary metal–oxide semiconductor (CMOS), shown in Figure 13.18. MOS transistors are of the field-effect type, called a metal–oxide semiconductor field-effect transistor. They are used both as power devices and as integrated-circuit MOS devices and function in essentially the same manner in both circuits. To explain their amplification process, we refer to Figure 13.17. Two *N*-type regions are formed in a *P*-type substrate, one being the source and the other the drain. The conductor or gate is the third component. In contrast to the bipolar transistor, no current flows into the MOS device. A potential is applied to the positive gate electrode and the source and draws electrons toward the gate, but the gate is insulated from the source, so the electrons cannot enter it. The concentration of electrons around the gate makes this region more conductive, so that a small positive potential at the gate will produce a large electron flow between the source and the drain, thereby amplifying the input. A negative potential on the gate will drive charge carriers out of the channel and reduce the conductivity. In essence a voltage applied to the gate electrode is used to set up an electric field that controls the conduction between the heavily doped *N*-type source and drain regions. By changing the input voltage between the source and gate, a larger output potential is achieved between the source and drain. The gain in metal–oxide semiconductor field-effect transistor devices is usually measured in terms of a voltage ratio instead of a current ratio as in the bipolar devices. The MOS devices are preferred by computer manufacturers because they are smaller, require less power, and are les expensive to produce.

The MOS process technology is based on an eight- to ten-mask sequence. The key to this process is the use of a polysilicon gate, which has the advantage of *self-alignment* and an added conductor layer. With the use of polysilicon as a gate material, the poly is grown on the gate oxide, patterned, and then used as part of the masking to define the source and drain areas. However, for high-performance devices, an additional layer of a refractory metal on the poly-silicon is used to reduce the resistance and increase the circuit speed.

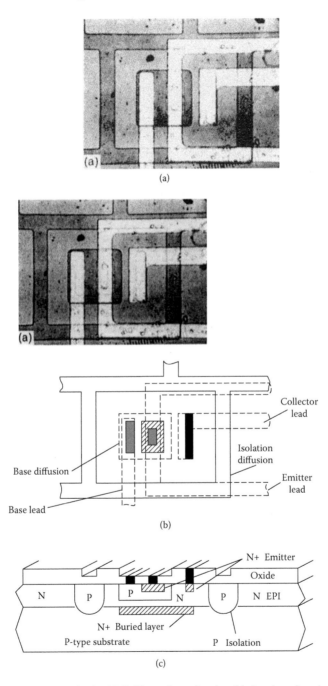

FIGURE 13.16 Typical bipolar *N–P–N* transistor showing (a) the plan-view device along with (b) the plan-view schematic and (c) the cross-section schematic. (Courtesy of DM Data, Inc.)

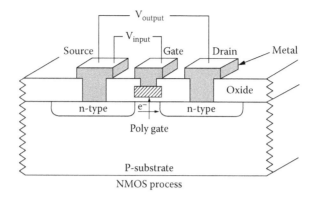

FIGURE 13.17 Cross section of an *N*-channel MOS field-effect transistor. A small input potential on the gate will produce a large output between the source and drain.

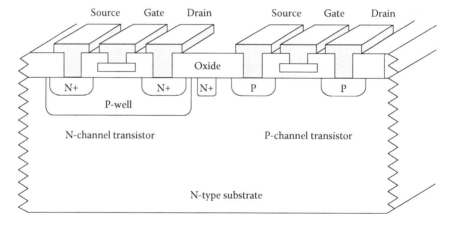

FIGURE 13.18 CMOS structure cross-section schematic. (Courtesy of DM Data, Inc.)

DEFINITIONS

Base: Electrode in a bipolar transistor that controls the flow of charge.

Bipolar: Type of transistor that has an emitter, a base, and a collector.

Bridgeman: A method of crystal growth whereby a molten zone is slowly moved along a charge that may be contained in a boat or suspended as a rod in a vertical fashion.

Chemical vapor deposition (CVD): The deposition of a material after first reacting it with another material to form a compound that is subsequently decomposed. Used for thin film deposition and for purification.

CMOS: Complementary MOS device that contains both *p* (MOS) and *n* (MOS) channels.

Collector: Electrode that collects electrons in a bipolar transistor.

Czochralski: A method of crystal growth in which a seed crystal is slowly pulled from a molten charge.

Doping: The injection of elements into a semiconductor in order to change and control the conductivity.

Drain: Electrode to which the electrons go in amplification devices such as MOS and bipolar devices.

Epitaxial: Growth of a thin single-crystal layer on a single-crystal substrate.

Feature size: Minimum width of pattern openings, gate size, or line spaces.

MOS: Metal oxide transistor.

MOSFET: Field effect MOS transistor.

Sputtering: A process for deposition of a thin film layer by dislodging atoms from a target material by bombardment with inert gas ions, the dislodged atoms depositing on a substrate.

Trichlorosilane: $SiHCl_3$ made by reacting HCl with crushed metallurgical grade silicon.

QUESTIONS AND PROBLEMS

13.1   Using a schematic of a phase diagram, such as that shown in Figure 13.13, show how the equilibrium coefficient K_0 is obtained. What other factors must be considered for the effective coefficient K_e?

13.2   What crystal growth method should be used for GaAs
    (a) If a low dislocation density is important? What applications require a low dislocation density?
    (b) If a large-diameter crystal is preferred?

13.3   Discuss the various techniques used for doping semiconductor crystals.

13.4   What are epitaxial layers?

(Questions 13.5 and 13.6 will require some library searching since these numbers are always changing with time.)

13.5   What are typical dislocation densities in silicon wafers?

13.6   What feature size is currently obtained in device deposition layers?

13.7   What is the orientation of the wafer plane obtained from a silicon crystal that has been grown such that the longitudinal direction is a [111] crystalline direction?

13.8   Trivia questions: When was the first transistor developed? What semiconductor material was used? Name the persons who were awarded the Nobel prize in physics for this achievement?

REFERENCES

1. Van Zant, P., *Microchip Fabrication*, 5th ed., McGraw-Hill, New York, 2004.
2. Barraclough, K. G., *The Chemistry of The Semiconductor Industry*, S. J. Moss and A. Ledwith, Eds., Paper by K. G. Barraclough, Chapman & Hall, New York, 1987.
3. Swaminathan, V. and Macrander, A. T., *Materials Aspects of GaAs and InP Based Structures*, Prentice Hall, Englewood Cliffs, NJ, 1991.
4. Campbell, S. A., *The Science and Engineering of Microelectronic Fabrication*, Oxford University Press, New York, 1996.
5. Nishi, Y. and Doering, R., Eds., *Handbook of Semiconductor Manufacturing Technology*, Marcel Dekker, New York, 2000.
6. Atherton, L. F. and Atherton, R. K., *Wafer Fabrication*, Kluwer Academic Publishers, Boston, 1995.

14 Environmental Degradation of Engineering Materials

14.1 INTRODUCTION

All materials react to some degree with certain environmental constituents, but there is a vast difference in the degree of reactivity of the three major groups of materials: metal, polymers, and ceramics.

Metals are by far the most reactive of these three groups. That is why metals are found in nature in the form of compounds. Metals are extracted from their ores where nature put them. The metal compounds in the ores, oxides, sulfides, nitrides, and the like are very stable compounds. When the metal is removed from these other elements, they want to go back to where they came from. We take metals from their ores and refine them so that they are formable and easily fabricated into many desirable shapes. Metals are very useful materials, as shown in Figure 14.1, but we pay a price for this convenience. On their way back to their natural habitat, they may react with many elements that exist in whatever environments they may become exposed to, including those that are not normally found in ambient atmospheres. These metal attackers include the acids; caustic bases; the gaseous hydrogen, oxygen, and nitrogen atmospheres; and many of the halides, chlorine and fluorine in particular. The metals industry spends billions of dollars extracting metals from their compounds as found in nature, and then billions more to protect them from reactions that occur in their new environments. Estimates of losses from corrosion of steels in U.S. industries fall in the range of $70 billion to $80 billion per year. However, some metals are not very reactive. These are the expensive precious metals. Gold does not have to be extracted from oxides and sulfides because it does not have to be extracted from such compounds. It is usually found (but not easily located) in nature as a pure elemental metal.

Metals are reactive because they have all those free electrons roaming around looking for something to which they can become attached, and in so doing lower the overall energy of the system. In polymers and ceramics the covalent and ionic bonds do not provide free electrons. Polymers are much more subject to environmental degradation then are ceramics. In polymers light, particularly in the ultraviolet-wavelength region, can break the C–C bonds, a process called *scission.* Neutron radiation can also cause scission. At elevated temperatures the bounds of the atoms attached to the side to the chain of the carbon backbone can be broken, causing *charring,* also called *carbonization.* Ozone degradation of elastomers is a serious problem for the rubber industry. One of the larger effects of the environment on polymers is that caused by organic solvents. Whereas metals are attacked by electrochemical reactions, which require electron conductivity, polymers are insulators. In polymers the solvents diffuse into the polymer body and, without breaking bonds, cause plastization and swelling effects, processes generally referred to as *physical corrosion.* Organic liquids such as cleaning fluids, detergents, gasoline, lubricants, and sealants lead to the deterioration of polymer properties. Chemical attack by these solvents can cause swelling, scission, and depolymerization (unzipping) of the chain molecules. Some chemicals can cause crazing and alteration of the mechanical properties of polymers. In short, polymers are affected by sunlight, nuclear radiation, heat, oxygen, humidity, and organic solvents. It must also be recognized that polymer degradation is a very useful process when the need arises for the destruction of waste materials. There is an increasing demand for biodegradable self-destructing polymers for packaging and agricultural and medical applications.

Ceramics are the materials least affected by their environment, but ceramics can be attacked by certain chemicals that lead to their degradation and loss of strength. The most obvious cases are the halides (e.g., NaCl), which are ceramics by most definitions and which are soluble in water and many other solvents up to certain concentrations. Some carbonates can be dissolved by weak acids. The carbides and nitrides are attacked by oxygen at elevated temperatures. Ceramic silicates are readily attacked by hydrofluoric acid, which is used to etch glasses, but the strongly bonded ceramics such as Al_2O_3 and Si_3N_4 are very stable and inert to attack by most aqueous solutions, including acids and alkaline solutions. A related ceramic corrosion problem is that facing highway engineers; reinforced concrete bridges deteriorate from deicing salts and marine environments. A January 1992 report by a committee of the National Academy of Sciences stated that NaCl is still the best way to clear ice and snow off roadways, even though its corrosive powers cost tens of millions of dollars to vehicles and bridges every year. The widespread use of salt substitutes, such as calcium magnesium acetate, is not warranted according to the 14-member panel of the National Research Council's transportation board because, even though the alternative deicers are less corrosive, they cost more than 20 times what NaCl costs and are less effective. It appears to the writer that this is a fruitful field for exploration by some of our entrepreneurial scientists and engineers.

14.2 CORROSION OF METALS

The reaction of metals with their environment can be divided into two major groups: reactions with aqueous environments via the establishment of galvanic cells, which we call electrochemical corrosion, and reactions with other reactants, such as oxygen (oxidation), acids, and a host of other media. Galvanic corrosion is by far the most prominent degradation mechanism for metals and is the subject of this section. The others are discussed briefly in a subsequent section.

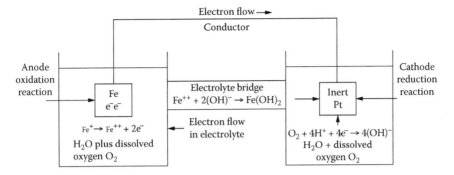

FIGURE 14.1 Laboratory galvanic cell.

14.2.1 LABORATORY GALVANIC CELL

Before examining the overall galvanic cell corrosion behavior let us construct a large galvanic (electrochemical) cell using a piece of iron plate that is a perfect single crystal (i.e., it contains no grain boundaries, dislocations, vacancies, or impurity atoms). When this piece of iron is immersed in a solution of pure distilled water that contains nothing more than dissolved oxygen, as depicted in the beaker on the left in Figure 14.1, the following oxidation reaction begins:

$$Fe^0 \rightarrow Fe^{2+} + 2e^- \tag{14.1}$$

Some of the neutral iron atoms go into solution as doubly charged positive ions, leaving electrons behind on the iron plate, but for every forward chemical reaction, there is also a backward reaction that involves the ions returning to the plate and recombining with the electrons. After a short period of time these two reactions are equal, as denoted by the two equal arrow lengths in the following:

$$Fe^0 \Leftrightarrow Fe^{2+} + 2e^- \tag{14.2}$$

At equilibrium there is no net loss of ions from the plate, so the corrosion process ceases. Galvanic corrosion is defined as the loss of metal by a galvanic reaction. For the reaction to proceed, the balance must be upset. This can be done by removal of either the iron ions or the electrons. In a galvanic cell the electrons are removed as illustrated in Figure 14.1, where we have connected the iron plate via a conducting wire to another metal, the latter also immersed in the same kind of pure water as was used for the iron crystal. We consider the second metal to be inert, such as the precious metals platinum and gold. No metals are entirely inert, but these metals do approach that state. Now the electrons are drawn off the iron plate in the left beaker and into the right beaker, thus permitting the reaction in the chemical Equation 14.1 to proceed. Now the corrosion process continues, but again not for long. In another short period of time the electron concentration on the inert plate will equal that produced on the iron plate. The reaction would again cease if it were not for the dissolved oxygen in the water. This permits the following reduction reaction to occur:

$$O_2 + 4H^+ + 4e^- \rightarrow 4(OH)^-$$ (14.3)

This is called the *cathode reaction*, and that occurring at the iron plate, as denoted by Equation 14.1, is the *anode reaction*. Electrons are produced at the anode and consumed at the cathode, so now the corrosion process can continue. Reaction 14.3 also has a backward reaction, and when the forward and backward reactions equilibrate, there will be no net consumption of electrons at the cathode, and again the corrosion process will cease. However, if we can mix the liquids in the beakers, for example, by constructing an electrolyte bridge between the two beakers, another reaction will take place, because now the positive iron ions will be attracted to the negative hydroxyl ions:

$$Fe^{2+} \rightarrow +2(OH)^- + Fe(OH)_2$$ (14.4)

We call these liquids *electrolytes*. We could have mixed the liquids in the same container without the electrolyte bridge, and the reactions would be the same. The electrolyte bridge was used only to illustrate that an electrolyte and the chemical reaction therein are required to complete the electric circuit that occurs within a galvanic cell. The overall reaction can be written as the sum of the equation above:

$$2Fe^0 + 2H_2O + O_2 \rightarrow 2Fe^{2+} + 4(OH)^- \rightarrow 2Fe(OH)_2$$ (14.5)

The compound formed is a white ferrous hydroxide. It is unstable and is further oxidized to ferric hydroxide, which has a reddish brown color called rust. The reaction is

$$2Fe(OH)_2 + H_2O + \tfrac{1}{2}O_2 \rightarrow 2Fe(OH)_3 \ (rust)$$ (14.6)

Now our galvanic cell is complete. We have an anode, a cathode, an electrolyte, and a conductor. All four are required to produce a galvanic cell, and they must from a complete circuit. The electrons flow from the iron anode through the conductor to the cathode. Many of the electrons will be consumed at the cathode according to Reaction 14.3, but some will flow through the electrolyte back to the anode. As long as more electrons are pumped out of the anode than are consumed at the cathode, the corrosion process will continue until all of the iron is consumed. With so many requirements for the corrosion reaction to proceed, the reader may wonder how so much corrosion can ever take place. We will pursue this further in a moment, but first note that what is depicted in Figure 14.1 is a battery. For many years, automobile batteries consisted of a Pb anode and a PbO_2 cathode in an acid electrolyte, the latter for the purpose of providing hydrogen ions. The reactions are listed as follows:

$$\text{anode}: Pb + SO_4^{2-} \rightarrow PbSO_4 + 2e^-$$

$$\text{cathode}: PbO_2 + 4H^+ + SO_4^{2-} + 2e^- \rightarrow PbSO_4 + 2H_2O$$

A typical automobile battery contains several of these cells in series. The Pb anode (which in your car battery is labeled as the positive terminal) corrodes and in so doing produces electrons and $PbSO_4$. A corrosion cell and a battery are the same thing, only differing in that one results in destruction and the other in producing usable energy.

Now let us consider a piece of real iron. As shown in Figure 14.2, it contains grain boundaries, dislocations, and impurity atoms, mostly in the form of pre-cipitates such as iron carbides or oxide inclusions left as a residue of the refining process. Any of these may become anodic or cathodic with respect to the main body of the iron. The atoms in the grain boundary, for example, can go into solution more easily than the atoms within the grain proper, because the former are more disorganized and not as tightly bound together as are those in the perfect lattice. The terms *anodic* and *cathodic*, or *anode* and *cathode*, are strictly relative. If more atoms go into solution at the grain boundary than from the lattice proper, more electrons will be generated there and this boundary will become the anode. These excess electrons will be consumed in the interior of the grain even though that region may be producing electrons by the dissolution of iron atoms, but at a lesser rate than the iron atoms at the grain boundary.

Thus, in a real piece of iron thousands of small galvanic cells are established all over the surface. The rust will appear uniform to the eye (uniform corrosion) because we cannot see each little cell. In addition, these negative and positive electrodes shift position with time because of the changes in the concentrations

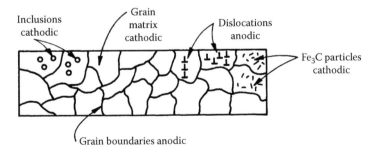

FIGURE 14.2 Possible location of anodes and cathodes on a piece of polycrystalline iron in a humid atmosphere.

of the chemical species involved in the reactions. This adds to the uniform corrosion appearance of the piece being destructed. A cold-worked iron or steel will corrode more than an annealed piece of iron because the dislocations are anodic with respect to the more perfect regions of the crystal. This characteristic of grain boundaries and dislocations being more reactive than the perfect crystal is used to reveal these defects by chemical attack with certain etchants. The grain boundaries are easily revealed by etching, and under certain controlled conditions, the dislocations will form etch pits when in contact with the etchant (Figure 5.8). Perhaps you have seen preferential rust formation at a bent or deformed region in a steel sheet caused by the higher dislocation density in the deformed regions.

It is not necessary to place the iron or steel in water to form rust. There is sufficient moisture in the air to provide a surface layer of electrolyte. Corrosion proceeds faster and is more extensive in humid atmospheres. The metal itself acts as the conductor, the humid atmosphere as the electrolyte, and the perfect lattice regions and defects become the cathodes and anodes, respectively. The ferric hydroxide rust flakes off the iron or steel surface owing to the large coherency strains set up as a result of a mismatch in their respective lattice parameters. This flaking provides for removal of the corrosion products. The rusting will continue until the iron disappears. The bridge shown in Figure 14.3 has been attacked extensively over its entire surface. It is a typical example of uniform corrosion.

14.2.2 ELECTROMOTIVE FORCE SERIES

Just as defects in iron are anodic with respect to the lattice proper, metals are anodic or cathodic with respect to each other. This is how battery cells are made: by putting two materials of widely different electrochemical characteristics into an electrolyte. One material is the anode, while the other is the cathode. The Pb-PbO_2 battery, mentioned previously, is the best example of one of the good guys: a galvanic cell that is put to work to generate power rather than to destroy the anode materials as in the case of a corrosion reaction. As the battery produces current, both plates are covered with lead sulfate as a result of the chemical reactions taking place in the electrolytes. Recharging this battery is accomplished by feeding current in the reverse direction, which decomposes the lead sulfate. The life of a battery is snuffed out when the amount of this lead sulfate and assorted corrosion products flake off the electrodes and fall in-between them. The battery can then no longer be recharged. It is not the consumption of the anode on which the life of the battery is dependent. The anode can always be restored by current reversal as long as lead sulfate or lead compounds are available to place the Pb back onto the anode during current reversal (recharging). Note also that the dry cell battery that we use in our flashlights is not really dry. In the common zinc anode–carbon cathode dry cell, the electrolyte is a wet paste containing moist aluminum chloride.

Returning now to metals, their corrosion tendency is determined by measuring the electrons that they generate vs. those of a standard electrode. The hydrogen electrode is universally used for measuring the standard electromotive force (emf)

FIGURE 14.3 Uniform corrosion on an unprotected bridge structure. (C. V. White.)

of metals. Since each metal is measured against a hydrogen electrode, they are written as half-cell values since the potential of the hydrogen standard electrode is taken arbitrarily as zero. If the metal electrode measured against the hydrogen electrode is negative, it has a greater tendency to give up electrons than does the hydrogen electrode, and this metal is anodic with respect to the hydrogen standard half-cell. Metals with high negative potentials (e.g., Li and Mg) are very reactive metals. These metals and most other metals characteristically give up electrons in what is called an *oxidation reaction*. Metals that measure a positive potential with respect to the standard hydrogen electrode give up fewer electrons than the standard cell and are more cathodic. These include the more noble precious metals (e.g., Au, Pt, and Pd). The hydrogen electrode (Figure 14.4) consists of an inert metal such as platinum immersed in a 1 M solution of H^+ ions into which hydrogen gas is bubbled. The metal to be measured is placed in a 1-M solution of its own ions. The platinum provides a surface for the oxidation reaction of hydrogen gas to hydrogen ion ($H_2 \rightarrow 2H^+ = + 2e^-$) to occur. The electrolytes of the two

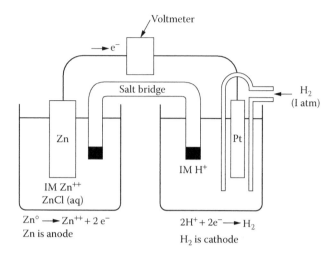

FIGURE 14.4 The hydrogen reference half-cell (on the right), better known as the hydrogen electrode, is being used to measure the emf of zinc.

half-cells are allowed to be in electrical contact but are not permitted to mix with one another. The voltage of each metal is measured vs. the hydrogen half-cell at 25°C. These voltages are usually ranked in descending order of positive values until the arbitrarily defined zero potential of hydrogen is reached. Those metals that generate fewer electrons than the hydrogen electrode are listed as positive, and those that generate more electrons are considered to be negative.

The electromotive force voltages resulting from these measurements are listed in Table 14.1, and this listing is known as the emf series for metals. Any metal below another metal in this listing is anodic with respect to the one above it. The anodic metal is also more reactive. If we form a cell (battery) between any two metals, the voltage generated will be the difference in the emf values of the two metals. For Li–Mn this voltage would be $-3.05 - (-1.18) = -1.22$ V. For Li–Cu the cell voltage would be $-3.05 - (+0.52) = -2.53$ V. Many emf series, including that of the *Handbook of Chemistry and Physics*, list Li at the top with -3.05 V and gold at the bottom with $+1.50$ V. This is the same listing in both sign and value that we show in Table 14.1 except that in the latter listing the more reactive metals are placed at the bottom of the list. Others, however, reverse the signs and put gold at the top with -1.50 V and Li at the bottom with $+3.05$ V. This difference is merely one of convention. Both conventions are still used, although this matter was supposedly settled at the International Union of Pure and Applied Chemistry Meeting in Stockholm in 1953. According to the Stockholm convention, a negative sign indicates a trend toward oxidation and corrosion in the presence of H^+ ions. Thus, Li has a negative sign when it is undergoing an oxidation reaction.

Corrosion engineers use the galvanic series, which includes alloys as well as pure metals, the latter seldom existing in most product designs. This series includes metals that form passive film and lists the most negative (anodic) material

TABLE 14.1
Electromotive Force Series of Metals

Electrode Reaction	Standard Potential at 25°C (77°F) (Volts vs. SHE)
$Au^{3+} + 3e^- \rightarrow Au$	1.50
$Pd^{2+} + 2e^- \rightarrow Pd$	0.987
$Hg^{2+} + 2e^- \rightarrow Hg$	0.854
$Ag^{2+} + e^- \rightarrow Ag$	0.800
$Hg_2^{2+} + 2e^- \rightarrow 2Hg$	0.789
$Cu^+ + e^- \rightarrow Cu$	0.521
$Cu^{2+} + 2e^- \rightarrow Cu$	0.337
$2H^+ + 2e^- \rightarrow H_2$	0.000
	(Reference)
$Pb^{2+} + 2e^- \rightarrow Pb$	−0.126
$Sn_2 + 2e^- \rightarrow Sn$	−0.136
$Ni^{2+} + 2e^- \rightarrow Ni$	−0.250
$Co^{2+} + 2e^- \rightarrow Co$	−0.277
$Tl^{2+} + e^- \rightarrow Tl$	−0.336
$In^{2+} + 3e^- \rightarrow In$	−0.342
$Cd^{2+} + 2e^- \rightarrow Cd$	−0.403
$Fe^{2+} + 2e^- \rightarrow Fe$	−0.440
$Ga^{3+} + 3e^- \rightarrow Ga$	−0.53
$Cr^{3+} + 3e^- \rightarrow Cr$	−0.74
$Cr^{2+} + 2e^- \rightarrow Cr$	−0.91
$Zn^{2+} + 2e^- \rightarrow Zn$	−0.763
$Mn^{2+} + 2e^- \rightarrow Mn$	−1.18
$Zr^{4+} + 4e^- \rightarrow Zr$	−1.53
$Ti^{2+} + 2e^- \rightarrow Ti$	−1.63
$Al^{3+} + 3e^- \rightarrow Al$	−1.66
$Hf^{2+} + 4e^- \rightarrow Hf$	−1.70
$U^{3+} + 3e^- \rightarrow U$	−1.80
$Be^{2+} + 2e^- \rightarrow Be$	−1.85
$Mg^{2+} + 2e^- \rightarrow Mg$	−2.37
$Na^+ + e^- \rightarrow Na$	−2.71
$Ca^{2+} + 2e^- \rightarrow Ca$	−2.87
$K^+ + e^- \rightarrow K$	−2.93
$Li^+ + e^- \rightarrow Li$	−3.05

at the top and the most noble (cathodic) at the bottom. The measurements are also usually conducted in about a 3.5% saltwater solution, comparable to seawater, and often measure the voltages, which are frequently not listed, against a graphite electrode. In practical corrosion problems galvanic coupling between metals in equilibrium with a 1-M solution of its own ions is seldom observed. The galvanic series is a more accurate prediction of galvanic reactions than is the emf series.

The galvanic series as determined by the LaQue Center for Corrosion Technology is given in Table 14.2. This series was obtained by measuring the specific metals listed against a control half-cell reference electrode, which is composed of mercury and calomel (mercurous chloride) in a solution of potassium chloride. The design engineer tries to avoid placing metals that are far apart in the series physically close together in an attempt to avoid large galvanic cells.

Most electrolytes in the real world are not 1 M but of a lesser concentration of ions. In this case the driving force to corrode the anode is greater since there are fewer of its ions in solution. The effect of ion concentration is given by the Nernst equation (after W. H. Nernst) as

$$E = E^{\circ} + \frac{0.0592}{n} \log C_{ion} \tag{14.7}$$

where

E = measure potential
E° = standard emf
n = number of electrons transferred per metal ion (valency)
C_{ion} = molar concentration of ions

Example 14.1   If 200 g of Mg^{2+} ions are present in 1000 g of water, what is the electrode potential of this half-cell?

Solution   A 1-M solution contains 1 mol of Mg^{2+} to 1000 g of water. The atomic mass of magnesium is 24.312 g/g-mol. The concentration for 200 g of Mg^{2+} becomes

$$C_{ion} = \frac{200}{24.312} = 8.22 \, M$$

$n = 2$ since the Mg ion has a +2 charge. E° is obtained from Table 14.1 as −2.37 V.

Thus,

$$E = -2.37 + \frac{0.0592}{2} \log(8.22) = -2.4 \, V$$

14.2.3 ELECTRODE REACTIONS AND THEIR ASSOCIATED GALVANIC CELLS

In our discussion of the corrosion of iron via electron production at the anode and electron consumption at the cathode, which were stated in Equations 14.1 and 14.3, respectively, the anode reaction is always the same, that is,

TABLE 14.2
Galvanic Series of Metals

Corrosion Potentials in Flowing Seawater (8 to 13 ft/sec) Temp Range 50–80°F
Votes: Saturated Calomel Half-Cell Reference Electrode

+0.2	0	-0.2	-0.4	-0.6	-0.8	-1.0	-1.2	-1.4	-1.6
									Magnesium
						Zinc			
						Beryllium			
						Aluminum Alloys			
					Cadmium				
					Mild Steel, Cast Iron				
				Low Alloy Steel					
					Austenitic Nickel Cast Iron				
				Aluminum Bronze					
					Naval Brass, Yellow Brass, Red Brass				
			Tin						
			Copper						
				Pb–Sn Solder (150/500)					
				Admiralty Brass, Aluminum Brass					
				Magnesium Bronze					
			Silicon Bronze						
				Tin Bronzes (G & M)					
					Stainless Steel—Types 410, 416				
			Nickel Silver						
				90–10 Copper-Nickel					
				80–20 Copper-Nickel					
					Stainless Steel—Type 430				
			Lead						
			70–30 Copper-Nickel						
				Nickel-Aluminum Bronze					
					Nickel–Chromium Alloy 600				
			Silver Braze Alloys						
			Nickel 200						
			Silver						
						Stainless Steel—Types 302, 304, 321, 347			
			Nickel–Copper Alloys 400, K-500						
					Stainless Steel—Types 316, 317				
			Alloy "20" Stainless Steels Cast and Wrought						
			Nickel–Iron–Chromium Alloy 825						
		Ni-Cr-Mo-Cu-Si Alloy B							
			Titanium						
			Ni-Cr-Mo Alloy C						
	Platinum								
	Graphite								

Source: Courtesy of the LaQue Center for Corrosion Technology, Inc., Wrightsville Beach, NC.

$$M^{\circ} \to M^{n+} + ne^{-} \tag{14.8}$$

where n is the valency of the metal. However, at the cathode several different reactions may occur, all being reduction reactions, the exact type being dependent on the solution or electrolyte involved. One of the more common cathode reduction reactions is that which occurs via the corrosion of metals in acid solutions, where a high concentration of hydrogen ions is present as represented by the following chemical equation:

$$\text{hydrogen generation: } 2H^{+} + 2e^{-} \to H_2 \uparrow \tag{14.9}$$

This reaction occurs when the cathode is another metal and the electrolyte is acidic. This reaction is sometimes referred to as the plating out of H_2, and the upward arrow indicates that the hydrogen escapes as a gas. Unfortunately, not all of this hydrogen escapes. When metals are electroplated, hydrogen is also plated out according to Equation 14.9 simultaneously with the metal at the plated part. Some of the hydrogen ions enter the plated metal. These ions have been the cause of many failed plated parts because hydrogen embrittles metals. Many other sources of hydrogen in metal processing lead to hydrogen embrittlement. This phenomenon is discussed more fully in Section 14.2.6.3.

$$\text{Water formation: } O_2 \, 4H^{+} + 4e^{-} \to 2(H_2O) \tag{14.10}$$

In acid solutions containing dissolved oxygen, water is produced at the cathode instead of the $(OH)^{-}$ of Equation 14.3, the later occurring when the water is neutral or basic. The reaction that predominates depends on the electrolyte and the ion concentration. As pointed out above, more than one reaction may occur at the same time, but the predominate or controlled reaction determines the type of cell involved in the corrosion process. Metal ions in solution may also be reduced, since these ions can exist in more than one valence state. The most common of these reactions in the corrosion process is the reduction of ferric to ferrous ions:

$$Fe^{3+} + e^{-} \to Fe^{2+} \tag{14.11}$$

14.2.3.1 Galvanic Cells between Two Metals

This is the type of cell that generates a large amount of corrosion when two metals that have a large separation between them in the galvanic or emf series are placed in direct contact within an electrolyte. This is also the type of cell that could be used as a battery. The Ni–Cd battery is another example of such a cell. The electrolyte is a solution of potassium hydroxide. Cadmium metal changes to cadmium oxide; the nickel changes to nickelous hydroxide on discharging and the reverse reaction occurs on charging. The reactions are very complex and still not fully understood. The potassium hydroxide solution does not change in composition and strength, and does not become involved in the reactions. It is said to be a nonvariant electrolyte. The Ni–Cd cell lasts longer than the Pb–PbO$_2$ acid battery and requires less care, but most large galvanic cells between two dissimilar metals result in unwanted corrosion. The connecting together of copper pipes with a silver braze alloy is a typical example. Note how far apart these two materials are in the galvanic series of Table 14.2.

14.2.3.2 Concentration Cells

Concentration cells occur when the concentrations of identical ions differ from one region to another in the system. One of the common cells of this type is the oxygen concentration cell. Consider the cell of Figure 14.5, where we have two iron electrodes in electrolytes of different oxygen concentrations. The electron-producing anode reaction and the cathode consumption reaction are the same as before:

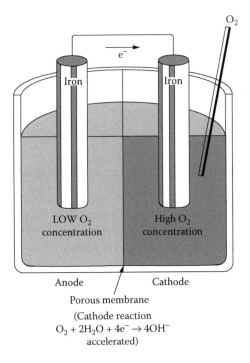

FIGURE 14.5 Oxygen concentration cell. The region of high oxygen concentration is cathodic to the region of low oxygen concentration. (After J. Wulff, Ed., *The Structure and Properties of Materials*, Vol. 2, Wiley, New York, 1964, p. 165. With permission.)

$$\text{anode: } Fe^0 \rightarrow Fe^{2+} + 2e^-$$
$$\text{cathode: } O^2 + 2H_2O + 4e^- \rightarrow 4(OH)^-$$

These reactions are the same ones that occurred in our laboratory cell, and the electrode reactions were expressed by Equations 14.1 and 14.3, but our electrodes are of the same metal now. Since the cathode reaction requires electrons and oxygen, it must have a higher concentration of oxygen than that at the anode. In an oxygen concentration cell the region low in oxygen will produce more electrons and become anodic with respect to the high-oxygen region. The metal can be a single piece. Two separate metal electrodes, as shown in Figure 14.1, are not needed. One example of this type of cell is that of waterline corrosion. Immediately above the waterline the oxygen in the air will be of a higher concentration than that of the dissolved oxygen in the water below. Thus, the latter region becomes anodic with respect to that on the same metal above the waterline. All of the requirements for a galvanic cell exist: the anode and cathode, the electrolyte, and the metal conductor. In this cell the corrosion is accentuated in the region of low oxygen concentration. Pitting often occurs immediately below the waterline in the oxygen-starved layer. Similar oxygen concentration cells can also form on metallic surfaces at localized spots where debris such as grease spots or dirt decrease the oxygen content underneath the accumulated junk. The same cell type could occur in cracks, where the oxygen concentration might be low. This type of accelerated localized corrosion can lead to pit formation.

14.2.3.2.1   Water Reduction

In the absence of all other reactions, water will be reduced as follows:

$$2H_2O + 2e^- \rightarrow H_2 + 2(OH)^- \tag{14.12}$$

which is equivalent to Equation 14.9, assuming dissociation of the water.

14.2.3.2   Electroplating

This reaction, although not considered a corrosion reaction, obeys the same principles and occurs when there is a high concentration of metallic ions, usually intentionally introduced to coat the cathode with metal atoms. The reactions is

$$M^0 + ne^- \rightarrow M^{n+} \tag{14.13}$$

14.2.4   Corrosion Rates

Michael Faraday (1791–1867) established many of the basic principles of electricity and magnetism but is probably best known for his work on ion migration in an electric field. *Faraday's law,* which is applicable to both electroplating and corrosion, is expressed as

$$W = \frac{ItM}{nF} \qquad (14.14)$$

where

W = weight of metal plated or corroded per unit of time t(g/s)
I = current flow (A)
M = atomic mass of metal (g/mol)
n = number of electrons produced or consumed
F = Faraday's constant = 96,500 C/mol or 96,500 A·s/mol

Example 14.2   For electroplating nickel from a sulfate solution, a current of 5 A was employed to plate 1-mm thickness of nickel onto a steel surface of 10 cm² area. How much time was required to complete this operation?

Solution   The quantity of Ni = 10 cm² × 0.1 cm = 1 cm³. Nickel has a density of 8.9 g/cm³. Thus, 8.9 g of Ni is required. In this reaction Ni has a valency of +2 since it goes into solution as follows: $Ni^\circ \rightarrow Ni^{2+} = +2e^-$.

$$W = \frac{ItM}{nF}$$

$$t = \frac{WnF}{IM} = \frac{8.9 \text{ g} \times 2 \times 96,500 \text{ A} \cdot \text{s/mol}}{5 \text{ A} \times 58.7 \text{ g/mol}} = 5097 \text{ s} = 1.4146 \text{ h}$$

14.2.4.1 Polarization

When metals are electropolated, the current density is controlled and the plate thickness for a specific time can be accurately predicted. During the corrosion reaction it is difficult to predict corrosion rates because of polarization, which may occur at either the anode or cathode and thereby alter the potential at these electrodes. As corrosion progresses, the corrosion rate will change owing to changes in the electrolyte and oxides or other surface films. These changes are termed *polarization*. When the electrode potentials change because of polarization, the corrosion rate changes. Electrochemical reactions at both electrodes depend on the availability of the electrons involved in the reactions. During metallic corrosion, in the absence of polarization, the rate of production of electrons at the anode equals the rate of consumption at the cathode. If electrons are made available to the cathode too fast, the cathode becomes more negative and cannot accommodate all of these electrons in the cathode reaction. This is called *cathodic polarization*. A deficiency in the electrons required to sustain dissolution of the anode, which leaves its electrons behind, is called *anodic polarization*. As the deficiency becomes greater, the driving force to produce more electrons becomes greater and the anodic dissolution increases. As the surface potential becomes more positive (i.e., fewer electrons available), the oxidizing power of the solution increases because of the larger anodic polarization. Polarization can be divided into two classes: *activation* and *concentration*.

Activation polarization is that related to the metal–electrolyte interface, where a sequence of steps is involved and the rate of one step in the process is the rate-controlling step of the entire process. The plating out of hydrogen is a good example to use here because it is also applicable to the hydrogen embrittlement problem to be discussed later. When hydrogen is evolved at the cathode, the reaction proceeds by the following steps:

1. Physical adsorption, where the gaseous molecules are attached to the surface by weak van der Waals forces.
2. Chemisorption, where the molecules become dissociated and are bound to the surface as atoms or radicals of the molecules (e.g., H^+ ions).
3. Electron transfer from the metal to form an H atom that combines with another H atom to form the H_2 molecule.
4. Evolution of H_2 gas as bubbles.

The slowest of these steps, probably step 1, determines the rate of the overall reaction. An activation energy barrier is associated with the rate of the limiting step. (Incidentally, many of the H^+ ions do not combine with electrons as in step 3 above but diffuse into the metal and cause hydrogen embrittlement.)

Concentration polarization exists when the reaction rate is limited by diffusion in the electrolyte. If the number of H^+ ions is small, the reduction rate may be controlled by the diffusion of H^+ ions to the metal surface. Concentration polarization is usually not a factor in anodic dissolution of metals. Increasing the agitation of the corrosive medium will increase the corrosion rate only if the cathodic reaction process is controlled by concentration polarization. When both cathodic and anodic reactions are controlled by activation polarization, agitation will have an influence.

When the surface area of the cathode in a galvanic cell is large in comparison to the anode, the anodic current density is large, and this anodic polarization leads to more pronounced anode dissolution. Conversely, a smaller cathode slows the galvanic action owing to the predominate cathodic polarization.

Passivation is a phenomenon that affects the corrosion rate significantly and always in a favorable direction. Most metals that have a high degree of passivity form tight adherent protective oxide layers on the surface of the metal, although it is believed that the passitivity mechanism is not simply that of an oxide barrier mechanically separating the metal from the electrolyte. Stainless steel is passive because of its chromium content. When this chromium content exceeds about 11 wt %, a complete surface oxide (Cr_2O_3) is formed in a few seconds on exposure to air. Similar oxide films form on aluminum (Al_2O_3), titanium (TiO_2), and silicon (SiO_2). Nickel and iron show some degree of passivity but not to the same extent as those mentioned above. There is some controversy as to the nature of this barrier, but it is believed by many to be some type of thin hydrated oxide layer, on the order of tens of angstroms in thickness.

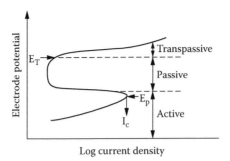

FIGURE 14.6 Schematic polarization curve for a metal that displays an active–passive transition.

Passivation is usually demonstrated by polarization curves, where the electrode potential is plotted vs. the log of the current density (Figure 14.6). For a normal active metal the current and hence the electrode dissolution rate increase exponentially with increasing positive potential on the electrode. This is termed the *active* state of the metal and, correspondingly, the active region of the polarization curve. However, the metals that possess the ability to become passive show a marked decrease in the corrosion rate, by a factor of 10^3 to 16^6, when the potential exceeds some critical value denoted as E_p on the polarization curve. The corresponding critical current is known as I_c. As the electrode potential exceeds the E_p value, the current suddenly decreases to a constant value that does not change with increasing electrode potential. This is the passive state of the metal and is noted on the polarization curve as the *passive* region. When the electrode potential attains a sufficiently high value (E_t), the metal again becomes active. This part of the curve is termed the *transpassive* region.

The state of passivity can be altered by the environment. Chloride ions are noted to be detrimental to most oxide films, particularly on aluminum. Breaks in this film lead to localized corrosion. The damage to bridge structures by NaCl is to some extent related to the destruction of passivity of the metal-reinforcing bars.

14.2.5 Types of Corrosion

The ability to identify the type of corrosion is of particular importance to selection of the kind of corrosion protection that could be employed. The types usually experienced can be identified by the appearance of the metal surface.

14.2.5.1 Uniform Corrosion

As the term *uniform corrosion* suggests, this type of corrosion appears visually the same over all of the surface, as shown in Figure 14.3 for the rusted bridge. Many small galvanic cells are on the surface, and these may shift position from time to time, causing the corrosion to spread over the entire surface. Weathering steels and some copper alloys generally show uniform corrosion (Figure 14.7).

FIGURE 14.7 Uniform atmospheric corrosion of a structural steel member from the Golden Gate Bridge after 50 years of service. The irregular-shaped holes are where the corrosion penetrated the entire thickness of the member. (Courtesy of R. H. Heidersbach, Cal Poly.)

The rusting of iron and steel and the formation of a patina on cooper occur readily in most atmospheres and are some of the best examples of uniform corrosion.

Uniform corrosion is one of the more common forms and can be controlled in a number of ways, such as by protective coatings, inhibitors, reduction in the moisture where possible, cathodic protection, or alloy selection. Small additions of copper, nickel, and phosphorus have proven to be very beneficial in retarding atmospheric corrosion in steels. Aqueous environments, including seawater, usually cause uniform corrosion on steel parts other than the stainless steels and aluminum, where pitting is more likely to be found. In certain industries chemical dissolution by acids, bases, and chelants frequently cause uniform corrosion and require alloys designed specifically for these environments. (*Note:* Uniform corrosion is sometimes referred to as *general corrosion.*)

14.2.5.2 Pitting Corrosion

Localized corrosion is generally classified as pitting corrosion when the diameter of the pit is less than the depth. At certain depths the pits may branch out into several directions. In pitting corrosion the anodic and cathodic sites do not shift but become fixed at certain points on the surface. Pitting often requires a certain amount of initiation time, but the pit grows faster with time as conditions within the pit become more aggressive. Oxygen concentration cells can initiate a pit, and the difference in oxygen concentration between the pit and the surface will accelerate pit growth rates. In sea water, positively charged anodes at the bottom of the pit attract negative chloride ions. The reaction at the bottom of the pit is the usual anode reaction of $M^\circ \rightarrow M^n + ne^-$, while the cathode reaction takes place at the surface forming $(OH)^-$ ions as in Equation 14.3 (i.e., $O_2 + 2H_2O + 4e^- \rightarrow 4[OH]^-$). The combination of these two reactions causes a buildup of acidic chlorides according to the following reactions:

$$M^+Cl^- + H_2O \rightarrow MOH + 2Cl^- + H^+ \qquad (14.15)$$

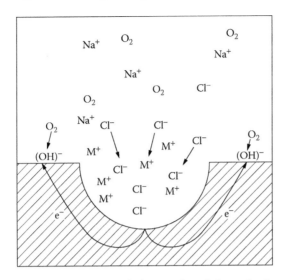

FIGURE 14.8 Schematic of an autocatalytic corrosion pit formation in seawater.

The acid buildup at the bottom of the pit increases the reaction rate, which now becomes an autocatalytic reaction (Figure 14.8). A 6% $FeCl_3$ solution is a common testing medium for pitting and crevice corrosion. Pits may be initiated at surface inclusions, breaks in the inhibitor, or passive film coverage and surface deposits. Examples of pitting corrosion in stainless steels are shown in Figure 14.9.

14.2.5.3 Crevice Corrosion

Crevice corrosion is a localized corrosion type that occurs in narrow gaps or small openings where two materials, either metal to metal or metal to nonmetal, come into contact. In the latter case the second material may be debris, such as grease or mud, gaskets, and packing materials, and as a result, this type of crevice corrosion has been called *deposite* or *gasket corrosion*. In the former situation cracks and seams may be the point of corrosion initiation even between two regions of the same metal. The mechanism varies from material to material and with the environment, but what is common in all crevice corrosion is the presence of a crevice where stagnant fluids collect. Furthermore, it is always a localized form of attack. Usually, there will exist some type of concentration cell such as that depicted in Figure 14.10, but the chemical reactions are believed to be more involved than those in simple concentration cells. Metal ions are released in the anodic crevice following Equation 14.8 when the potential in the crevice exceeds some critical value. The cathodic reaction in the regions surrounding the crevice occurs according to Equation 14.3. However, for iron alloys and the presence of chloride ions, the following reaction may occur:

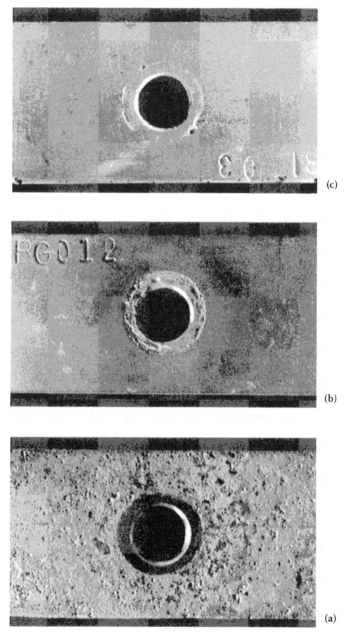

FIGURE 14.9 Variation in stainless steel pitting corrosion resistance in a model SO_2 scrubber environment; (a) type 304 in acid condensate, (b) type 316 in acid condensate, (c) type 304 in a limestone slurry zone. (Courtesy of the LaQue Center for Corrosion Technology, Inc., Wrightsville Beach, NC.)

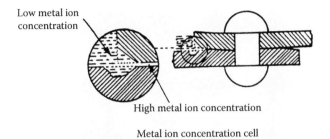

Metal ion concentration cell

Oxygen concentration cell

FIGURE 14.10 Crevice corrosion owing to concentration cells. (From P. A. Schweitzer, Ed., *Corrosion and Corrosion Protection Handbook*, 2nd ed., Marcel Dekker, New York, 1989, p. 11. With permission.)

$$Fe^{2+} + 2H_2O + 2Cl^- \rightarrow Fe(OH)_2 + 2HCl \qquad (14.16)$$

Such crevice corrosion will proceed autocatalytically as more chloride is attracted to the crevice, much as in pitting corrosion.

For stainless steels and the passive alloys in general, the concentration of acid and chloride ions becomes sufficiently aggressive to break down passive films. Examples of crevice corrosion are shown in Figure 14.11. The occurrence of crevice corrosion in copper alloys is believed to be of the more simple metal ion concentration type.

A Teflon or similar sealant can be used around mechanical joints to keep out the electrolyte. However, one must be careful to avoid crevices between metallic and nonmetallic parts such as gaskets and seals. The avoidance of regions where stagnant solutions will accumulate is a must. Increasing the alloy content with Cr, Ni, Mo, and nitrogen increases the resistance to both crevice and pitting corrosion.

14.2.5.4 Stray-Current Corrosion

This type of corrosion occurs in the presence of some external direct current, most commonly encountered in soils containing water. It is more frequently a

(a)

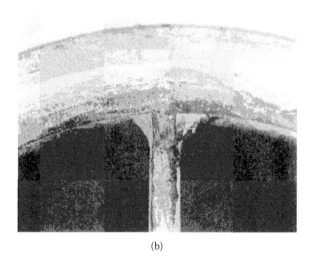

(b)

FIGURE 14.11 Examples of crevice corrosion: (a) metal-to-metal crevice site formed between components of a type 304 stainless steel fastener in seawater; (b) at a nonmetal gasket site on an alloy 825 heat exchanger. (Courtesy of the LaQue Center for Corrosion Technology, Inc., Wrightsville Beach, NC.)

problem in underground structures than in underground piping. Protective coatings and insulation are two possible preventive methods for this type of corrosion.

14.2.5.5   Intergranular Corrosion

Intergranular corrosion consists of selective attack at the grain boundaries. The fact that chemical etchants reveal grain boundaries is evidence of their susceptibility to chemical attack and dissolution at these boundaries.

One of the most common examples of grain boundary corrosion is that termed *sensitization*. It occurs in austenitic stainless steels, some superalloys, and perhaps in other high-chromium alloys that have been heated to certain temperature ranges. For the austenitic stainless steels this range is from 500 to 600°C (950 to 1450°F). The chromium in this particular temperature range precipitates out of the solid solution as chromium carbides. The chromium is no longer available to form the passive chromium oxide (Cr_2O_3) protective film. The alloy is said to be in the *sensitized condition* or, more simply, is now susceptible to galvanic corrosion and the related rust formation. The austenitic stainless steels alloys 321 and 347 contain small amounts of Ti and Nb, respectively, which are introduced to tie up the carbon and reduce the precipitation of chromium carbides, but this is not a cure-all. 316L and 304L austenitic stainless steels of low carbon content have also been used, particularly where welding is involved. Welding operations often heat the adjacent metal sections into the critical temperature range for the chromium carbides to precipitate. A typical microstructure of a sensitized alloy is shown in Figure 14.12.

The other major types of intergranular corrosion—stress corrosion cracking and hydrogen-assisted cracking—are treated in a later section.

14.2.5.6 Selective Leaching

This type of corrosion occurs when a particular element in an alloy is preferentially removed with respect to the major element. Dezincification of brass is a typical example, where zinc is removed from brass in certain environments. Similar processes have been observed for other alloys in which Al, Fe, Co, and Cr are preferentially removed.

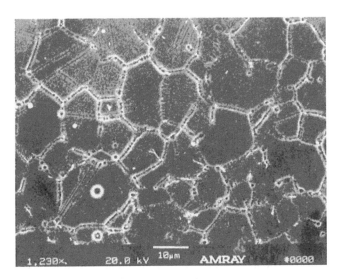

FIGURE 14.12 Example of sensitization in X750 Inconel after 2 h at 1400°F plus 5 h at 1100°F. Note the carbides at the grain boundaries. (Courtesy of Roxanne Hicks, Cal Poly.)

14.2.5.7 Erosion–Corrosion

Erosion–corrosion describes a situation where metal removal occurs by the combined action of corrosion (chemical attack) and erosion (the wearing away of the metal by fluid motion). Soft metals are sensitive to this form of attack. Grooves and gullies will appear. Corrosion tests conducted under static conditions are not valid for predicting corrosion rates when a moving fluid is involved.

14.2.5.8 Corrosion Fatigue

Corrosion fatigue is a subject that, as its name suggests, involves two mechanisms, both of them very complex. We can simplify this phenomenon somewhat by separating the two mechanisms prior to joining them together, but first let us define corrosion fatigue as that type of fatigue failure that occurs during application of a cyclic stress in the presence of a corrosive environment or corrosive action. In Chapter 2 we described how fatigue crack nucleation occurs by the to-and-fro movement of dislocations but that most fatigue cracks propagate from surface defects. One of the primary roles of corrosion in the corrosion fatigue process is that of providing surface defects for crack formation and propagation. A corrosion pit, for example, can concentrate stresses by factors of 10 to 100 or more. There is little doubt that fatigue cracks frequently start at corrosion pits. Crack propagation can also be enhanced by a corrosive process in some alloys whereby the energy required to break atom bonds and provide two new surfaces is reduced by the corrosive media. The breaking of surface films may also be involved. The interrelation of crack propagation in the presence of cyclic stress and corrosive environments is far beyond the scope of this book. The ASM *Atlas* (see the Suggested Reading) presents a considerable amount of data and insight on this phenomenon. For the present we leave this subject to specialists in this field.

14.2.6 STRESS CORROSION CRACKING AND HYDROGEN DAMAGE

When discussing stress corrosion cracking and hydrogen damage, we encounter some difficulties, in part owing to semantics and in part to the controversy that exists in the mechanisms involved in each process. People from different industries label the same embrittlement process by different names, so let us begin with the semantics problem by listing the various names that one could encounter in the literature for these embrittlement phenomena. First it must be mentioned that as far back in 1959, A. R. Troiano and coworkers concluded that much of what had been previously called stress corrosion cracking (SCC) was related to hydrogen effects, commonly referred to at that time as hydrogen embrittlement (HE). Today, it is generally considered that one form of HE is a mechanism of SCC. Other names are:

Stress corrosion cracking (SCC): A phrase developed early in the twentieth century that has generally been defined as the result of the combined action of static stresses and corrosion, the stresses being either residual or externally applied.

Hydrogen-assisted cracking (HAC): A phrase, mostly used by metallurgists, to describe the principal type of HE, where a few parts per million of hydrogen in the form of H^+ ions cause extreme embrittlement of metals, resulting in loss of ductility and often strength. HAC is the form of HE mentioned above that is now considered to be one mechanism of SCC.

Hydrogen-induced cracking (HIC): A phrase that to many engineers means the identical process as HAC. However, in the petroleum industry this phrase is used to describe blister formation on the surface of steels, where the hydrogen, in the form of H_2 gaseous molecules, builds up sufficient internal pressure at points near the surface to cause the formation of bumps (blisters). A 1989 article in the journal *Corrosion* states, "HIC normally observed in pipeline steels includes not only hydrogen blistering but also internal stepwise cracking parallel to the pipeline" [1]. It appears that even those in the petroleum industry do not always agree on the meaning of this term.

Sulfide stress cracking (SSC): A type of cracking identical to HAC, but the term was coined by the petroleum industry to describe cracking when hydrogen sulfide environments generate the hydrogen, in contrast to the term HAC used in other industries, where the source of hydrogen is not defined.

Sulfide stress corrosion (SSCC): The same process as SSC.

Hydrogen embrittlement (HE): A term used synonymously with HAC by many metallurgists but by others to include all forms of hydrogen damage.

Despite the confusion in semantics, certain forms of SCC can, we believe, be separated from hydrogen effects, and vice versa. It will assist in clarifying the picture if SCC is considered as a phenomenon that consists of two mechanisms, one that we label as anodic dissolution, which occurs independently of hydrogen, and the other that is some type of HAC mechanism, where only a few parts per million of hydrogen ions is required for embrittlement.

14.2.6.1 Stress Corrosion Cracking

The first recognition of SCC occurred early in the twentieth century when cartridge shells made of 70% Cu–30% Zn were found to crack over a period of time. It was later realized that ammonia from decaying organic matter in conjunction with the residual stresses in the brass was responsible for the cracking of these shells. The presence of high humidity also promoted SCC. This phenomenon was termed *season cracking* because it was more prominent in warm, moist climates (or seasons). Stress-relief annealing of the brass reduces the residual stresses without significantly reducing the strength and at the same time reduces the susceptibility of the brass to SCC. SCC of certain nonferrous metals and the responsible environments are listed in Table 14.3. Anodic dissolution is the most likely cause of this type of SCC. Steels have been omitted from this table intentionally, because in steels the cracking mechanism of the SCC phenomenon is probably due to hydrogen effects.

The mechanisms of the forms of SCC in Table 14.3 are not completely understood, although anodic dissolution of the metal by a localized galvanic cell accompanied by tensile stresses is responsible for crack growth. Crack nucleation may begin at surface pits or similar discontinuities. The crack grows in a plane perpendicular to the applied stress (this is also the frequent mode of crack propagation in the absence of corrosion). It has been stated *that if either the corrosion or the stress is removed, the crack stops growing,* provided that the crack is below a certain critical length for the stress intensity present at the crack tip. This is the stress intensity that is a function of the applied stress and the crack length as described in Chapter 2 (Equation 2.17) and becomes the K_{1c} material constant at the critical crack length for propagation to fracture. In corrosion we often see the term K_{1SCC} used for the critical stress intensity to propagate a crack in the presence of a corrosive medium. This K_{1SCC} is not a true material constant but is a useful measure of the relative susceptibility of materials to cracking in certain environments. The role of the tensile stresses is not completely understood but is believed to be associated with the rupture of surface films, which promotes further anodic dissolution and an increase in stress concentration as the crack grows via the anodic dissolution process. A tensile stress is necessary for propagation of the cracks in addition to the rupturing of surface films. SCC that occurs by anodic dissolution often occurs along grain boundaries and results in intergranular cracking, although considerable transgranular cracking has also been observed.

TABLE 14.3
Environments That Produce Stress Corrosion Cracking in Metals (Partial Listing)

Nonferrous Metals That Show SCC	Environment
Copper alloys	Ammonia
Aluminum alloys	Most halide ion solutions
Magnesium alloys	Chloride solutions
Titanium alloys	Most halidium alloys
Zirconium alloys	$FeCl_3$ or $CuCl_2$ solutions
High-nickel alloys	Aerated hydrofluoric or hydrofluorosilicia acid vapor

14.2.6.2 SCC by Hydrogen

In this book the SCC by hydrogen (HAC) type of hydrogen embrittlement is reserved for that commonly encountered in steels, although it has also been observed in Al, Ti, Ni, and possibly other nonferrous alloys, where only a few parts per million of hydrogen can cause embrittlement. This is also the type of cracking referred to above as a subclass of SCC. In many cases it can occur without residual or applied stresses. Applied stresses after introduction of the hydrogen are usually necessary for the embrittlement effect, which changes the fracture from a ductile to a brittle type as a result of the small hydrogen content. In a few cases a ductile failure has been observed accompanied by a reduction in ductility. Some have speculated that an applied stress, subsequent to or during the introduction of hydrogen, allows dislocations to move, and during their movement they collect hydrogen and dump it at interfaces such as grain boundaries and phase boundaries, and as such, stress is required. Stress may also enhance the diffusion of hydrogen ions. The belief that stress was required is probably the result of early tests where stress applied in hydrogen atmospheres produced curves such as that shown in Figure 14.13. This curve shows stress vs. time to fracture, and where the stress levels off, the material appears to be immune to the effects of hydrogen. Because of the close similarity to the *S–N* curves of fatigue (Figures 2.19 and 2.20), this phenomenon has been labeled *delayed fatigue*. This term is misleading and should eventually disappear from the literature.

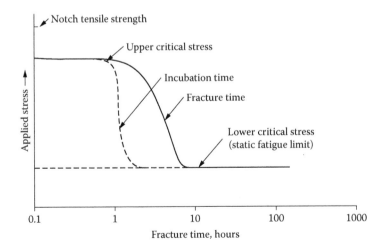

FIGURE 14.13 Schematic representation of delayed failure characteristics of a hydrogenated high-strength steel. (From A. R. Trioano, *Trans. ASM*, 52, 54, 1960. With permission.)

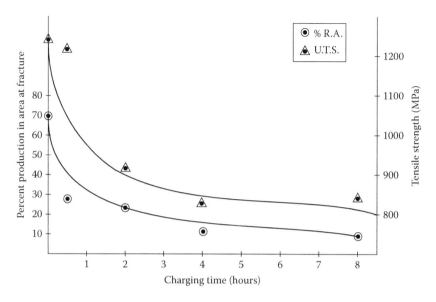

FIGURE 14.14 Effect of charging time on strength and ductility of a PH 13-8 Mo stainless steel (H 1050 condition). (From G. T. Murray et al., *Corrosion*, 40, 147, 1984. With permission.)

There is ample evidence, however, that in many cases stress is not required for HAC, other than that for observing the effect. A few isolated cases have been reported where cracking occurred from the introduction of the hydrogen ions alone (not H_2 gas), and many results can be found in the literature where cracking and fracture of previously hydrogen-charged specimens occurred when stresses as low as 30% of the yield stress were subsequently applied to hydrogen-charged specimens (Figure 14.14). Thus, the hydrogen ions have drastically reduced the fracture stress. It is unlikely that stress-induced hydrogen migration plays a significant role at such low stresses. For low hydrogen contents the ductility is reduced, while at higher hydrogen contents both strength and ductility are severely reduced. Figure 14.15 shows a brittle ring in a PH stainless steel tensile test specimen that had been electrolytically charged with a few parts per million of hydrogen prior to testing to failure. The boundary between the brittle and ductile failure regions in Figure 14.15 is that where the concentration of hydrogen was below some very small critical level.

Scanning eletron microscope (SEM) fractographs (Figure 14.16) show brittle intergranular failure in the brittle ring and a dimpled ductile microvoid coalescence core in the center of the specimen. In a few cases the brittle fracture was transgranular, possibly owing to hydrogen trapping at carbide–matrix interfaces. The crack, of course, will follow the path of least resistance, and this is often the grain boundary. In many cases it is difficult to distinguish the anodic dissolution form of SCC from HAC since both often show branched cracking (Figure 14.17) and both can be either intergranular or transgranular. The classic form of anodic

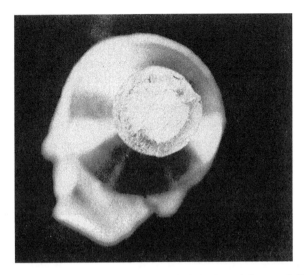

FIGURE 14.15 Brittle ring on the exterior of PH 13-8 Mo stainless steel (H1050 condition) that was fractured in a tensile test after being electrolytically charged with hydrogen at a current density of 9 mA/cm^2 for 4 h (10×). (From G. T. Murray et al., *Corrosion*, 40, 147, 1984. With permission.)

dissolution SCC, such as that observed in brass, often shows more branched cracking than does that caused by hydrogen. The separation of the two mechanisms is further complicated by our inability to ascertain what chemical reactions may be occurring at the crack tip. The chemistry at the crack tip may be entirely different from that taking place on the external surface of the metal part.

There are many sources of hydrogen that can cause HAC, including H_2S, water (particularly salt water), surface cleaning acids, electroplating, welding, and contact with hydrogen gas. Surface protection, particularly oxides on steel, will retard hydrogen penetration. Another preventive measure is to determine the critical stress level and ensure that this stress is not exceeded. Heating the metal after hydrogen exposure (baking) will remove a large percentage of the hydrogen, but in many cases this operation will not prevent HAC.

14.2.6.3 Hydrogen Damage (Other Than Stress Corrosion Cracking [SCC])

Certain forms of hydrogen damage are definitely distinguishable from SCC. These hydrogen effects are:

> *Hydrogen attack:* This type of embrittlement is a result of hydrogen combining with carbon to form a gas, usually methane. The gas pressure builds up to the point at which it can cause cracking of the metal. Surface decarburization by the hydrogen occurs at temperatures above 540°C, while internal decarburization can occur at temperatures as low as 200°C.

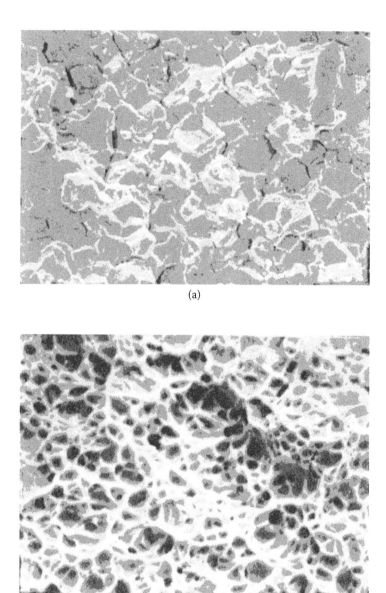

(a)

(b)

FIGURE 14.16 SEM fractographs of (a) brittle hydrogen failure taken from the brittle ring in the specimen of Figure 14.15 and (b) ductile failure taken from the central portion of the specimen in Figure 14.15. (From G. T. Murray et al., *Corrosion*, 40, 148, 1984. With permission.)

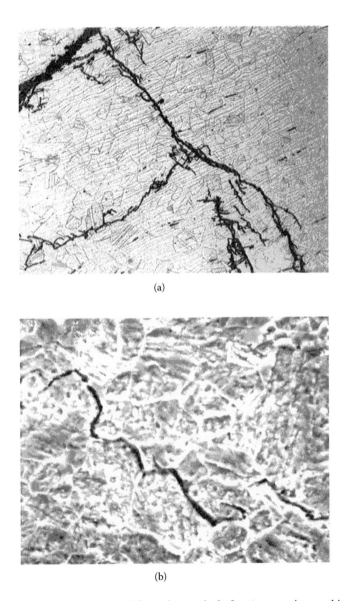

(a)

(b)

FIGURE 14.17 (a) Branched crack formation, typical of stress corrosion cracking, found in an austenitic stainless steel bracket on a sailboat. Both transgranular and intergranular cracking are evident. (Courtesy of Eric Willis, Cal Poly). (b) Intergranular cracking in a hydrogen-charged PH stainless steel.

Closely related to this effect is the decreasing solubility of H_2 in molten metal with decreasing temperature. Upon solidification, H_2 pressure builds up sufficiently to cause cracks to form. These cracks were observed many years ago, and because of their shiny fracture surfaces, visible without microscopes, were called *fisheyes, flakes*, and an assortment of similar names.

Hydrogen blistering: This term covers a multitude of causes. Hydrogen gas blistering has been found to occur in plated surfaces and in the petroleum industry, where it is caused by H_2S reacting with the metal. In the latter case it would be called HIC.

Hydride formation: Hydrogen cracking via brittle hydride formation occurs in metals that readily form hydrides (e.g., Ta, Ti, Zr, and Nb). Hydrogen concentrations greater than 100 ppm are usually required. The cracks form in the brittle hydrides rather than the metal matrix.

In summary, there are many ways in which hydrogen can enter and embrittle or cause cracks, blisters, and so on to form in metals. The vocabulary of hydrogen damage is large and sometimes overlapping and confusing. Strain rates and temperature have an effect. Some mechanisms are better understood than others. The HE (HAC) caused by a few parts per million hydrogen content is believed to be related to some type of decohesion (i.e., the energy to form two new surfaces is reduced by the presence of hydrogen ions).

14.2.7 MICROBIOLOGICAL CORROSION

This subject has been given more attention recently, although its intrusion into the field does not present any real new form of corrosion process nor does it affect any of the fundamental electrochemical equations stated earlier. It plays a role in two ways: (1) by the introduction into the system of new chemical entities such as acids, alkalides, sulfides, and other aggressive ions and (2) by entering into one or more of the electrochemical reactions at the surface of a part, thereby affecting the kinetics of the reaction. Hydrogen ingress has been reported to occur at some microbiological deposits. Microorganisms covering a wide range of species are encountered in corrosion problems, but in the common ones, sulfur and its compounds play a vital role. Sulfate-reducing bacteria reduce sulfates to sulfides, which appear as curved rods. They are aerobic; that is, they require oxygen for growth. The most serious corrosion problems involving microorganisms occur on iron alloys in the absence of oxygen (anaerobic conditions). Also, some rather serious corrosion of concrete pipes, cooling towers, and building stones have been associated with this type of corrosion. Prevention methods include inhibitors, which kill the organisms; elimination of stagnant conditions where growth proceeds at a rapid rate; and protective coatings. A typical form of this corrosion is shown in Figure 14.18.

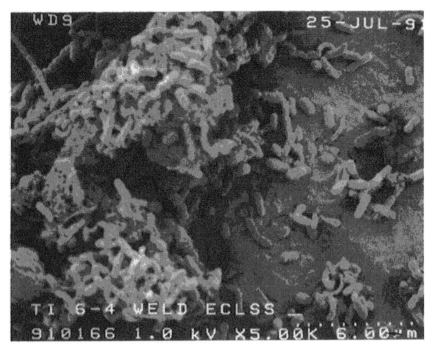

FIGURE 14.18 Microbiological corrosion: deposits of rod-shaped sulfate-reducing bacteria on titanium after exposure to tap water. (Courtesy of Dan Walsh, Cal Poly.)

14.2.8 MANAGEMENT AND PREVENTION OF CORROSION

Corrosion in the context of an engineering design is not something that is totally preventable if the device is to satisfy its intended use. Sometimes a strategy to manage the corrosion is preferred to total elimination for economic and engineering reasons. Corrosion management/prevention strategies fall into four general categories.

1. Change the Material
2. Change the Design
3. Change the Environment
4. Protect the Structure

A simple example using a galvanic cell can illustrate how each of the above can be implemented in Figure 14.20.

1. The driving force for a galvanic cell is the potential difference between different metals. By selecting material nearer to each other on the electromotive series thus providing a smaller electrochemical difference that can be used to reduce the rate of corrosion.

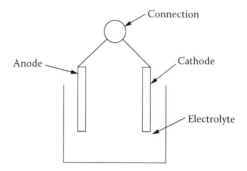

FIGURE 14.19 Simple galvanic cell.

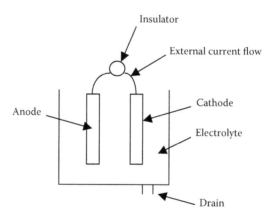

FIGURE 14.20 Schematic of simple galvanic cell.

2. A design change could be to elimination or cutting off the flow of current by insulating the contact between dissimilar metals.

3. Adding an inhibitor to the electrolyte or draining the tank can change the environment.

4. Coating the anode and cathode with a nonconductive material (paint) will again disconnect the system from the corrodible metals.

These examples are used to illustrate the concept but by no means are the only approaches to take.

Substituting or sacrificing a metallic anode in place of the structure is a common method of protecting many systems. This scheme is called *cathodic protection* and is used on buried structures such as pipelines, ship hulls, home water heaters, and automotive trim.

Cathodic protection is accomplished by attaching an extremely anodic (active) metal such as zinc or magnesium on the metal surface at close intervals. Cathodic protection uses the more anodic material to supply electrons to the metal structure to be protected. These electrons flow to the cathode, allowing the electrical circuit to be completed without drawing electrons from the anodic metal that is being protected. Cathodic protection of a buried steel pipeline using a magnesium *sacrificial anode* is depleted in Figure 14.19. In this pipeline the magnesium corrodes preferentially to the steel. Cathodic protection using sacrificial anodes has been applied to ship structures and hot-water tanks, to name a few applications. To determine the size and spacings of the sacrificial anode, the energy content and anode efficiency must be known. For magnesium, the most common sacrificial anode metal (approximately 10 million pounds per year), the theoretical energy content is 100 A-h/lb with 50% efficiency. The number of pounds of Mg required to provide 1 A for a year can be computed as follows:

$$\text{lb Mg/A-yr} = (8760 \text{ h/yr})/(500 \text{ A-h/lb}) = 17.52$$

Equations have been developed for protection of underground structures based on soil resistivity, number of anodes, anode spacing, and size (see the book *Corrosion and Corrosion Protection* by Schweitzer in Suggested Reading). The resistance of a system in terms of the variables above is expressed by the Sunde equation, as follows:

$$R = \frac{0.005P}{NL}\left[2.3\log\left[\frac{8L}{d-1}\right] + \frac{2.3}{S(\log 0.656N)}\right] \qquad (14.17)$$

where
  R = resistance of system (Ω)
  P = soil resistivity ($\Omega \cdot$ cm)
  N = number of anodes
  L = anode length (ft)
  d = diameter of anode (ft)
  S = anode spacing (ft)

Another protection technique, somewhat analogous to the sacrificial anode method, is to impress a current with an external dc power supply connected to an underground tank or pipeline. The negative terminal of the power supply is connected to the metal to be protected, and the positive terminal, to an inert anode such as graphite, located some distance away from the structures to be protected. For underground structures, the resistivity of the soil will determine the required applied voltage. If the leads are properly insulated, current passes to the metallic structures and suppresses the corrosion reaction that would otherwise take place by metallic ions of the protected metal going into solution and leaving electrons behind. This reverses the reactions stated in Equation 14.1. To provide cathodic protection, a current density of a few milliamperes per square foot is required. Insulating coatings are also helpful. One magnesium anode is capable of protecting about 100 ft of bare pipeline compared to about 5 miles of coated pipeline.

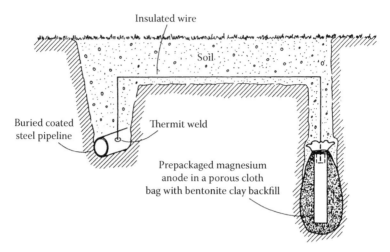

FIGURE 14.21 Sacrificial anode of magnesium used to protect a buried steel pipeline. (Courtesy of R. H. Heidersbach, Cal Poly.)

For prevention of giant galvanic corrosion cells where large pieces of dissimilar metals are placed in contact with each other, a number of possible approaches exist. First, try to avoid dissimilar metal contact by design, or at least minimize it by choosing two metals or alloys that are close together in the galvanic series. When it is not possible or practical to avoid high anodic–cathodic combinations, try to find a way to insulate these two metals (i.e., prevent their contact). Polymer washers, couplings, and bolt heads have been used for this purpose. Where metallic junctions cannot be avoided, cover the junction with paint or some other protective medium that will assist in preventing the electrolyte from contacting the dissimilar junction.

Crevice corrosion must be approached primarily from a design perspective. Avoiding the creating of a device is impossible for some means of fabrication of components. In the automotive industry, the hem flange at the bottom of the door panel is a classic example. The inner and outer door panels are folded over on themselves to form a mechanical bond along the bottom edge of the door. This configuration creates a water catching lip at the top edge of the fold, which, if not properly sealed with mastic or a paint film, can develop into a cosmetically unattractive line of corrosion visible when the door is open.

Some simple approaches to avoiding crevice problems can be seen in Figure 14.22A and B; we see where a simple rearrangement of a lap joint can reduce the potential for trapping liquid in a restricted (crevice) area. In Figure 14.22C liquid dripping down the side of the tank (rainwater) can collect in the region between the tank bottom and the concrete pad resulting in a crevice type zone. In Figure 14.22D a smaller pad or a pad which slopes away from the tank bottom will eliminate the collection of rainwater and eliminate the crevice zone.

It is often more efficient to provide for good drainage rather than creating a perfect seal.

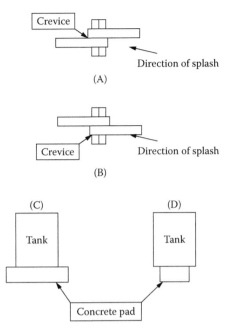

FIGURE 14.22 Example of crevice corrosion prevention design strategies.

Pitting corrosion prevention must be approached in the same manner as that for crevice corrosion. Some materials pit more than others (e.g., austenitic stainless steels, aluminum, and titanium), so material selection is important. Aluminum is particularly susceptible to pitting in chloride environments. The Cl⁻ ion tends to break down the protective passive film. Both hydrogen and chloride ions promote dissolution of metals and may start a pit that feeds on itself by collecting more H⁺ than Cl⁻ ions. These environments must not come into contact with the bare metal. The selection of the duplex stainless steels is a possible solution that should be seriously considered for pitting environments.

Sensitization of stainless steels can be minimized by selection of a stabilized 321 or 347 or a low-carbon stainless steel. For other forms, consult tables (see the Suggested Reading) that show the metal and the media in which intergranular attack can be expected to occur.

For many years the most publicized and touted method for prevention of SCC has been a *stress-relief anneal*, which is simply an anneal or heat treatment of a cold-worked metal that relieves much of the elastic stresses resulting from the cold work without sacrificing most of the strength intentionally introduced by the cold-working operation. In theory this causes a realignment or disentanglement of the dislocations introduced by, and responsible for, the increase in strength caused by the cold-working operation. Whether or not the theory is correct, a stress-relief anneal does appear to reduce the susceptibility of the material to the SCC of brass. In the case of the HE form of stress corrosion cracking, it has been observed that high dislocation density promotes cracking.

In stress corrosion SCC, the two major components are a tensile stress and a material/environment combination that favors cracking. The most successful approach is to avoid metals and environment combinations which are known to be suscrptible, as it is often more difficult to redesign a component to be in neutral or compressive loading.

Inhibitors reduce the rate of anodic oxidation or cathodic reduction, or both processes. Inhibitors are physically or chemically absorbed, the latter involving transfer or sharing of charge. Inhibitors can be broken down as passivation, organic, or precipitation inhibitors. Passivation inhibitors are the most effective and usually consist of chromate or nitrite substances that are absorbed into metal surfaces in the presence of dissolved oxygen. Chromates are inexpensive and widely used in automobile radiators, engine blocks, and cooling towers. A wide variety of organic inhibitors (e.g., sodium benzoate, ethanoline phosphate [used in ethylene glycol cooling systems]) are available, which are absorbed onto the surface and provide a protective film, thus retarding dissolution of the metal in the electrolyte. Precipitation inhibitors provide a protective film in the form of a solid precipitate in the metallic surface. Sodium phosphate will cause precipitation of a calcium or magnesium orthophosphate onto a metal surface.

14.2.9 OTHER TYPES OF METALLIC DEGRADATION

Metals are attacked by many environments that result in deterioration of their desirable properties, even in the absence of galvanic cell formation. The most notable of these effects are those caused by oxygen. Oxidation can occur by the oxygen penetrating a porous, previously formed oxide film such that the oxidation reaction at the metal–oxide interface continues. Oxygen may also diffuse atomically through a nonporous oxide film and react at the metal–oxide interface, a somewhat slower process of oxidation. Similarly, the metallic cations may also diffuse through the nonporous oxide layer and react with the air–oxide or oxygen–oxide outer surface.

The oxide may form a thin protective layer, as in the passivation process, or continue to grow in thickness until it flakes off the metal, exposing a new fresh metal surface, a process that continues until the metal is entirely consumed. The Pilling–Bedworth ratio R, which was first presented in the *Journal of the Institute of Metals* in 1923, can be used to predict the tendency to form a protective oxide coating. R is expressed as follows:

$$R = \frac{Md}{amD} \qquad (14.18)$$

where
  M = molecular weight of oxide of formula M_aO_b
  d = metal density
  m = metal atomic weight

For a value of $R > 1$, the oxide tends to be protective. In addition to the passive metals mentioned earlier, Cu, Mn, Be, and Co tend to form protective oxides according to the Pilling–Bedworth criterion.

The reactivity of metals to other environments is far too broad a subject to be treated here. The interested reader should consult the ASM *Atlas* listed in the Suggested Reading.

14.3 ENVIRONMENTAL DEGRADATION OF POLYMERS

Polymers are inert in many aqueous solutions and as such are frequently used to replace metal parts. Typical examples include ABS polypropylene, and PVC as piping, tanks, scrubbers, and columns in process equipment. Teflon has been used in heat exchangers, and nylons and acetals for small parts and gears. In the chemical industry polymers have been used alone and as metallic coatings, the latter being used where more strength is desired. Because of all of these applications of polymers, their reactions with certain environments must be considered.

Chemical attack of polymers can be broadly classified as follows:

- Degradation of a physical nature such as permeation or solvent action
- Oxidation, the breaking of bonds
- Hydrolysis–bond splitting by the addition of water (e.g., attack of ester linkages)
- Radiation

One could include thermal degradation to this list, but since this is not a chemical attack, it will not be included in this chapter (see Chapter 7). The theory involved in all such attacks is beyond the scope of this book. Suffice it to say that softening, embrittlement, crazing, dissolution, and blistering are typical results. There is a significant, and in some cases vast, difference in the reaction of polymers to specific environments.

The most chemically resistant polymer developed to date is tetrafluoroethylene (Teflon or Halon). It is practically unaffected by acids, alkalies, and organics at temperatures to 260°C (500°F). It is an excellent material for gaskets, O-rings, and seals. It has also been used as linings for tanks and ductwork. Polyethylene is a low-cost polymer that has excellent resistance to a wide variety of chemicals. Because of its low strength and low cost, it is used in large quantities as thin sheet liners in metal containers and for agricultural purposes. It is temperature limited and is used primarily at ambient temperatures. PVC materials have excellent resistance to oxidizing acids, other than nitric and sulfuric, and all nonoxidizing acids. Many of the synthetic rubbers have good resistance to nonoxidizing weak acids but not to oxidizing acids. Epoxy resins have excellent resistance to weak acids and to alkaline materials but poor resistance to strong oxidizing acids. Many polyesters, particularly bisphenol polyester, have good general chemical resistance. The bisphenol polyesters have been the workhorses of the chemical industry. These materials are easy to fabricate and have superior acid resistance and moderate resistance to alkaline and most solvents, such as the alcohol, gasoline, and ethylene glycol. The chemical resistance and applications of the more common and commercially used polymers are listed in Table 14.4.

14.4 ENVIRONMENTAL DEGRADATION OF CERAMIC MATERIALS

Since ceramic materials are basically insulators, they do not experience the type of uniform corrosion experienced in metals because of the lack of a conductor, one of the four requirements for the formation and existence of a galvanic cell. However, other forms of chemical attack are possible. Almost all materials can be attacked by some environments. Hydrofluoric acid is used to etch glass and other ceramic materials. The degradation of ceramics by their environment has not been categorized to the same extent as that for metals and polymers. Many ceramic materials decompose at high temperatures, and in many cases this decomposition may be accelerated or altered by the environment to which they are subjected.

It is more common in ceramics to group corrosion and other environmental effects as *dissolution corrosion,* which may include some of the same metallic characteristics of grain boundary or stress corrosion and even galvanic cell corrosion. Ceramics contain many components (e.g., bonding agents, mineralizers, grain growth inhibitors, etc.), many of which are capable of providing a conductive path for electrons to flow in a galvanic cell reaction. Corrosive attack often centers around minute quantities of such constituents. Hot pressing, reaction sintering, and chemical vapor deposition have removed the need for many of these minor constituents and, hence, there is less concern about corrosive effects.

Considerable work has been done on the galvanic corrosion of refractories containing a glassy phase. Refractories that establish an electrical potential relative to glass of around 0.5 V are fairly resistant to this type of corrosion, but when this potential exceeds 1.0 V, the corrosion resistance is poor. In many ceramics a glassy phase surrounds all other phases. Refractories that have a negative potential relative to glass should not be used.

The corrosion of glasses by atmospheric conditions is referred to as *weathering.* It is basically a result of a water vapor reaction and is believed to be related to tensile stresses set up by an ion exchange of the alkali by hydrogen ions. Soda–lime silica glasses lose considerable strength owing to atmospheric corrosion by water vapor. Glass corrosion by liquids is much more frequent than that by vapors. The release of PbO and other toxins is currently of much interest and is being studied intensely. Low-pH solutions release Pb more slowly than the higher-pH liquids. Glasses are soluble in liquids over a wide range of pH values from acids to bases, including water to a slight extent. The high-silicate glasses such as the borosilicates and alumina silicates of about 98% SiO_2 content have excellent corrosion resistance compared to the somewhat water-soluble sodium silicates. In the soda–lime glasses the H^+ ion from the water replaces an alkali ion that goes into solution while the $(OH)^-$ ion destroys the Si–O–Si bonds forming nonbridging oxygens. The reaction feeds on itself since the nonbridging oxygen reacts with the H_2O molecule forming another nonbridging oxygen–hydrogen bond plus another $(OH)^-$ ion. The hydrolysis of glasses determines the service life from weathering and corrosion. The rate of hydrolysis of the alkali glasses increases according to the modifier elements in the following order: Cs > K > Na > Li.

Glass fibers are much more subject to corrosion than are the bulk glasses because of the higher surface-to-volume ratio. Humid environments lower the strength of E-glass fibers. For this reason, many glass fibers are produced with a polymer coating.

The resistance of ceramics to chemical attack can be related to bond strength. Whereas weakly bonded ionic salts, nitrates, oxalates, chlorides, and sulfates are dissolved in water and weak acids, the more strongly bonded Al_2O_3, Si_3N_4, and $ZrSO_4$ are resistant to many types of chemicals, including aqueous solutions, strong acids, bases, and liquid metals and hence can be used as crucible materials. Many of the oxides are used for furnace linings, metal refining, and glass processing. These oxides are also resistant to acids and bases and are used either to replace metals or as a coating for them.

Elevated-temperature environmental reactions of ceramics are of more concern now since those materials are being considered seriously for use in gas turbines. Oxides and silicates are more stable in high-temperature oxide environments than are the carbides, nitrides, and borides. Oxygen pressures of 1 mmHg will form a protective layer of SiO_2 on SiC and Si_3N_4, rendering them passive materials. It is essential that the latter be free of impurities, or else low-melting silicates many be formed. Porosity is also a factor since oxidation can occur along interconnected pore channels. For heat engines, Si_3N_4 and SiC appear to be the leading candidate materials since they possess good stability, thermal conductivity, low thermal expansion, and good high-temperature strength. In addition, they can be protected to some extent by SiO_2 layers.

14.5 CORROSION OF COMPOSITES

Steel rusts, polymers swell in moisture and change color and strength in sunlight, rubber ages faster under the effects of ozone, wood splits because of loss of water, and ceramics are attacked by a host of chemicals. Since the properties of composites are a function of their constituents, one could say that the environmental effects would follow the rule of mixtures. To a large extent, this is true. Any of the environmental effects mentioned previously on the individual constituents would be expected to have similar effects on these respective constituents, provided that there exists a path whereby the damaging ingredients can reach the individual constituents. For example, it is now generally accepted that steel in concrete is protected by the passivity provided by the highly alkaline nature of

TABLE 14.4
Summary of the Chemical Resistance of the Common Polymers and Their Applications

Type of Plastic	Normal Upper Service Limit		Typical Areas for Use
	°C	°F	
Polyvinyl chloride	60	140	Piping: water, gas drain, vent, conduit, oxidizing acids, bases, salts; ducts (breaks down into HCl at high temperatures); windows plus accessory parts and machine equipment in chemical plants; liners with FRP overlay
Chlorinated PVC	82	180	Similar to PVC, but upper-temperature limit is increased
Polyethylene	60	140	Tubing, instrumentation (laboratory): air, gases, potable water, utilities, irrigation pipe, natural gas; tanks to 12-ft diameter
High-density polyethylene	82	180–220	Chemical plant sewers, sewer liners, resistant to wide variety of acids, bases, and salts; generally carbon filled; highly abrasion-resistant; can be overlaid with FRP for further strengthening
ABS	60	140	Pipe and fittings; transportation, appliances, recreation; pipe and fittings are mostly drain, waste, vent; electrical conduit; resistant to aliphatic hydrocarbons but not resistant to aromatic and chlorinated hydrocarbons; formulations with higher heat resistance have been introduced
Polypropylene	82–104	180–220	Piping and as a composite material overlaid with FRP in duct systems; useful in most inorganic acids other than halogens; fuming nitric and other highly oxidizing environments; chlorinated hydrocarbons cause softening at high temperatures; resistant to stress cracking and excellent with detergents; flame-retardant formulations make it useful in duct systems; further reinforced with glass fibers for stiffening, increases the flex modulus to 10^6 and deformation up to 148°C (300°F)
Polybutylene	104	220	Possesses excellent abrasion and corrosion resistance; useful for fly ash and bottom ash lines or any lines containing abrasive slurries; can be overlaid with fiberglass for further strengthening when required

TABLE 14.4 (Continued)
Summary of the Chemical Resistance of the Common Polymers and Their Applications

Type of Plastic	Normal Upper Service Limit		Typical Areas for Use
	°C	°F	
General-purpose polyesters	50	120	Made in a wide variety of formulations to suit the end-product requirements; used in the boat industry, tub-showers, automobiles, aircraft, building panels
Isophthalic polyesters	70	150	Increased chemical resistance; used extensively in chemical plant waste and cooling tower systems plus gasoline tanks; liners for sour crude tanks
s	120–150	250–300	Includes the families of bisphenol, hydrogenated bisphenols, brominated, and chlorendic brominated, and chlorendic types; a wide range of chemical resistance, predominantly to oxidizing environments; not resistant to H_2SO_4 above 78%; can operate continually in gas streams at 148°C (300°F); end uses include scrubbers, ducts, stacks, tanks, and hoods
Epoxies (wet)	150	300	More difficult to formulate than the polyesters; more alkaline resistant than the polyesters, with less oxidizing resistance; highly resistant to solvents, especially when postcured; often used in filament-wound structures for piping; not commonly used in ducts; more expensive than polyesters
Vinyl esters (wet, dry)	93–140	200–280	Especially resistant to bleaching compounds in chlorine-plus-alkaline environments; wide range of resistance to chemicals, similar to the polyesters; used extensively in piping, tanks, and scrubbers; modifications developed to operate continuously up to 140°C (280°F)
Furans (wet)	150–200	300–400	Excel in solvent resistance and combinations of solvents with oxidizing chemicals; one of the two resins to pass the 50 smoke rating and 25 fire-spread rating; carries about a 30–50% cost premium over the polyesters; excellent for piping, tanks, and special chemical equipment; does not possess the impact resistance of polyesters or epoxies

Source: J. H. Mallison, in *Corrosion and Corrosion Protection Handbook*, P. A. Schweitzer, Ed., 2nd ed., Marcel Dekker, New York, 1989, pp. 348–350. (With permission.)

the porewater and persists as a result of the presence of the water and oxygen. However, if the concrete fails to provide a barrier to the ingress of certain substances, the reinforcing steel bars can corrode. Carbon dioxide and chloride can cause depassivation of the steel. One of the problems is preventing crack formation in the concrete that allows such substances to enter.

In the more sophisticated composites such as the organic matrix–fiber composites, other problems arise. The most commonly feared is the deterioration and breaking of interfacial bonds. Water can cause swelling and plastization of resins and thereby permit the ingress of water or other liquids by capillary action along the fiber–matrix interface. Deterioration of PMCs by moisture or chemical absorption is of great concern. The strength of these composites can be reduced if the chemical species in the environment attack the polymer, the reinforcement material, or the interface between the two. To some extent these problems can be circumvented by judicious selection of the materials used in the composite construction. Similar problems can arise in metal and ceramic matrix composites as well. MMCs are not as susceptible to moisture deterioration as are PMCs. Of more concern for MMC applications is the temperature stability of the matrix and reinforcement constituents in the environment of interest. Damage from the external environment can be minimized if the reinforcement material is not exposed. Galvanic corrosion has been observed in some MMCs, particularly Al–B, Al–SiC, and Mg–graphite. The coupling of electrochemically dissimilar materials can occur in these composites when electrolytes are available. The compatibility of certain alloys and intermetallic compounds with hydrogen atmospheres is also of concern. The state of the art in many of these composites has not advanced to the point where many corrosion data are available. For the resin composites, a number of studies have been made, but for now the corrosion of composite materials will be considered beyond the scope of this book.

DEFINITIONS

Anode: Electrode in a *galvanic cell* that is attacked by the electrolyte, thereby removing ions and producing electrons.

Cathode: Electrode in a galvanic cell that consumes electrons by a variety of chemical reactions.

Composition cell: Galvanic cell composed of two different materials.

Concentration cell: Type of galvanic cell that establishes anodes and cathodes because of a difference in concentration of ions on or near the metal surface. The *oxygen concentration cell* is the best known of this type.

Crevice corrosion: Corrosion occurring where stagnant fluids collect in crevices. The corrosion involves a number of reactions depending on the materials and the environment.

Electrolytes: Liquids in galvanic cells that act as media for the combination of the ions resulting from anode and cathode reactions.

Electromotive force series (emf):   Ranking of the elements according to their tendency to corrode.

Erosion–corrosion Combination of corrosion and fluid flow, where the latter may accelerate the former.

Galvanic series: Ranking of metals and alloys according to their corrosion tendency.

Hydrogen attack: Type of hydrogen damage where a gas, often methane, is generated via a chemical reaction of hydrogen with Fe_3C. The gas builds up sufficient pressure to cause cracking.

Hydrogen electrode: Electrode against which metals are measured to establish their position in the emf series. It consists of the bubbling of hydrogen into 1-M solution of H^+ ions. Its voltage is arbitrarily set at zero.

Hydrogen embrittlement: Form of hydrogen damage and SCC where a few parts per million of hydrogen cause decohesion. Sometimes called *hydrogen-assisted cracking* (HAC).

Hydrogen-induced cracking: Type of cracking usually associated with H_2S in petroleum products that produces hydrogen gas sufficient to cause blistering.

Inhibitors: Complex ion formation on the surface of a metal that protects it in some way from the electrolyte. (*See also* Passivation.)

Intergranular corrosion: Type of anodic dissolution of the grain boundary. A form of *stress corrosion cracking.*

Nernst equation: Equation that expresses the effect of metal ion concentration on the standard emf of a metal half-cell, the latter being set at a 1-M solution of the metal ions.

Passivation: State existing on the metal surface where the corrosion rate does not change with increasing electrode potential. It is a characteristic of metal surfaces that usually occurs when thin adherent metal oxides are formed, although inhibitors are also believed to affect passivity by being absorbed into the metal surface.

Pitting corrosion: Localized corrosion that forms a pit where the surface diameter is less than the pit depth.

Polarization: *Activation polarization* affects the corrosion rate by a rate-controlling reaction at the metal–electrolyte interface. *Concentration polarization* affects the corrosion rate by a rate-controlling step in diffusion within the electrolyte.

Sacrificial anode: Highly anodic metal, such as Mg, that is attached to a less anodic metal and corrodes in preference to this metal.

Selective leading: Preferential removal of one element from a solid alloy by a corrosion process.

Sensitization: Precipitation of chromium carbides at the grain boundaries in austenitic stainless steels and other high-chromium-content alloys that render them sensitive to galvanic corrosion.

Stray corrosion currents: Corrosion by current flow other than the intended path; often found in soils.

Sulfide stress cracking: Type of hydrogen embrittlement in which the H_2S is the principal source of hydrogen.

QUESTIONS AND PROBLEMS

14.1 What is the maximum voltage that one could obtain from any two pure metals arranged as two electrodes in a galvanic cell?

14.2 What chemical reaction always occurs at a metallic anode in a galvanic cell?

14.3 Write balanced chemical equations for three types of reactions that could occur at the cathode of a galvanic cell. What types of electrolytes are involved in each case?

14.4 What is the difference between the electromotive force and the galvanic series?

14.5 What are the four requirements for the operation of a galvanic cell?

14.6 Describe and give examples of the following corrosion types: (a) uniform corrosion; (b) crevice corrosion; (c) pitting corrosion.

14.7 Determine the electrode potential with respect to hydrogen for a copper electrode in a solution containing 10 g of Cu^+ ions/liter.

14.8 A TiO_2 layer that forms on a sheet of titanium passivates the metal. How thick should this layer be to limit the current density to 10^{-4} A in a cell that produces 4 V? Assume the electrical resistivity of TiO_2 to be $10^{18}\ \Omega \cdot m$.

14.9 Why is concentration polarization seldom rate controlling for oxidation reactions?

14.10 Gold and copper both have a valency of +1 and both are good conductors of electricity. Why is copper so much more aggressively attacked in ambient environments, whereas gold is essentially stable? Would iron be more reactive than copper?

14.11 Suppose it was necessary in a certain construction to connect a 1020 steel to a Ni–Co Monel K alloy. Suggest a method to prevent the formation of a galvanic composition cell.

14.12 What is meant by the term *sensitization* of austenitic stainless steels? How can this phenomenon be avoided in the welding of such steels?

14.13 Cd shows –0.403 V and nickel –0.250 V on the emf series. Why is the voltage of this galvanic cell +0.146 V rather than –0.403 – (–0.250) = –0.153 V?

14.14 What is the corrosion rate in grams per second for a Zn plate in combination with a Cu plate of the same area in a galvanic cell that is producing 1 A of current flow?

14.15 What is the meaning of a standard half-cell oxidation–reduction potential?

14.16 Copper concentrations of 0.05 and 0.1 M occur in an electrolyte at 25°C at opposite ends of the copper wire. Which end will corrode?

14.17 Within a mild steel in a humid atmosphere, what regions or constituents will become anodic with respect to the matrix material? What regions or constituents will be cathodic with respect to the matrix?

14.18 What is the purpose of galvanizing steel? With respect to corrosion prevention, how does galvanizing differ from tin plating?

14.19 Define the term *passivity* and how it is measured. Name four passive-type metals and the film that causes their passivity.

14.20 Do inhibitors causes metals to become passive? Explain how they differ from oxide films.

14.21 List and describe four types of hydrogen damage.

14.22 In what ways are SSC and hydrogen effects similar? In what ways are they different?

REFERENCE

1. *Corrosion*, American Society for Metals (ASM), *Metals Handbook*, Vol. 13, ed., ASM International, Metals Park, OH, 1998.

SUGGESTED READING

American Society for Metals, *Metals Handbook*, Vol. 13, 9th ed., ASM International, Metals Park, OH, 1987.

Fontana, M. G., *Corrosion Engineering*, 3rd ed., McGraw-Hill, New York, 1986.

Jones, D. A., *Principles and Prevention of Corrosion,* Macmillan, New York, 1992.

McEvily, A. J., Jr., Ed., *Atlas of Stress-Corrosion and Corrosion Fatigue Curves,* ASM International, Metals Park, OH, 1990.

Schweitzer, P. A., Ed., *Corrosion and Corrosion Protection Handbook,* 2nd ed., Marcel Dekker, New York, 1989.

Schweitzer, P. A., Ed., *Corrosion Resistant Tables,* 3rd ed., Marcel Dekker, New York, 1991.

15 Materials and Process Selection

15.1 INTRODUCTION

Engineers apply the principles of science to benefit mankind. The most common way to do this is through design. Materials selection is key to design. The purpose of this chapter is to present some basic methods that you can apply throughout your career. It is not possible to give you a "one-size-fits-all" approach; every situation is somewhat different. For this reason only generic examples will be presented. The material selection decisions presented in this chapter therefore are not to be considered as recommendations for design decisions. Material property data presented in this chapter should not be used for reference. The process by which the design decisions are made is the focus of the chapter.

There are two approaches one can take regarding materials selection and design. The first is to design with specific materials. This method is inadvertently presented in many of the introductory engineering courses. It is not very practical. Consider that one is designing a simple tensile support and the following two design criteria (assume that all factors of safety are included) are critical:

- It is required to support a load of 13.8 kN.
- The length of the support before load is 125 cm, and the maximum length under load is 125.8 cm.

With these criteria, any material can be used. The required cross-sectional area will vary and may be impractical for many materials, but with these two criteria any material can be used.

The minimum cross-sectional area based on strength is

$$\sigma \equiv \frac{F}{A_0};$$

$$A_0 \equiv \frac{F}{\sigma} \tag{15.1}$$

The minimum cross-sectional area based on modulus is

$$\varepsilon \equiv \frac{\Delta L}{L_0};$$

$$\Delta L = L_0 \varepsilon;$$

$$\Delta L = L_0 \frac{\sigma}{E};$$ (15.2)

$$\Delta L = L_0 \frac{F}{A_0 E};$$

$$A_0 = \frac{F L_0}{E \Delta L}$$

The required cross-sectional area will depend upon both the yield strength and the modulus of the material. One can use a spreadsheet (shown in Table 15.1) to determine the suitability of various materials. The spreadsheet was set up to do the following:

- Calculate the required cross-sectional area based on strength according to Equation 15.1.
- Calculate the required cross-sectional area based on modulus according to Equation 15.2.
- Determine the maximum of the two required cross-sectional areas and based on this determine the required diameter.

Once the minimum dimensions are calculated, one would need to determine which materials are suitable. For example, if the maximum diameter of the tensile rod were 2.5 cm (approximately 1 in), then both polymers could be eliminated from consideration.

15.2 GENERAL PHILOSOPHIES

The method used to determine the best material for the tensile rod application was inefficient. Rather than determining the required dimension, it would have been more practical to determine required material properties. Assume that the maximum diameter is 2.5 cm (the corresponding cross-sectional area is 4.9 cm²). One can then determine the required strength and modulus of the material.

The required strength is

$$\sigma = \frac{F}{A_0} = \frac{13.8 \times 10^3 \, N}{4.9 \times 10^{-4} \, m^2} = 28.1 MPa$$ (15.3)

TABLE 15.1

Example of a Spreadsheet Used to Determine the Suitability of Various Metals

Material	Yield Strength (MPa)	Modulus (GPa)	Required Diameter (cm)
1020 Steel (annealed)	395	197	0.667
1050 Steel (annealed)	636	197	0.526
1050 Steel (540°C Temper)	876	197	0.448
1050-O Aluminum	28	62	2.505
1050-H16 Aluminum	125	62	1.186
2014-O Aluminum	97	62	1.346
2014-T4 Aluminum	290	62	0.778
Alumina	172	310	1.011
Silicon Nitride	207	165	0.921
ABS Polymer	30	2	3.70
PBT polymer	25	0.5	7.410
ABS/Alumina Composite	115.2	186.8	1.235
ABS/Si$_3$N$_4$ Composite	136.2	99.8	1.136
PBT/Alumina Composite	113.2	186.2	1.246
PBT/Si$_3$N$_4$ Composite	134.2	99.2	1.144

The required modulus can be calculated from Equation 15.2:

$$E = \frac{FL_0}{A_0 \Delta L} = \frac{(13.8 \times 10^3 \, N)(125 \times 10^{-2} m)}{(4.9 \times 10^{-4} m^2)(0.8 \times 10^{-2} m)} = 4.4 GPa \qquad (15.4)$$

Based on the values determined in Equation 15.3 and Equation 15.4, PBT polymer and ABS polymer can be eliminated from consideration. 1050-Aluminum can still be considered because it was assumed in this example that the design criteria included all factors of safety.

It is not yet possible to make a materials selection decision. It is very unlikely that only two material properties will dictate a design decision. Even if this were the rare exception, cost must be considered. It is also very unlikely that only 11 candidate materials exist. In this case three steel alloys, four aluminum alloys, two ceramics, and two polymers. An engineer has thousands of materials to consider. The purpose of this section is to give you some strategies that can be employed to better select materials.

Consider that instead of designing a tensile rod support, one is designing a simple beam to support a point load at its center as shown in Figure 15.1. The following primary design criteria apply:

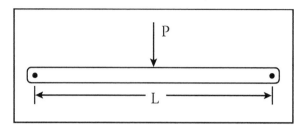

FIGURE 15.1 Simple beam support.

- The beam must support a load of 75 kN.
- The beam must be between 1.2 and 1.45 m in length.
- The deflection under its own weight must be minimal.
- The width of the beam must be between 7.5 and 10 cm, and the thickness of the beam must be between 1.8 and 2.5 cm.

We will use a combination of techniques to identify the best materials from the list considered earlier.

15.2.1 Ashby's Method

A method advanced by Michael Ashby allows for one to screen based on a set of two or more primary criteria. This is most commonly presented as a graphical method. Three material properties are of interest: strength, modulus, and density. The greater the strength, the greater the load that can be supported. To consider deflection under the beam's own weight, one must consider that deflection will decrease as the ratio of modulus to density increases. Thus, the second parameter of interest is, $\frac{E}{\rho}$ where E is the elastic modulus, and ρ is the density. This term is sometimes called the specific modulus and is considered a *lumped parameter*, as it contains more than one material property. Ashby's approach is to plot each of the two parameters on a graph. Such a plot is shown in Figure 15.2. Note that there are three lines shown on the plot. The upper line assumes that the modulus-to-density ratio is more important than the specific modulus, while the lower line assumes the opposite. The middle line assumes that both will be considered equally. If we consider both criteria equally, then nine materials remain to be considered. They are shown in Table 15.2.

This method is good for initial screenings. Note that there is a limitation in that only two material properties (or combinations thereof) are considered. In this example we did not consider specific material properties. We could have done this by only considering a specific region of Figure 15.2.

This is done by establishing some criteria, in this example based on yield strength and specific modulus. Consider the following criteria mentioned for the beam:

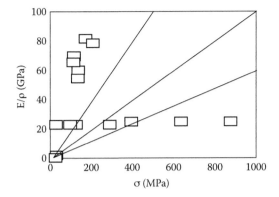

FIGURE 15.2 Ashby style plot for materials under consideration for the beam shown in Figure 15.1.

TABLE 15.2
Example of a Spreadsheet Based on Figure 15.2

Material	Yield Strength (MPa)	Specific Modulus (GPa)
1020 Steel (annealed)	395	25
1050 Steel (annealed)	636	25
1050 Steel (540°C Temper)	876	25
1050-H16 Aluminum	125	23
2014-T4 Aluminum	290	23
Alumina	172	82
Silicon Nitride	207	79
ABS/Si$_3$N$_4$ Composite	136.2	60
PBT/Si$_3$N$_4$ Composite	134.2	55

- First, the beam must support a load of 75kN.
- Second, the beam must be between 1.2 and 1.45 m in length.
- Third, the deflection under its own weight must be minimal.

By minimizing the length of the beam, the deflection will be minimized and the load-bearing capability will increase. Recall that the bending stress will increase if the beam is longer. Thus, in this example, one could specify a length of 1.2 m.

- The fourth criteria, the width of the beam must be between 7.5 and 10 cm, and the thickness of the beam must be between 1.8 and 2.5 cm, can be applied as well.

The bending stress and the deflection are inversely proportional to the moment of inertia, I. This means that I should be maximized. For a rectangular beam,

$$I = \frac{1}{12} bh^3 \qquad (15.5)$$

where b is the width of the rectangle and h is the height. According to Equation 15.5, both base and height should be maximized. This means that we can design the beam so that the width is 10 cm and the thickness is 2.5 cm. Based on this, the required yield strength is 60 MPa. According to Figure 15.2, we eliminated any material with a yield strength less than 125 MPa. Materials with a specific modulus less than 25 GPa were also eliminated. If one further specifies a deflection under weight to be 1 mm, then a specific modulus of 5.65 GPa is required. According to Figure 15.2, we eliminated any material with a specific modulus less than 25 GPa.

15.2.2 DECISION MATRICES

The candidate materials identified through the graphical method would exceed the design criteria. While the best-performing materials are identified through this method, given cost and other constraints, many of the materials identified would not be appropriate. Also note that only two criteria were considered. A decision matrix is the basis of another materials selection method.

A decision matrix is used to assign a numerical rank or figure of merit to each candidate material. However, prior to assigning a numerical rank for the material, a set of go/no-go criteria should be established. The establishment of go/no-go criteria allows one to eliminate any material from consideration which would fail to meet a required design constraint. Therefore, the first step in forming a decision matrix is to determine which properties will be considered. In the example being discussed, strength and specific modulus have already been considered. The design could require, for simplicity, that two additional criteria be considered:

1. Assume there is an advantage to using a light material. A light beam may be less expensive to install. Density will therefore be a factor.
2. Assume that the beam will be outside. The temperature during the course of a year will vary by 50°C and therefore the linear thermal expansion coefficient will be a factor.

Go/no-go criteria must be established for these properties. To minimize mass (or weight), assume that the maximum density is 5 g/cm^3, this corresponds to a maximum mass of 15 kg or 30 lbs. Thus, in this design it may be preferable for the beam to be able to be lifted by one person. To minimize the strain associated with thermal expansion, assume that the maximum thermal expansion coefficient is $2.5 \times 10^{-5}\,C^{-1}$. This corresponds to a net thermal elongation of approximately 3 cm. Thus, the four go/no-go criteria are:

1. The strength must be at least 125 MPa.
2. The specific modulus must be at least 12 GPa.
3. The density must be less than 5 g/cm^3.
4. The thermal expansion coefficient must be less than $2.5 \times 10^{-5}\,C^{-1}$.

The original list of candidate materials is shown in Table 15.3. A ~~strikeout~~ indicates that the go/no-go criteria were not met. As a result of this analysis, only four candidate materials remain: two grades of aluminum and two ceramic materials. Were we making a materials selection decision, these would have to be further investigated.

The candidate materials can be ranked for performance using a decision matrix. The first step in creating a decision matrix is to assign a point value for each material property. The maximum point value is aarbitrarily set, for this example it will be "10." To determine the "points" for the strength of each material one needs to set a value of strength corresponding to "0" points and a value of strength corresponding to "10" points. The go/no-go criteria can be used to identify the "0" point value, in this case 125 MPa. The value of strength corresponding to "10" points is set byt determining the strength above for which there is no further benefit. For this example it is arbitrarily set at 250 MPa. This is

TABLE 15.3
Original List of Candidate Materials

Material	Yield Strength (MPa)	Specific Modulus (GPa)	Density (g/cc)	Thermal Expansion Coeff. (C^{-1})
1020 Steel (annealed)	395	25	~~7.9~~	1.18×10^{-5}
1050 Steel (annealed)	636	25	~~7.9~~	1.18×10^{-5}
1050 Steel (540°C Temper)	876	25	~~7.9~~	1.18×10^{-5}
1050-O Aluminum	~~28~~	23	2.7	2.36×10^{-5}
1050-H16 Aluminum	125	23	2.7	2.36×10^{-5}
2014-O Aluminum	~~97~~	23	2.7	2.36×10^{-5}
2014-T4 Aluminum	290	23	2.7	2.36×10^{-5}
Alumina	172	82	3.8	7×10^{-6}
Silicon Nitride	207	79	2.1	3.00×10^{-6}
ABS Polymer	~~30~~	~~2~~	1	~~9.50×10^{-5}~~
PBT Polymer	~~25~~	~~0~~	1.4	~~7.00×10^{-5}~~
ABS/Alumina Composite	~~115.2~~	70	2.68	~~4.22×10^{-5}~~
ABS/Si$_3$N$_4$ Composite	136.2	60	1.66	~~3.98×10^{-5}~~
PBT/Alumina Composite	~~113.2~~	66	2.84	~~3.22×10^{-5}~~
PBT/Si$_3$N$_4$ Composite	134.2	55	1.82	~~2.98×10^{-5}~~

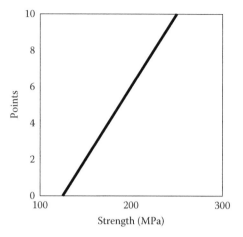

FIGURE 15.3 Determination of "Points" based on "Strength" for decision matrix.

TABLE 15.4
Example of a Spreadsheet Showing
Points Assigned Each Material
Based on Strength

Material	Yield Strength (MPa)	Material Score
1050-H16 Aluminum	125	0
2014-T4 Aluminum	290	10
Alumina	172	3.76
Silicon Nitride	207	6.56

shown graphically in Figure 15.3. The points assigned to each material based on strength are shown in Table 15.4.

Note that in Table 15.4 the material which barely met the go/no-go criteria for strength, 1050-H16 Aluminum has a point score of "0." The material which has a strength greater than 250 MPa, 2014-T4 Aluminum has a point value of "10." This can be done for all the other material properties as well.

To assign point values for the other properties, ranges must be set up. The "0" point value will be based on the go/no-go criteria, and the "10" point value will be established. For the four material properties, these willbe assigned as follows.

- For strength the "0" point value will correspond to 125 MPa, and the "10" point value to 250 MPa.

- For specific modulus the "0" point value will correspond to 12 GPa, and the "10" point value to 50 GPa.
- For density the "0" point value will correspond to 5 g/cc, and the "10" point value to 2 g/cc.
- For thermal expansion coefficient the "0" point value will correspond to 2.5×10 C, and the "10" point value to 1.0×10 C.

Note for the density and thermal expansion coefficient a lower value is generally better. The points assigned to each material property is shown in Table 15.5. If, and this is rarely the case, each property affects performance of the design equally, then one can simply add up the points for each property and total the score for each material. Typically a weighted total is more appropriate. One would then assign a "weighting" to each material property. In this case strength, which affects load-bearing capability, and density, which affects installation, would be more important than the other material properties. Thus, we can "weight" each property as shown:

- The load-bearing ability, which is determined by strength, is extremely important. Thus, the overall weighting of strength will be 40%, or 0.4.
- The density, which will determine how easy the beam is to install, is also important. Thus, the overall weighting of density will be 30%, or 0.3.
- The other two material properties will be weighted 15%, or 0.15.

The overall material ranking is shown in Table 15.6. The materials are now ranked in descending order of performance. Note that to determine this ranking, one had to make a series of decisions based on engineering judgment. The importance of this cannot be understated. The engineer making these decisions must make decisions. These must be based first and foremost on safety. They also must be based on listening to the customer's needs and making engineering

---

TABLE 15.5
Example of a Spreadsheet Showing the Points Assigned to Each Material Property

Material	Yield Strength (MPa)	Specific Modulus Points	Density Points	Expansion Coeff. Points
1050-H16 Aluminum	0.0	2.9	7.7	0.9
2014-T4 Aluminum	10.0	2.9	7.7	0.9
Alumina	3.8	10.0	4.0	10.0
Silicon Nitride	6.6	10.0	9.7	10.0

TABLE 15.6
Example of a Spreadsheet
Showing the Overall Material
Ranking

Material	Material Score
Silicon Nitride	8.5
2014-T4 Aluminum	6.9
Alumina	5.7
1050-H16 Aluminum	2.9

decisions. Also note that this example was simplified. In general, ten or eleven criteria may be considered, and the initial list of materials will be much longer.

15.2.3 COST–BENEFIT ANALYSIS

Having identified and rank-ordered a list of candidate materials, one needs to determine if the cost of each material is justified and assess the benefit to the company. This is the last step in the materials selection process. Once it has been determined that a design is safe and meets customers' needs, a cost analysis is appropriate. To determine cost one has to consider not only the raw cost of materials but the cost of material handling, environmental compliance, installation, warranty, and repair. This list is not exhaustive but illustrates that all costs must be considered. It can easily be shown that one would need 6.3 kg of silicon nitride, 8.1 kg of aluminum, and 11.4 kg of alumina to make the beam. The ceramics are likely to cost more (on a mass basis) than the two grades of aluminum. Therefore, using alumina, the third-ranked material, would likely be the most expensive choice, and alumina can be eliminated from consideration.

A beam made of silicon nitride would be more expensive than a beam made of either grade of aluminum. The questions that must be answered are:

• Is the increased cost worth it?
• Will the customer pay more for a beam made of silicon nitride?
• Will it enable the company to expand its market?
• Will the lighter weight of the beam reduce installation costs and save the company money?
• How significant is a mass reduction of 10.8 kg?

In this case for a simple beam it is probably not worth the extra expense. Therefore, the first-ranked material can be eliminated for cost–benefit reasons. The added cost does not justify the benefit.

The next decision is whether to use 2014-T4 Aluminum or 1050-H14 Aluminum. Assuming the costs are comparable, you are now deciding whether to use a thermal-treated aluminum copper alloy or a work-hardened material. Based on performance, the 2014-T4 Aluminum would be selected.

Finally, you must be prepared to justify your decision. Explaining the process by which the material was selected is important. When making a recommendation such as this, one needs to be ready to justify and explain the implications of the recommendation. In this case one might be asked to explain why all aluminum is not the same. The person making this decision would also have to be prepared to explain the implications of using 2014-T4 Aluminum. Subjecting this material to elevated temperature for extended periods of time would reduce the strength.

15.3 GENERAL CRITERIA

When making materials selection decisions, one needs to determine what material properties should be considered and what material properties are required. The examples presented earlier are oversimplified, as the focus of this chapter is on the procedure used to make the decision, not the decision itself.

15.3.1 PHYSICAL CRITERIA

The density, thermal expansion coefficient, heat capacity (or specific heat), heat of fusion, melting temperature, color, and transparency are all physical properties of a material. The energy required to prepare a casting will have a direct impact on the cost of the material.

15.3.2 MECHANICAL CRITERIA

The mechanical properties discerned from a tensile test are the most familiar. The yield strength, fracture strength, ultimate tensile strength, ductility, modulus, and toughness all can be considered. Fracture toughness, which determines the resistance to brittle failure, is often a key parameter used in design decisions. The fatigue and creep resistance of a material will have long-term use implications. Hardness will affect processability and could impact a material selection decision.

The temperature dependence of material properties must be considered. When using polymers, dramatic changes can occur with slight variations in ambient temperature.

15.3.3 ELECTRICAL CRITERIA

Electrical resistivity or conductivity can impact material selection decisions. The dielectric constant and breakdown potential of an insulator can also impact design decisions. The band-gap characteristics of a semiconductor will determine the light-emitting properties of a light-emitting diode.

15.3.4 Chemical Criteria

Corrosion is a leading cause of failure. Chemical interactions with the environment must be considered. Whenever two metals are in contact with each other and a conducting medium (such as water or wet sand) corrosion can occur. Diffusion of one material into another is a problem for small devices at elevated (based on melting point) temperatures. Toxicity and environmental impact must be considered.

15.4 CONCLUDING REMARKS

It cannot be overemphasized that the purpose of this chapter is to present methodologies that can be used to make material selection decisions. The examples have been simplified and therefore should not be considered design recommendations. However, it is hoped that by following the examples, one can more efficiently make material selection decisions.

SUGGESTED READING

Ashby, M. F., *Materials Selection in Mechanical Design*, Pergamon Press, New York, 1992.

Dieter, G. E., *Engineering Design: A Materials and Processing Approach*, 2nd ed., McGraw-Hill, New York, 1991.

Murray, G. T., *Handbook of Materials Selection for Engineering Applications*, Marcel Dekker, New York, 1997.

16 Ferrous and Nonferrous Metals for Special Applications

16.1 INTRODUCTION

A high percentage of the metal products produced in the world are made of steel. The reason is obvious: it is a low-cost, readily available, easily fabricated material that is well understood in the industrial world. If an engineering designer faces a special application that requires corrosion resistance, low- or high-temperature service applications, superior heat or electrical transfer properties, low or high density, or a nonmagnetic characteristic, the options available are more costly but the property management protocols are similar within the boundary conditions of a particular metal family. In this chapter some specialty steels and common nonferrous options are discussed. More complete details for a given material can be found in the references.

16.2 SPECIAL FERROUS-BASED MATERIALS

16.2.1 HIGH-STRENGTH LOW-ALLOY (HSLA) STEELS: MICROALLOYED STEELS

These steels have a lower alloy content and less strength than the alloy steels discussed in the preceding chapter. They are primarily structural steels for large buildings, bridges, ships, and oil and gas pipelines. Today, they are also finding their way into automotive applications. The microalloyed steels are a relative newcomer to this group but are rapidly replacing many of the older structural steels. The American Society for Metals (ASM) *Metals Handbook* separates these steels into four groups, two of which are heat treatable. The other two groups, which account for the major use of structural steels, fall into the categories of carbon–manganese and microalloyed steels. The typical structural steel is a carbon–manganese steel of about 0.18% carbon and 1.3% manganese and has the American Society for Testing and Materials (ASTM) designation of A36. The microalloyed steels, of which the ASTM A572 group is the most common, contain small amounts of niobium, titanium, or vanadium, all of which are strong carbide formers. The strength is achieved by a fine dispersion of these carbides plus a carefully controlled hot-rolling process to produce a fine grain size. The small carbides restrict grain growth. Both of these strengthening mechanisms produce

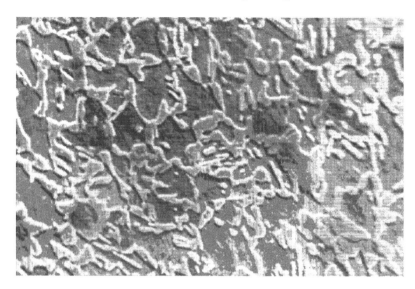

FIGURE 16.1 Microstructure of a dual phase quenched from 730°C that contains approximately 28% martensite (light areas) (× 2000). (Courtesy of Barbara Blumenthal, Cal. Poly.)

structural steels of up to 550 MPa (80 ksi) yield strength, with an ultimate tensile strength in excess of 690 MPa (100 ksi). Because these steels are strengthened by the strong carbide formers, most of the carbon is removed from the ferrite solid solution. In welding processes the steel is at in the γ-phase region some point during the cooling process. At moderately fast cooling rates the undesirable brittle martensite phase could form. The microalloyed steels have little carbon available for martensite formation, another plus for these steels.

Another, more recent, addition to the HSLA family of steels is the dual-phase steels. They are now classified as HSLA steels, although when they were first introduced around 1980, there was some confusion concerning their classification. Their microstructure consists of regions of hard martensite in a much softer α-ferrite phase matrix (Figure 16.1) and is achieved by quenching from the $\alpha + \gamma$ region of the iron–iron carbide phase diagram. The γ phase transforms to martensite on quenching, while the α phase remains as such. Their carbon contents are about 0.1%, so the major phase is α ferrite, while the martensite comprises only about 20%. They have strengths of about 621 MPa (90 ksi), with good formability and a very high rate of strain hardening. With carbon contents in the range of 0.2 to 0.4, strengths of 1040 MPa (140 ksi) have been attained after cold forming. The uses of these steels to date have been primarily in the automotive industry.

The treatment, properties, and applications of some typical Society of Automobile Engineers (SAE) steels are summarized in Table 16.1.

TABLE 16.1
Properties and Applications of Plain Carbon, Alloy Steels, and HSLA Steels

Alloy Type and Condition	Ultimate Tensile Strength [MPa (ksi)]	Elongation (%)	Applications
Plain Carbon Steels			
Low carbon <0.2% C hot rolled	290 (42) to 414 (60)	40 to 25	Nails, stampings, cans, common sheet steel products, auto bodies, consumer goods, deep-drawn products
Medium carbon 0.2 to 0.5% hot rolled	414 (60) to 552 (80)	25 to 40	Machine parts, rivets, carburized gears, fasteners, cams, camshafts, forgings, general-purpose heat-treated parts
Medium-high carbon 0.5 to 0.9% C hot rolled	552 (80) to 828 (120)	20 to 8	Oil-hardening gears, set and socket screws, spring steel, ball bearings
High carbon 0.9 to 1.5% hot rolled	828 (120) to 966 (140)	8 to 1	Cutting tools, files, saws, knives, boring tools (often quenched or quenched and tempered)
Low-Alloy Steels			
Cr–Ni–Mo A736 A517 A542 A543	760 (118) to 930 (135)	18 (min)	Automotive and other machinery, shafts of many types, steel, pressure vessels
Cr-MO A387 A382	380 (55) to 690 (100)	18 (min)	Oil and gas industries, nuclear power plants, fossil fuel plants
Ni–Mo A302 A533 A645	550 (80) to 860 (125)	18 (min)	Pressure vessels
HSLA and Structural Carbon Steels			
Microalloyed A572 A588	480 (70) to 345 (50) min	18 min	High-strength niobium–vanadium steels of good structural quality
Structural carbon A36	400 (50) min	20	General structural steel

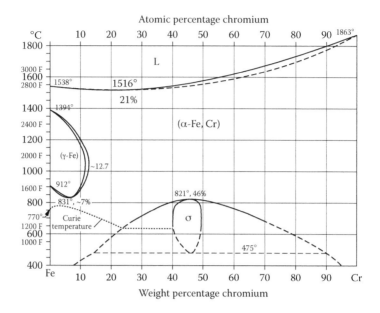

FIGURE 16.2 Fe–Cr phase diagram. (From *Metals Handbook*, Vol. 8, 8th ed., ASM International, Metals Park, OH, 1973, p. 291. With permission.)

16.3 SPECIALTY STEELS

16.3.1 STAINLESS STEELS

The stainless steels account for the majority of what we will call specialty steels. The stainless variety can be separated into five groups: the austenitic, which account for well over half of the use of all stainless steels; martensite; ferritic; precipitation hardening; and duplex stainless steels.

Stainless steels are stainless because of their chromium content. Chromium is a passive metal. Like aluminum and titanium, it forms a thin oxide layer, Cr_2O_3, on immediate exposure to air. When the chromium is in solid solution with iron or nickel, about 10.5 wt % Cr is required to make the alloy passive. The iron–chromium phase diagram is shown in Figure 16.2. The important feature is the γ (austenite) loop on the left side of the diagram. Chromium and α-iron are very similar metals with respect to atom size, valency, and crystal structure. Thus, the body-centered cubic (BCC) iron can contain lots of BCC chromium in a substitutional solid solution, but the γ–face-centered cubic (FCC) field can dissolve only about 12.7% Cr. Also, the γ region is restricted to temperatures between about 830 and 1390°C.

16.3.1.1 Ferritic Stainless Steels

Ferritic stainless steels are Fe–Cr alloys containing between 12 and 30% Cr. The upper limit is determined by the σ phase. This phase, located in the central portion of the diagram, is very brittle and undesirable and should be avoided. Ferritic steels on the high end of the Cr content scale have good corrosion resistance but relatively low strength compared to the other stainless steels. Since they do not contain nickel, they are less expensive than the austenitic steels. The 409 and 430 are common ferritic stainless steels.

16.3.1.2 Martensitic Stainless Steels

Martensitic stainless steels become martensitic just like the other steels (i.e., by quenching from the γ region). The Fe–Cr phase diagram suggests that the maximum Cr content would be about 12.7%, but the carbon content expands the γ region to the extent that larger chromium contents are possible. Common alloys are 410, containing 12% Cr and low carbon, and alloy 440 of 17% Cr with a high carbon content. The martensitic steels are the strongest of all stainless steels, having strengths to 1897 MPa (275 ksi), but at such high strength levels they lack ductility. They are used for cutlery, surgical instruments, springs, and applications that in general involve high strength and corrosion resistance.

16.3.1.3 Austenitic Stainless Steels

These steels are a good compromise between the ferritic and martensitic stainless steels, which explains their widespread usage. These steels are ternary alloys and as such require a ternary phase diagram for a complete understanding of the phase relations. By the addition of nickel in quantities of about 8%, the γ transformation temperature, which in pure iron is 916°C, can be depressed to the extent that the γ can be retained at room temperature. The γ phase is metastable at room temperature, but the tendency to transform is so sluggish that for all practical purposes it can be considered to be stable. Higher nickel contents (e.g., 10%) can ensure γ stability.

Most alloys of austenitic stainless steels contain 16 to 25% Cr, 7 to 20% Ni, and the balance Fe. The chief advantage is their greater ductility and hence formability that goes with the FCC structure. Their strengths are in-between those of the ferritic and martensitic steels, but via cold working and the related strain hardening, strengths of around 759 MPa (110 ksi) with ductilities of 50% Reduction in Area (R.A.) are attainable. The austenitic steels are the 300 series of steels, with the 301 and 304 series of roughly 18% Cr–8% Ni being the most widely used.

Most austenitic steels are susceptible to *sensitization* when heated in the range of about 425 to 870°C or slowly cooled through this temperature range. The chromium becomes tied up as chromium carbide and is not available to form the passive Cr_2O_3 surface film. In this state they are *sensitive* to corrosive media. The chromium carbide frequently forms first in or near the grain boundaries, and hence corrosion begins in the grain boundary regions. Two steels have been

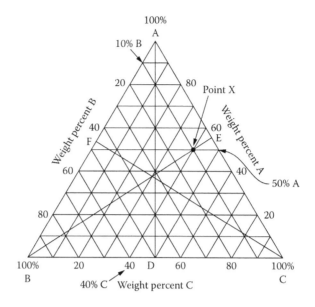

FIGURE 16.3 Schematic of a three-element phase diagram.

developed to minimize sensitization: 347, containing about 0.6% niobium, and
321, containing similar amounts of titanium. These elements readily form carbides, leaving the chromium free to form Cr_2O_3. This is not a cure-all, but it
helps. The 321 and 347 steels are called *stabilized* stainless steels. The best
solution to the problem is to avoid this temperature. During the welding of 321
and 347 steel the niobium and titanium carbides go back into solution near the
weld and make this region again subject to sensitization. A 304 L steel, L meaning
"low carbon," was developed for situations where welding is required.

16.3.1.4 Ternary Phase Diagrams

The austenitic stainless steel discussion is a good place to introduce the subject
of ternary phase diagrams. The metallurgist often tries to circumvent these diagrams by one means or another because of their complexity. Also, in many cases
in important alloy systems they may not exist. However, these diagrams are very
commonplace in ceramic systems. Let us examine their concept and see how it
works in the iron–chromium–nickel system.

In the ternary diagram the composition of the three components are defined
in the form of a triangle, as depicted in Figure 16.3. The binary alloy composition
AB, BC, and *AC* is represented on the three edges of the triangle. The temperature
is a constant throughout this triangle. To find the composition of an alloy denoted
by point *X* in the diagram, a line is constructed from each corner representing
the pure metals to the side of the triangle opposite that corner and perpendicular
to these sides. At the point where the line from corner *A,* for example, intersects
line *BC,* the % *A* is zero. The total length of the line from *D* to *A* represents

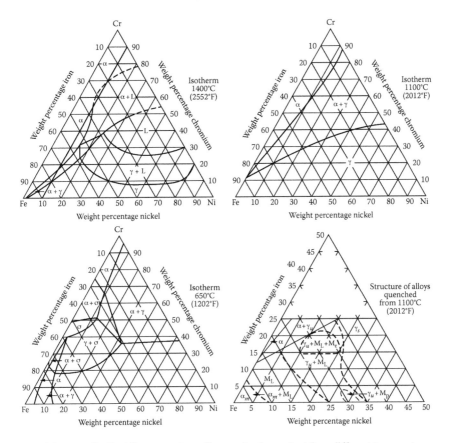

FIGURE 16.4 Cr–Fe–Ni ternary phase diagram isothermals at four different temperatures. (From *Metals Handbook*, Vol. 8, 8th ed, ASM International, Metals Park, OH, 1973. With permission.)

100% *A*. Point *X* is on an isocomposition line at 50% *A*; thus, the percentage *A* in the alloy is 50%. Note that the line from the point of interest *X* to line *AC* is perpendicular to line *AD*. Similarly, the % *B* is read from line *BE*, which is constructed from 100% *B* perpendicular to line *AC*, and which in this case turns out to be 10% *B*. The balance must be the % *C*, which can be verified by constructing line *CF*. This is the composition for only one temperature. To know the phases that exist at any given temperature, the composition triangle must be known for that particular temperature. These are called *composition isotherms*. The isotherms for different temperatures for the Fe–Cr–Ni system are shown in Figure 16.4.

16.3.1.5 Precipitation-Hardening (PH) Stainless Steels

These steels can attain very high strength levels, in excess of 1380 MPa (200 ksi) with good ductilities (50 to 70% R.A.). They attain their high strengths by precipitation of intermetallic compounds via the same mechanism as that found in aluminum alloys. The compounds usually are formed from Fe or Ni with Ti, Al, Mo, and Cu. Typical compounds are Ni_3Al, Ni_3Ti, and Ni_3Mo. Chromium contents are in the range of 13 to 17%. These steels have been around for several decades but are now being recognized as an alternative to the other stainless steels. They have the good characteristics of the austenitic steels plus strengths approaching the martensitic steels. One of the early problems centered around forging difficulties, but these problems have been overcome to some extent. They are of two types: martensitic and semiaustenitic, the latter being able to be converted to martensite by special heat treatments. The most popular are the 13-8, 15-5, and 17-4 martensitic types, where the numbers refer to their Cr and Ni contents, respectively. They are susceptible to hydrogen embrittlement, but this too can be overcome to some extent by overaging to a strength level in the vicinity of 1173 (170 ksi).

These steels are finding a myriad of uses in small forged parts and even in larger support members in aircraft designs. They have been considered for landing gears. Many golf club heads are made from these steels by investment casting techniques, and the manufacturers proudly advertise these clubs as being made from 17-4 stainless steel, although golfers do not have the least idea what the numbers mean. The 17-4 is the most popular, although the 13-8 can attain higher strength levels. The precipitate in an aged 13-8 steel is shown in Figure 16.5.

16.3.1.6 Duplex Stainless Steels

The duplex stainless steels contain both austenite (γ) and ferrite (α) phases. These steels can be heat-treated to strengths of 828 MPa (120 ksi) with 35% elongation. The chief advantage of these steels is their good corrosion resistance in salt water, making them candidates for marine applications. Chromium contents are about 25%. They have trade names such as Ferrallium 255, U50, and 7-Mo plus. They contain about 50% austenite–50% ferrite. They have a relatively high solubility for molybdenum, and it is the Mo content of about 2% that results in a good corrosion resistance to chloride environments. The microstructure of a duplex stainless steel is shown in Figure 16.6.

The properties of the various stainless steels and some of their applications are listed in Table 16.2. Many stainless steels are used at elevated temperatures, and for this reason they are sometimes classified in the category of *heat-resistant alloys*. The maximum operating temperatures of the austenitic, ferritic, and martensitic grades are listed in Table 16.3, and their creep and stress-rupture properties are shown in Figure 16.7.

16.3.2 Maraging Steels

The maraging steels, as a group, are the highest-strength steels currently available. Their yield strengths range from 1030 MPa (150 ksi) to 2420 MPa (350 ksi). The lower-strength maraging steels are on the same order of strength as the low-alloy

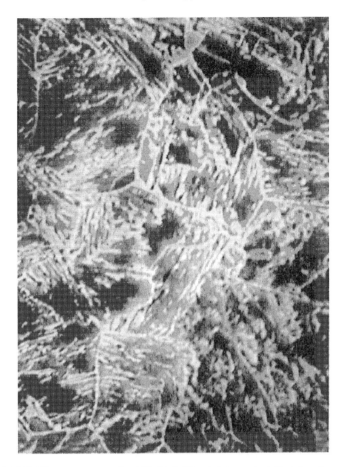

FIGURE 16.5 Microstructure of a PH 13-8 Mo stainless steel that has been solution annealed and aged for 4 h at 1050°C. (Courtesy T. Mousel, Cal. Poly.)

high-hardenability steels such as 4130, 4340, and 8640, but the higher-strength maraging steels exceed the strength levels of all other steels.

The maraging steels are hardened by the precipitation of intermetallic compounds (e.g., Ni_3Mo and Ni_3Ti), and in this sense their physical metallurgy is similar to that of the PH martensitic stainless steels. The latter, however, have a high chromium content for good corrosion resistance. The maraging steels do contain sufficient carbon to form a low-carbon iron–nickel soft lathe martensite, but the intermetallic compounds account for their strength. These steels typically contain high nickel (18%), cobalt (~9%), and molybdenum (~5%). Because of their low carbon content, their weldability is excellent, and the high alloy content provides good hardenability. They also have good fracture toughness, and except for the very high-strength alloys, adequate ductility. The properties in the aged condition of two of these steels are listed in Table 16.4.

FIGURE 16.6 Microstructure of Ferrallium 225 duplex stainless steel quenched from 1120°C showing islands of γ in an α-ferrite matrix (×200). (Courtesy of Kerri Page, Cal. Poly.)

Probably their chief disadvantage is their high cost and somewhat lower ductility than most low-alloy steels. They have been used as missile cases, aircraft, structural forgings, springs, shafts, bolts, and punches. A cobalt-free maraging steel has been developed for the nuclear industry, and a few stainless grades have been produced.

16.3.3 Tool Steels

Any steel used to make metals by cutting, forming, or shaping is a tool steel. For simplicity and brevity we will classify tool steels into the two broad categories of cold- and hot-worked tool steels. The ASM *Metals Handbook*, Vol. 1, 10th ed., pp. 757–792, lists compositions and properties in much more detail than will be presented here.

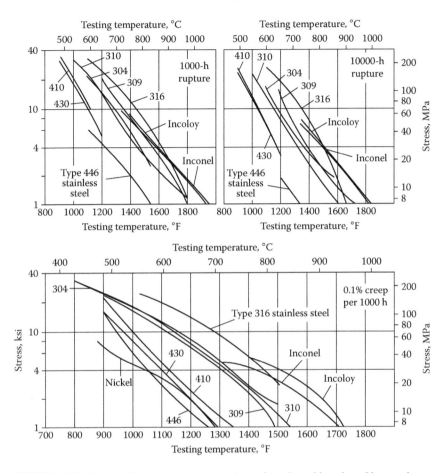

FIGURE 16.7 Creep and stress rupture comparisons for selected iron-based heat resistant alloys. (From *Metals Handbook*, Vol. 3, 9th ed., ASM International, Metals Park, OH, 1980, p. 195. With permission.)

16.3.3.1   Cold-Worked Tool Steels

Most of the cold-worked tool steels are subclassified according to the quenching media employed in the heat-treating process. *Water-hardening* tool steels, known as the W type, contain from about 0.6 to 1.4% carbon. They are essentially plain carbon steels that are inexpensive, have low hardenabilities, and as such must be used in thin sections. Typical applications include chisels, drills, lathe tools, and files. The *oil-hardening* (O) types have carbon contents in the range of 0.9 to 1.5%, with some small additions (0.5 to 1.0%) of chromium, tungsten, and manganese for hardenability. Thus, they can be quenched in larger sections at slower rates and still possess a high hardness value. The quenched structure consists of martensite, bainite, some retained γ, and undissolved carbides (carbides that do not go into the γ phase during the austenizing anneal).

TABLE 16.2
Room-Temperature Properties and Applications of Stainless Steels

Type	Condition	Minimum Y.S. MPa (ksi) U.T.S. [MPa (ksi)][a]	R.A. %	Applications
Martensitic 410	Quenched and tempered, 500°F	800 (116) 1000 (150)	60	High-strength and elevated-temperature applications; not suitable for cryogenic uses
Martensitic 431	Quenched and tempered, 500°F	750 (109) 1000 (150)	65	Similar to 410; both have acceptable resistance to stress corrosion cracking (SCC) at hardness below Rc 22
Ferritic 430	Annealed bar	205 (30)	45	Popular general-purpose steel where welding is not required; low toughness after welding
Ferritic 409	Annealed plate	205 (30) 380 (55)	—	General-purpose construction steels for automotive exhaust systems and light applications
Ferritic 405	Annealed bar	205 (30) 415 (60)	45	Toughness satisfactory after welding
Austenitic 304	Annealed bar	205 (30) 515 (75)	50	Good formability; use 304L (low carbon) for welding structures; general-purpose, corrosion resistant
	Cold-worked bar, $\frac{1}{4}$ hard	515 (75) 860 (125)	—	High-strength, corrosion-resistant structural parts; railway passenger cars, truck trailers, missiles, cryogenic applications
Austenitic 316	Annealed bar	205 (30) 515 (75)	5	Mo added to increase pitting, general corrosion resistance to chloride environments
Austenitic 347	Annealed bar	205 (30) 515 (75)	5	Stabilized with Nb and Ta to reduce sensitization in range 480°C (900°F) to 813°C (1500°C)

TABLE 16.2 (Continued)
Room-Temperature Properties and Applications of Stainless Steels

Type	Condition	Minimum Y.S. MPa (ksi) U.T.S. [MPa (ksi)][a]	R.A. %	Applications
PH 17-4 martensitic	Quenched and aged, 1025°C	1000 (145) 1070 (155)	45	Weldable, strong, corrosion resistance, varied applications for small forged precision cast parts; aerospace, hydrogen embrittles
PH 13-8 Mo martensitic	Quenched and aged, 1025°C	1210 (175) 1275 (185)	45	Highest-strength stainless steels; aerospace, fasteners, cooler sections of aircraft engines, susceptible to hydrogen embrittlement
Duplex ferralium 225	Annealed	550 (80) 760 (110)	65	Good corrosion resistance to chlorides; marine applications

[a] Y.S., yield strength; U.T.S., ultimate tensile strength.

Generally, these steels are tempered slightly to reduce the brittleness of martensite. They are the most widely used tool steels. They have good toughness in the tempered condition and are subject to less distortion on quenching than are the W-type tool steels. They are used for shear blades and a wide variety of cutting tools. The *air hardening* (A) types contain 1 to 2% C and larger quantities of CR, Mn, Mo, V, or Ni to achieve high hardenability. The as-quenched martensite is later tempered, but these steels tend to retain some γ. Sometimes at room temperature this retained γ transforms to martensite, which could result in dimensional changes. Applications for the A-type steels include punches, forming dies, blanking, and trimming dies.

The *shock-resistant* or S types have good toughness owing to a relatively low carbon content of about 0.5%. They also contain small amounts of alloying elements, giving a medium range of hardenability. Most are oil quenched. These steels are used for repetitive-wear applications, such as rivet sets, shear blades, punches, pipe cutters, and concrete drills.

There is one more cold-worked tool steel, called the D type, for dies. They have high carbon (1.5 to 2.3%) and high chromium (12%) contents and were originally developed for high-temperature applications. However, they have been replaced by other high-temperature steels and are now used primarily for wear resistance, such as in long-run drawing dies, rolls, deep drawing dies, thread rolling, and slitters.

TABLE 16.3
Maximum Service Temperatures for the
Major Stainless Steels

AISI Type	Maximum Service Temperature (°C)	
	Continuous	Intermittent
Austentic Grades		
201	815	845
202	815	845
301	840	900
302	870	925
304	870	925
308	925	980
309	980	1025
310	1035	1150
316	870	925
317	870	925
330	1035	1150
347	870	925
Ferritic Grades		
405	705	815
406	815	1035
430	815	870
442	980	1035
446	1095	1175
Martensitic Grades		
410	705	815
416	675	760
420	620	735
440	760	815

Source: *Metals Handbook*, Vol. 1, 10th ed., ASM
International, Metals Park, OH, 1990, p. 878. (With
permission.)

16.3.3.2 Hot-Worked Tool Steels

All hot-worked tool steels have one thing in common, and that is referred to as
secondary hardening. This can be better explained by Figure 16.8, where the
hardness vs. tempering temperature is shown for molybdenum-containing hot-
worked tool steels. The hot-worked steels show a hump in the tempering curve
that is due to secondary hardening brought about by carbide precipitation, usually
of Mo, W, or Cr carbides. These are more stable than Fe carbides. The hot-worked
tool steels are used for extrusion and forging dies, die cast dies, and high cutting
rates of hard materials where the tool becomes very hot.

TABLE 16.4
Properties of Two Maraging Steels

Grade	Ultimate Tensile Strength		Yield Strength		R.A. (%)	Fracture Toughness	
	MPa	ksi	MPa	ksi		$Mpa\sqrt{m}$	$ksi\sqrt{in}$
18 Ni (250)	1800	260	1700	247	55	120	110
18 Ni (300)	2050	297	2000	290	40	80	73

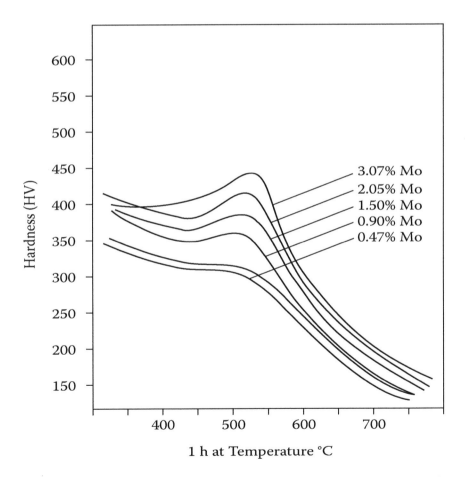

FIGURE 16.8 Secondary hardening in Mo-containing tool steels. (From K. J. Irvine and F. B. Pickering, *J. Iron Steel Inst.*, 194, 137, 1960. With permission.)

TABLE 16.5
H-Type Hot-Worked Tool Steels

Type	Alloy	
H 1 to H 19		Cr alloying element
H 20 to H 39	W	Cr alloying element
H 40 to H 59	Mo	Cr alloying element

The hot-worked tool steels are classified as H, M, T, or W type. This sounds very confusing, but such classifications are often based on the chronological order in which the alloys were developed.

The H type is most often used as extrusion and forging dies. There are three types, based on the alloying element used to provide secondary hardening. They are listed in Table 16.5. They all exhibit secondary hardening effects. The Cr steels have the higher hardenability characteristics, while the Mo-containing steels have a very marked secondary hardening peak. The W-containing steels, also used for extrusion and forging dies and mandrels, have a secondary hardening peak similar to the Mo steels but of less magnitude and at lower temperatures.

Finally, we have high-speed tool steels of the T and M types, containing tungsten and molybdenum, respectively. These are somewhat different from the H-type steels, in that the molybdenum contents are much higher than in the M-type steels of the H 40 to H 59 group of hot-working tool steels. However, in the T type the tungsten contents are only slightly higher than the tungsten contents of the H 20 to H 39 group, the latter containing from 8 to 19% W, while the T types contain about 12 to 19% W. Carbon contents are about 0.75 to 1.5 in both of these tungsten-containing tool steels. The M steels are much more popular than the T steels, primarily because of the approximately 40% lower cost of Mo compared to W. The M steels were developed after the T type, when large deposits of molybdenum were found in Colorado in the late 1930s. However, England still continued to use the T steels during World War II and later, perhaps because of their familiarity and some better temperature control on the older T-type steels. The compositions and applications of the various types of tool steels are listed in Table 16.6.

16.4 CAST IRONS

Cast irons are iron–carbon alloys that contain carbon in amounts larger than the steels, varying from 2 to 4 wt % C. As such, they are brittle alloys and must be cast rather than formed to their final shapes. Cast irons do not have the strength levels of most steels, except for perhaps the low-carbon steels in the annealed or hot-formed condition. Cast irons are easily melted and have a high fluidity and thus can easily be cast to sizes and shapes that are difficult to form by other means. Their competition, from a metal-forming viewpoint, is forgings. Many

TABLE 16.6
Compositions and Applications of Various Types of Tool Steels

Type	Typical Compositions	Applications
Cold-Worked Tool Steels		
W: Water hardening	0.7–1.4% C	Chisels, drills, files, lathe cutting tools
O: Oil hardening	0.8–1.5% C	Shear blades, wide variety of room-
	0.5–10% Cr, W, Mn	temperature cutting tools
A: Air hardening	1–2% C, Cr, Mn, Mo,	Punches, forming dies, blanking and
	V, and Ni to approx.	trimming dies
	3% total	
S: Shock resistant	0.5% C, 0.5–1.8% Cr,	Repetitive-wear applications, rivet sets,
	0.1–1.5% Mo	shear blades, punches
Hot-Worked Tool Steels		
H Type		
H10–H19	0.3–0.5%, 3–5% Cr	Shearing knives
H20–H39	0.2–0.5% C, 8–19% W	Extrusion and forging dies for brass
H40–H59	0.5–0.7% C, 4–6% Mo	Hot-worked steel tools
High-speed T	0.65–1.6% C	High-speed tool steels, both T and M
	0.9–5% V	types, used for heavy-duty work calling
	12–19% W	for high hot hardness and abrasion
		resistance; hot forging and extrusion dies
		and mandrels, cutting tools
High-speed M	0.7–1.5% C	
	1–6% W	
	4–11% Mo	

products that are cast can also be forged to near-final shape. The decision must take into consideration properties and costs, the latter frequently being the determining factor. In large structures, such as machine tool bases, the cast irons have the ability to absorb vibrations more than other metals. Also, most large structures cannot be forged.

If we examine the iron–iron carbide phase diagram again (Figure 16.25), we find that the cast irons solidify with some eutectic component. Furthermore, at high carbon contents they can solidify with the Fe–carbon (graphite) structure rather than the Fe–Fe_3C steel structures. The cast irons have wide variations in mechanical properties, depending on whether they form the stable austenite (γ)–graphite (C) or the metastable austenite–Fe_3C eutectic. Fe_3C is a metastable phase, even in low-carbon steels, but in the latter steels the tendency to decompose is virtually nil, so we treat the iron carbide as a stable phase. This is not so in the cast irons, particularly the high-carbon alloys and those containing silicon.

There are four types of cast irons: gray, white, malleable, and nodular. Their properties are dependent on heat treatment and the quantities of carbon and silicon. Microstructures of the four types are presented in Figure 16.9.

(b)

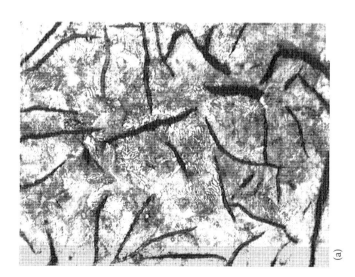

(a)

FIGURE 16.9 Typical microstructures of cast irons: (a) gray, 400×; (b) malleable, 200×; (c) nodular, 200×; and (d) white, 480×. (Courtesy of John Haruff and Roxanne Hicks, Cal. Poly.)

(d)

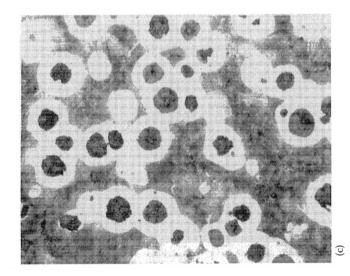

(c)

FIGURE 16.9 (Continued.)

16.4.1 GRAY CAST IRON

In gray iron the carbon is in the form of graphite flakes. Silicon additions assist in making the Fe_3C unstable. As the metal slowly cools in the mold, the Fe_3C decomposes to graphite. Its name comes from the gray appearance of a fractured surface. The gray irons are noted for their good machinability and high damping capacity.

16.4.2 WHITE CAST IRON

In white cast iron the silicon and carbon contents are relatively low and the cooling rates moderately high. The carbon is thus primarily in the form of Fe_3C, which results in a very hard, white-looking microstructure. White cast iron is used when good wear resistance is important. Most white cast iron consists of pearlite and massive Fe_3C regions. From examination of the phase diagram, an alloy containing about 3% C will solidify with some eutectic plus the γ phase. At lower temperatures the γ will be of the eutectoid composition and thus convert to pearlite on cooling through the eutectoid temperature. Very fast cooling can prevent pearlite from forming, and this product, called *chilled cast iron*, will contain considerable amounts of martensite. Ni, Cr, and Mo additions promote the formation of martensite.

16.4.3 MALLEABLE CAST IRON

This iron is made by heating white cast iron for long periods of time at about 950°C, causing the Fe_3C to decompose to α and graphite, the graphite being in the form of small particles rather than flakes. At the eutectoid temperature, on slow cooling, more Fe_3C decomposes, yielding more carbon. This carbon is called *temper carbon*. This malleable iron has slight ductility and varies considerably in properties, depending on the cooling rate from the graphitization temperature. If the alloy is fast-cooled from the 950°C anneal to about 750°C, and then slowly cooled to room temperature, the γ transforms to α + C, with the graphite depositing on the existing temper carbon. This ferrite matrix cast iron is called ferritic malleable iron. Cooling very slowly to 870°C, followed by air cooling, allows the γ to transform to pearlite. The microstructure then consists of temper carbon nodules in a pearlite matrix, known as pearlitic malleable iron. By cooling to around 850°C, holding for about 30 min to homogenize the γ, and then oil-quenching to room temperature, a martensite matrix is produced. This martensite is subsequently tempered, resulting in a final microstructure consisting of carbon nodules in a tempered martensite matrix. In summary, malleable iron can consist of either a ferritic, pearlitic, or martensitic matrix containing graphite nodules, and the properties will vary accordingly.

16.4.4 NODULAR CAST IRON

In this iron small additions of magnesium and cerium cause the graphite to form spherical particles instead of the flake form found in gray iron. The flake form has better machinability, but the spheroidal form yields much higher strengths and ductility. As in the case of the malleable iron, the matrix can be ferritic, pearlitic, or martensitic, depending on the heat-treatment process. The most common form is that shown in Figure 16.9c, in which the graphite nodules are surrounded by white ferrite, all in a pearlite matrix.

From the brief presentation above, one can see that cast iron metallurgy is just as varied and complex as that of steels. By understanding the Fe–Fe$_3$C phase diagrams and the associated heat treatments, it is relatively easy to comprehend the processing of cast irons. The nonmaterials engineer, however, is more interested in properties than in processing. The properties and some applications are summarized in Table 16.7. The nonmaterials engineer, by being familiar with the microstructures, has a better understanding of the advantages and disadvantages of the variety of cast irons available.

In recent years an austempered ductile iron has appeared on the market that can have a matrix varying from ferrite, to ferrite plus pearlite, to pearlite, to bainite, and even to martensite, depending on the austempering heat-treatment cycle. Yield strengths can vary from 550 to 1300 MPa (80 to 188 ksi) and elongations from 1 to 10%. It has excellent wear properties and is used for gears, crankshafts, chain sprockets, and high-impact and high-fatigue applications.

16.5 ALUMINUM ALLOYS

Aluminum ranks second in global metal usage. Its principal advantage is its low density (about one-third that of steel) and high corrosion resistance. The use of aluminum in automobile and truck manufacturing has seen a significant increase as car makers strive to make more fuel-efficient vehicles. Aluminum is used in both a cast and wrought form, exhibiting high strength and ductility in both. In the aluminum–aluminum alloy system's properties are commonly managed in three basic formats:

1. Alloy additions are used to increase the properties of the aluminum by solid solution strengthening.
2. Mechanical working, principally cold work, is used to increase the strength and reduce the ductility of the alloys.
3. Precipitation hardening is used as discussed in Section 16.5.

Aluminum and its alloys are commercially designated by a numerical code system (Table 16.8). The code signifies the major alloying elements. In each of the alloy families the major alloying element is designated by the code number. Other alloys are added in smaller quantities to improve specific properties.

TABLE 16.7
Types, Properties, and Applications of Cast Irons

Type	Composition (wt %)	Yield Strength [MPa (psi)]	Elongation %	Application
		Gray		
UNS No. F10006	3.1–3.14% C, 0.6–0.9% Mn, 2.3–1.9% Si	207 (30)	—	Cylinder blocks, pistons, brake drums, clutches
UNS No. F10008	3.0–3.3% C, 0.6–0.9% Mn, 2.2–1.8% Si	276 (40)	—	Diesel engine castings, liners, cylinders, pistons, camshafts
		Ductile (nodular)		
UNS No. F32800 (ferrite matrix)	3.5–3.8% C, 2.0–2.8% Si	276 (40)	18	Shock-resistant parts, valves and pump bodies
UNS No. F34800	3.5–3.8% C, 2.0–2.8% Si	483 (70) (mostly pearlite	3	Best combination of strength and wear resistance; can be surface hardened; crank
		Malleable		
UNS No. F22200, ASTM A338	2.3–2.7% C, 1.0–1.7% Si	224 (32)	10	Low-stress parts required; good machinability: steering-gear housings, brackets, carriers
ASTM A602 M5503	2.2–2.9% C, 0.9–1.9% Si	379 (55) (tempered martensite)	3	For machinability and response to induction hardening
		White		
Class I, D	2.5–3.6% C, 1.3% Mn, 1.0–2.2% Si, 5–7% Ni, 7–11% Cr	Hardness[a] 500–600 Brinell, tempered martensite	—	For abrasion resistance

[a] Hardness only reported for white cast irons.

Source: Adapted from *Metals Handbook*, Vol. 1, 10th ed., ASM International, Metals Park, OH, 1990. (With permission.)

TABLE 16.8
Aluminum Alloy Commercial Code System

Series Designation Number	Major Alloying Elements
Wrought Alloys	
1000	Commercially pure, Al 99%
2000	Copper
3000	Manganese
4000	Silicon
5000	Magnesium
6000	Mg and Si
7000	Zinc
8000	Other elements
9000	Unused series
Cast Alloys	
1XX.X	Unalloyed
2XX.X	Copper
3XX.X	Silicon +Cu and Mg
4XX.X	Binary Al–Si
5XX.X	Manganese
6XX.X	Unused
7XX.X	Zinc
8XX.X	Tin
9XX.X	Unused

In addition to the four-digit alloy code, a condition code is added as a suffix. The suffix is a combination of numbers and letters that designate treatment. They are:

- F: as fabricated
- O: annealed
- H: strain hardened
- T: heat treated
- W: solution treated

For alloys that are not heat treatable, strengthening is accomplished by strain hardening, and the degree of strain hardening is specified by a number from 1 to 9. An example would be 3003 H 14. The H signifies strain hardening, the 1 is a code for the specific process, and the 4 indicates a condition halfway to the maximum tensile at numerical designation 8. The number 9 is given to a condition approximately 10 MPa greater than the H8 condition. The specific values of strength are per alloy grade.

TABLE 16.9

Typical Mechanical Properties of Representative Nonheat-Treatable Wrought Aluminum Alloys

Alloy	Nominal Composition	Temper	Tensile Strength (ksi)	Yield Strength (ksi)	Elongation (% in 2 in)	Hardness (Brinell hardness number [Bhn])
1199	99.99+% Al	0	6.5	1.5	50	—
		H18	17	16	5	—
1100	99+% Al	0	13	5	35	23
		H14	18	17	9	32
		H18	24	22	5	44
3003	1.2% Mn	0	16	6	30	28
		H14	22	21	8	40
		H18	29	27	4	55
5005	0.8% Mg	0	18	6	36	30
		H14	23	22	6	41
		H18	29	28	4	51
3004	1.2% Mn	0	26	10	20	45
	1.0% Mg	H34	35	29	9	63
		H38	41	36	5	77
5052	2.5% Mg	0	28	13	26	47
		H34	38	31	19	68
		H38	42	37	7	77
5456	5.1 % Mg	0	45	23	24	70
	0.8% Mn	H343	56	43	2	94

For heat-treatable alloys, a T suffix is used with a number from 1 to 10, where each number designates a specific process. More details for each of the systems can be found in the ASM aluminum and aluminum alloys handbook. Examples of property management techniques are shown in Table 16.9 and Table 16.10.

16.6 OTHER NONFERROUS ALLOYS

16.6.1 SUPERALLOYS (NICKEL AND COBALT BASE)

Nickel-based alloys, although not used in the quantities of the copper and aluminum alloys, still play a major role in industrial applications, particularly at elevated temperatures. *Superalloys* have been defined as those possessing good high-temperature strength and oxidation resistance. They are alloys of nickel, cobalt, and iron that contain large amounts of chromium (e.g., 25 to 30%) for oxidation resistance. They are generally classified as iron–nickel-, nickel-, and cobalt-based alloys. For many years cobalt-based superalloys held the lead, but owing to the precarious availability of cobalt, primarily from South Africa, the

TABLE 16.10
Typical Mechanical Properties of Representative Heat-Treatable
Wrought Aluminum Alloys

Alloy	Nominal Composition	Temper	Tensile Strength (ksi)	Yield Strength (ksi)	Elongation (% in 2 in)	Hardness (Bhn)
2219	6.3% Cu	0	25	10	20	—
	0.3% Mn	T37	57	466	11	117
		T87	69	57	10	130
2024	4.4% Cu	0	27	11	20	47
	1.5% Mg	T4	68	47	20	120
	0.6% Mn	T6	69	57	10	125
		T86	75	71	6	135
4032	12.2% Si	T6	55	46	9	120
6061	1.0% Mg	0	18	8	25	30
	0.6% Si	T4	35	21	22	65
		T6	45	40	12	95
7005	4.6% Zn	0	28	12	20	—
	1.4% Mg	T6	51	42	13	—
7075	5.6% Zn	0	33	15	17	60
	2.5% Mg	T6	83	73	11	150
	1.6% Cu	T73	73	63	13	—
7001	7.4% Zn	0	37	22	14	60
	3.0% Mg	T6	98	91	9	160
	2.1% Cu					

nickel-based superalloys have now been studied more extensively and have replaced many of the cobalt-based alloys.

The physical metallurgy of these alloys is due basically to the precipitation of a very fine distribution of small particles, primarily Ni_3Al and Ni_3Ti, that have the generic name *gamma prime* in a gamma matrix (Figure 16.10). The nickel–iron-based alloys also have a γ phase in the form of a Ni_3Nb compound. The superalloys are *dispersion-hardened* alloys because they achieve their strength by a fine dispersion of these compounds. Although these compound particles are most often obtained via an aging or precipitation heat treatment, they do not develop the coherency strains that the true precipitation hardening alloys do. These particles still resist dislocation motion and thereby strengthen the base metal, but their beauty is that they resist growth (coalescence) at elevated temperatures. Thus, they have found their niche in the turbines and hot components of jet aircraft. Dispersion-hardening alloys do not over-age as readily as precipitation-hardening alloys. This is their basic difference. Furthermore, dispersion-hardening alloys can be made by processes other than the conventional precipitation reaction. Oxide dispersion-hardening alloys are manufactured by mixing oxide powders with metallic powders by a mechanical alloying process

FIGURE 16.10 Gamma prime in s gamma matrix in a Mar 247 alloy. (Courtesy of T. Devaney, Cal. Poly.)

that involves the repeated fracturing and rewelding of a mixture of powder particles in highly energetic ball mills. Nickel alloys do have a numbering system that in some ways attempts to relate the alloy numbers to the trade names, but most alloys are still known by their trade names. The letters MA preceding the number signify that the alloy was produced by the mechanical alloying process.

It should be pointed out here that the time to fracture at a certain stress and temperature is more important for superalloys than the decrease in strength as a function of temperature. The stress for rupture after 100 h versus temperature for superalloys is summarized in Figure 16.11. For life at 1,000 and 10,000 h, see the reference for Figure 16.11. Some typical stress-rupture data are listed in Table 16.11. Results are also included in the table for the single-crystal turbine blades.

In Chapter 2 one of the mechanisms of creep and rupture discussed was that caused by the grains sliding past one another at high temperatures under only moderate stresses. In the 1970s, single-crystal superalloy turbine blades were developed (Figure 16.12) to eliminate grain boundaries and their sliding. These

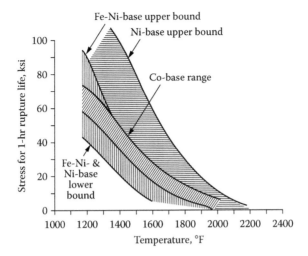

FIGURE 16.11 Stress-rupture vs. temperature data for a 100 h life of superalloys. (From M. J. Donachie, *Metals Handbook*, Desk ed., ASM International, Metals Park, OH, 1985, p. 165. With permission.)

TABLE 16.11
Stress-Rupture Data for Selected Superalloys

Alloy	Major Element Composition (wt %)	1000-h Stress-Rupture	Temperature (0°C)
Inconel 718	52.5 Ni, 19.0 Cr, 18.5 Fe	225 (37)	750
Inconel × 750	73.0 Ni, 19.0 Cr, 7.0 Fe	125 (18)	800
Udimet 720	55 Ni, 17.9 Cr, 14.7 Co	340 (49)	800
Incaloy MA 754	78 Ni, 20 Cr, 0.5 Ti, 1.0 Fe, 0.6 oxide	149 (21.6)	760
Incoloy MA 956	75 Fe, 20 Cr, 0.5 Ti, 4.5 Al, 0.5 oxide	63 (9.1)	982
Haynes 188	37 Co, 22 Cr, 22 Ni, 14 W	165 (24)	750
Incoloy 901	13 Cr, 40 Ni, 45 Fe	240 (35)	750
Mar M 246 cast	60.0 Ni, 10.0 Co, 9 Cr, 10 W, 5.5 Al	470 (68)	800
CMSX 2 single crystal	64 Ni, 8 Cr, 5.0 Co, 8.0	325 (47)	870

single-crystal alloys are now used in many jet engines, especially in high-speed, high-temperature military aircraft. Ceramic and composite turbine blades are the materials predicted at the time to replace the single-crystal superalloy blades, but these materials will require considerable improvement and testing before their reliability is established.

Not all nickel-based alloys are used for high-temperature applications, the Monels and some solid-solution Inconels being the most notable exceptions. The properties of some nickel-based alloys are listed in Table 16.1. For more extensive

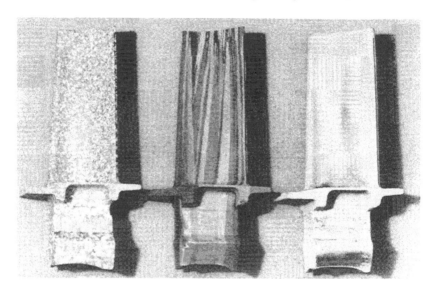

FIGURE 16.12 Single-crystal superalloy turbine blade. The left blade is a polycrystalline, the center one has long columnar grains, and the right one is a single crystal. (From D. H. Maxwell and T. A. Kolakowski, *Quest: New Technology at TRW*, Autumn 1980, p. 48. With permission.)

coverage, see Table 16.5, pages 437–440, of Vol. 2 of ASM *Metals Handbook*, 10th ed.

16.6.2 TITANIUM AND ZIRCONIUM ALLOYS

Titanium was described as the *wonder metal* during its early development period of the late 1940s. It possesses a density of 4.54 g/cm^3, which is much lower than iron, but light enough and strong enough to be used in aircraft structural members. One of the first alloys developed, Ti-6% Al-4% V (UNS R56400), has a higher strength-to-weight ratio than that of either aluminum or steel, or of any other commercial alloy. This alloy still accounts for about 55% of titanium use. This alloy has tensile strengths on the order of 1140 MPa (160 ksi), with ductilities as high as 60% R. A.

Titanium is polymorphic. Below 883°C, it has the hexagonal close-packed (HCP) structure called α-titanium but transforms to the BCC beta structure when it is heated to 883°C. A pseudo-binary-phase diagram, which shows how the Ti + 6% Al alloy phases vary with temperature and vanadium content, is shown in Figure 16.13. This type of phase diagram avoids the complexity of ternary diagrams. When titanium and some of its alloys are water quenched, a type of martensite is formed, but it is not a hard martensite like that obtained in the steels, partially because it is not supersaturated with interstitial elements. The Ms line is shown in Figure 16.13 for this alloy. By a variety of heat treatments, such as quenching, aging, and so on, the quantities of the α and β phases can be altered

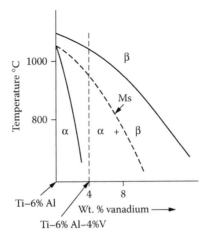

FIGURE 16.13 Psudo-binary-phase diagram showing how the phases of the Ti–6% Al alloy varies with temperature and vanadium content.

as well as the distribution of these phases. Ti-6% Al-4% V is thus known as an α–β alloy. The heat treatments on this and other titanium alloys are quite varied. Polymorphic alloys have the unique ability to change the quantities and distribution of the high- and low-temperature phases via the heat treatment process. This is somewhat different from the control of phase distribution by quenching and aging that was described in Section 16.4 for the precipitation of small particles. The latter dealt with a rather small quantity of the second phase (compound), whereas the α–β titanium alloys we are dealing with have more massive quantities of the two phases. In both cases, of course, a wide range of properties are available, dependent on the specific heat treatment selected.

More recently, some all-β and β plus small quantities of α alloys have been developed with somewhat higher strengths than the α–β alloys, but with less ductility. One of the more promising β alloys is the one designated as Ti 1023, of composition 85% Ti, 10% V, 2% Fe, and 3% Al.

Because of its high strength-to-weight ratio, titanium has been used mostly in the aircraft industry as forged support components and as the fuselage for aircraft flying beyond Mach 1 speeds. The Concorde and some military fighter aircraft have fuselage skin temperatures that are too high for the weaker aluminum alloys.

Titanium has not quite lived up to its initial hopes and predictions. For aircraft usage it is limited to temperatures of about 500°C because of its high reactivity with oxygen and nitrogen at higher temperatures. Unless protective coatings can be developed, titanium cannot be used in the hot regions of jet engines. Furthermore, as these operating temperatures increase, Ti alloys will not possess the required high-temperature strength. The melting points of titanium alloys are not much greater than those of the superalloys.

While these aerospace developments were proceeding, other applications for titanium have been realized. Titanium is a passive metal. Just like aluminum and chromium, an adherent oxide film (TiO_2) forms a protective surface layer almost immediately upon exposure to air. It can be anodized to control the oxide thickness. It is now used for surgical implants and prosthetic devices, for which good corrosion resistance, strength, and ductility are required. Chemical processing equipment, parts exposed to salt water, and similar corrosion-resistant applications have used some titanium alloys. Even in the automotive industry, some limited usage has been found in valves, springs, gears, and so on. Probably the most significant factor limiting titanium usage is its high cost, being more than five times that of aluminum on a weight basis. This cost is slowly being reduced by increased consumption and cost-reduction processing, but barring a major breakthrough, it will probably never compete on a cost basis with aluminum and steel.

Zirconium, a sister metal to titanium, was first used extensively in the nuclear industry because of its transparency to neutrons, but zirconium also has excellent corrosion resistance in a variety of media. Accordingly, it has been found to be of some use in the chemical industry. Zirconium is polymorphic, and like titanium, heat-treatable alloys have been developed with strengths as high as 587 MPa (85 ksi). It is not a light metal, however (6.5 g/cm^3 density), and because of its high cost and only moderate strength is unlikely to be used in many structural applications.

16.6.3 Low-Melting-Point Metals

Zinc is probably better known as the galvanized coating on steel than as a structural member. Zinc castings, particularly die castings, which are of rather odd shapes and difficult or expensive to manufacture by other means, are one case where zinc is used. Die castings are usually limited to low-melting-point metals (419°C for zinc). Zinc castings are used in the automotive industry as carburetors, wiper parts, speedometer frames, and a host of similar small parts, primarily because of the ease with which zinc can be cast into intricate shapes. Zinc castings are also used extensively in general hardware, electronic and electrical fittings, and domestic appliances. Zinc is also rolled, extruded, forged, and drawn. These wrought products include roof flashing, dry-cell battery cans, ferrules, gaskets, and solar collectors, to name a few.

Tin is used as a protective coating for steel (tinplate). It is a low-melting-point, low-strength metal. Its most popular application is as a solder when alloyed with other low-melting-point metals, such as lead, antimony, or silver. Since lead has been banned for internal household plumbing, silver is now frequently used, a 95 wt % Sn–5 wt % Ag being a popular composition.

Lead has found limited usage as lead–tin babbitt alloys (bearings). Pure lead is used widely for nuclear shielding and in $Pb–PbO_2$ batteries.

Indium, another low-melting-point metal, is used in solders. In the electronic industry it has been the most prominent solder, essentially as pure indium, since it makes low-ohmic contacts to silicon and gallium arsenide substrates. However, pure indium is expensive (about $500 per pound) and is often sold by the troy ounce.

16.6.4 PRECIOUS METALS

Precious metals are considered to be those high-priced, fairly inert metals that include gold, platinum, palladium, and silver. Silver is relatively inexpensive compared to the others and as such is more widely used, particularly as silverware. It is less inert than the others. The precious metals over the years have been used primarily as jewelry, art objects, and dental alloys (also as corrosion-resistant platings). The electronics industry is now a large user of these metals in both thin- and thick-film forms such as conductors, capacitors, and protective layers.

16.6.5 REFRACTORY METALS

Because of their high melting points (2468 to 3387°C), the refractory metals niobium, tantalum, molybdenum, tungsten, and rhenium have been widely studied for use in the higher-temperature regions of jet engines. Protective coatings, however, are needed since these metals are very reactive with hot gases. The best course of action is still being pursued. The strengths at high temperatures for some refractory metal alloys are given in Table 16.12.

Tungsten, of course, has been used for many years as light-bulb filaments. Tantalum has good corrosion resistance and accordingly is used to some extent in chemical process equipment. The largest use of tantalum is in electrolytic and foil capacitors. Molybdenum is an alloying element in steel (<3%). Molybdenum-based alloys have found some applications in the missile industry as high-temperature structural parts such as nozzles, leading edges of control surfaces, supports, vanes, and struts. Niobium, the lowest-melting-point refractory metal, has also found some limited use in hypersonic flight vehicles. It is easier to fabricate than molybdenum and competes with it for high-temperature applications. Niobium also has a low-neutron-absorption cross section and competes to some extent

TABLE 16.12
High-Temperature Strength of Some Refractory Metal Alloys

Alloy	Composition (wt %)	Temperature (°C)	Strength [MPa ksi)]
FS 85	Nb, 28 Ta, 10 W, 1 Zr	1000	414 (60)
HWM	Nb, 25 Mo, 25 W, 1 Hf	1000	1352 (196)
W–1% Re		1400	862 (125)
Ta–10% W		1000	304 (44)

with zirconium for nuclear applications. Niobium–tin and niobium–titanium are superconducting alloys at temperatures in the vicinity of 20 K. Protective coatings for niobium are further advanced than those for the other refractory metals and thus have a good possibility for more extensive high-temperature use.

16.7   FORMING BY MECHANICAL MEANS

In metalworking applications, over 200 processes have been developed to convert raw material to a final product; these processes may be separated into categories listed as follows (after G. E. Dieter): (a) direct compression, (b) indirect compression, (c) tension-type processes, (d) bending processes, and (e) shearing processes. These basic processes are depicted in Figure 16.37. Many variations and combinations of these basic processes are in use today, and references to specific processes can be found on the Internet or in the technical literature.

In *rolling and forging* operations, direct compression forces are used to move metal at right angles to the compression direction. Indirect compression includes *drawing* and *extrusion,* where the metal flows under the action of combined stresses that include a high compressive stress. Seamless tubes are made by extruding the metal over a mandrel.

The forging operation includes breaking down the as-cast structure of an ingot or cast bar shape (open-die forging or rolling) to refine the grain size, or processing a net or near-net shape in a closed-die or extrusion processes. In the former the ingot is often reduced to size by repeated blows from a drop hammer, whereas in the latter a single squeezing action is employed, causing the metal to be slowly squeezed by pressure into a mold of the final shape desired. Rolling includes both sheet and rod shape, the latter not being shown in Figure 16.14. In rod or round rolling the matching roll surfaces contain grooves, starting at one side of the roll with a large diameter and progressing in steps to the other side of the roll, which contains the smallest groove. The smallest diameter produced in the round rolling operation is about 1/8 in. Further processing is done by drawing the 1/8-in rod through successive wire drawing dies — a very fast, inexpensive, continuous operation. There also exists the less frequently used *swaging* operation (Figure 16.15), which is employed for small runs to form rod and large-diameter wire. In this process the rod is usually fed through die sections that when placed together form a circular shape. There are two- and four-die-section swagers. The metal diameter is reduced by a hammering action created by a set of revolving hammers, which alternately strike the die sections as the hammers revolve through 360°. Gun barrels are made this way by predrilling a hole in the rod and hammering the metal down onto an inserted mandrel of the finished diameter size. Swaging provides a compressive stress and is useful for metals that cannot withstand the tensile force of a drawing operation.

Superplasticity is a unique forming process that can be used on metals and alloys that exhibit superplastic behavior. Superplasticity occurs by grain boundary sliding and is accompanied by grain rotation. It occurs in certain metals at high temperatures (e.g., 0.7 Tmelting [Tm]) and at low strain rates on the order

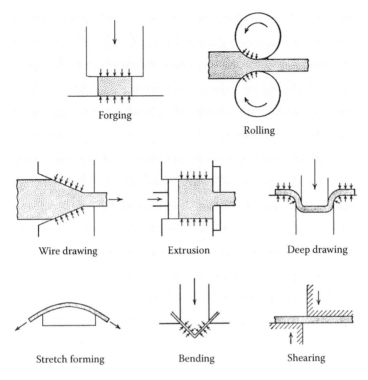

Forging

Rolling

Wire drawing

Extrusion

Deep drawing

Stretch forming

Bending

Shearing

FIGURE 16.14 Mechanical metal-forming processes. (From G. E. Dieter, *Metallurgy,* 3rd ed., McGraw-Hill, New York, 1986, p. 504. With permission.)

FIGURE 16.15 Rotary four-die swaging machine. (Courtesy of Penn Mfg. Co.)

of 10^{-3} per second. Tensile elongations as high as 4800% have been obtained in some superplastic alloys. A fine grain size, <10 μm, is required for this behavior. The grain shape is preserved during deformation, contrary to other mechanical forming methods. Superplastic forming of commercial aluminum-, titanium-, and nickel-based high-temperature alloys is done routinely, but not a large number of alloys exhibit this behavior. Another drawback to superplastic forming is the low strain rates required, which result in low production rates and corresponding high costs.

Metal injection molding is a near-net-shape process that has been used for years in the polymer industry but has only recently been considered seriously for metal processing. The starting material consists of fine metal powder mixed with up to 50 vol % of a polymer binder material. After being injected into mold cavities, the binder is removed by either solvent extraction or thermal treatment. The metal powder forms are then sintered to densities of around 95%. For small parts this process competes with investment and die casting, powder metallurgy methods, and machining processes.

16.8 FORMING BY CASTING METHODS

Casting involves pouring or injecting a molten metal into a mold of the shape of the desired product. Casting methods range from very basic sand casting with low upfront cost, fast production cycles, and near net shape products to investment casting with long, involved precast processing with an end product that is ready to be used directly in an application.

16.8.1   Sand Casting

This casting method is the oldest of all methods, and foundries have produced medium- to large-size parts by this method for centuries. A two-piece mold is formed by packing sand around a pattern of the final desired shape. The second mold piece filled with packed sand fits over the first one containing the pattern. They are called cope and drag sections, respectively. Passage ways called gates and runners are provided in the mold for filling of the mold cavity (mold plumbing). The pattern is a reusable item and can be made of metal, wood, plastic, or foam, depending on the production quantities needed. Sand castings are used for many components such as engine blocks, break parts, machine bases, and cooking pots. This process provides a low-cost shape that requires additional finishing prior to use.

16.8.2   Investment Casting

Other terms for this process include precision casting and the lost wax process. The one-time use pattern is usually made from wax or a low-melting, easily formed polymer. The pattern is surrounded by ceramic slurry and an aggregate mixture in layers to develop the desired mold wall thickness. After the slurry–aggregate mixture is dried, the wax is melted out and the ceramic is hardened by firing at an elevated temperature. Molten metal is then poured into the resulting cavity. Whereas in sand casting the pattern is reusable but the mold is expendable, in investment casting a reusable metal pattern die is used to produce expendable wax patterns, which in turn are used to produce expendable ceramic molds. The majority of investment castings are gravity cast, but other modifications to the process have included centrifugal or vacuum-assisted casting methods to improve the speed and reliability of the casting process. Investment casting has been used for over 2000 years to produce intricate detailed shapes, such as jewelry and rings. Modern applications include dental alloy fixtures such as crowns and inlays, prosthetic hardware such as hip joints, golf club heads, and turbine blades for jet engines. Investment castings offer precise dimensions, very fine surface finish, and excellent properties, but at a premium cost.

16.8.3   Die Casting

Die casting is very similar to injection molding of plastics except that liquid metal is forced into a permanent mold, often made of steel. It is very useful for long runs of low-melting-point metals such as small parts of zinc, magnesium, and aluminum alloys. The split die is closed and the molten metal flows into a warm piston pump chamber where a plunger drives the metal into the die while the air escapes through vents. The die is then opened, the part is ejected, and the die is rapidly water cooled, closed (locked), and ready for the next cycle. Thousands of pieces can be made from each die before rework or replacement is required. It is a fast process. The equipment is expensive, and large production runs are needed to justify the costs. Products made by die casting include zipper teeth, automotive door hinges, transmission parts, and aluminum engine castings.

Several modern adaptations of the casting processes include Thixocasting® and semisolid metal casting. In each of these processes a specific property feature is exploited to address a market. In Thixocasting a specially prepared magnesium ribbon is fed into a modified injection molding machine and a semimolten metal is forced into the casting cavity. Very thin (1 mm) parts can be formed this way. Applications for this process include cell phone housings and portable CD player decks.

In the semisolid metal casting process, a specially shear-formed ingot of aluminum is heated to a point in the liquid–solid range of the phase diagram and pressed into shape. This process allows the formation of thin aluminum parts with low production pressures and temperatures with properties greater than achieved by normal die casting processes. Typical commercial applications include fuel rails and transmission parts.

16.9   POWDER METALLURGY METHODS

Powder metallurgy dates back to the years prior to World War II when the process was used on a limited scale that involved the pressing of metal powders, usually

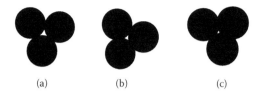

(a)   (b)   (c)

FIGURE 16.16 Three-particle sintering model.

for small parts that were difficult to cast or machine, and the subsequent sintering of these powder compacts to bind the powders together. The sintering process is illustrated in Figure 16.16. Here we consider three small spherical powder balls that have been pressed with sufficient plastic deformation that they are in contact along a short segment of their interface. When they are heated above about 0.5 T_m, diffusion of the atoms occurs sufficient to form a bridge between the particles and thereby decrease the pore size that existed among the three particles. As diffusion continues, the pore size decreases further, as shown in Figure 16.16c. Ideally, one would like to have zero porosity, but this is probably never achieved with pressing and sintering operations. Densities of the order of 95 to 97% of theoretical are considered quite good.

Some time in the 1950s, hot pressing, at least on an experimental level, began to be used. Pressing and sintering simultaneously caused much more plastic deformation to occur and thus provided a much larger interface contact area over which diffusion could take place. Densities in excess of 95% were common with the hot pressing process. Today, connecting rods for most production cars are a powder metal compact that is forged to its final density and dimensions.

Room-temperature isostatic pressing was the next development, which, via the equal application of a hydrostatic pressure in all directions on the part, yielded a more uniform density. This process, commonly referred to as *cold isostatic pressing*, uses high fluid pressures to densify powders in rubber molds. After pressing and vacuum sintering, uniform densities in excess of 95% of theoretical are quite common.

Hot isostatic pressing (HIP) has been developed over the past 40 years and is now an established but expensive process used in the aerospace and defense, automotive, marine, medical, and electronic industries. The HIP process involves the application of high pressures via inert gases to pressure-tight containers in which the powders or castings are enclosed. After pressing at the required high temperatures, the metal container is removed. Uniform densities approaching 100% of theoretical have been achieved. This process has been used for near-net-shape fabrication of titanium alloys, superalloys, stainless steels, and aluminum alloys. A 16-in-diameter HIP unit that attains temperatures to 1343°C (2450°F) and pressures to 310 MPa (45 ksi) is shown in Figure 16.17.

Although not envisioned originally, a significant beneficial aspect of the HIP process has been found to be densification of castings. The minimization or elimination of porosity in castings has been a long-sought goal. HIP has been

FIGURE 16.17 Sixteen-inch-diameter HIP Press. (Courtesy of International Materials Technology, Inc.)

described as a "defect-healing" process for investment castings of titanium- and nickel-based superalloys. HIP of previously cold isostatic pressed parts has been found to be a good two-step process to achieve high densities and net or near-net shapes. This process, sometimes referred to as the CHIP process, has been used for titanium, nickel, iron, tantalum, and molybdenum alloys. Some typical parts are shown in Figure 16.18. The HIP and CHIP processes have also been extended to the forming of near-net-shape ceramic parts and metal matrix composites and to high-pressure impregnation of carbon to create carbon–carbon composites.

Despite all the new processes developed or being explored, the press and sinter method is still the backbone of the powder metallurgy industry, primarily because of cost. The production of iron and steel powder compacts by the press and sinter route is used by the cost-conscious auto industry as well as by other industries involved in large quantities and low costs per part.

16.10 METAL JOINING

16.10.1 WELDING

The American Welding Society defines a weld as a "localized coalescence of materials, wherein coalescence is produced by the heating to suitable temperatures,

FIGURE 16.18 Number of high-temperature alloys that have been produced in a variety of shapes by the CHIP process. (Courtesty of Dynamet Technology, Inc.)

with or without the application of pressure and with or without the application of a filler material." Note that no mention is made of melting. Furthermore, there is no mention of the type of heat source used. The various processes include (a) ultrasonic welding, where vibration of the part produces the atomic closeness and cleanliness requisite for bonding on the atomic scale; (b) low-temperature "forge welding" operations, where heat and pressure are superimposed to produce joining; (c) diffusion bonding, where tower pressures and higher homologous temperatures (still below melting) are used; (d) transient liquid processes; (e) resistance welding; and (f) the common arc and beam processes, where melting of the materials produces closeness and cleanliness on the atomic scale.

The most common arc welding processes are shielded metal arc welding, gas metal arc welding, gas tungsten arc welding (or Till), plasma arc welding, and submerged arc welding. Typically, shielded metal arc welding is manual (travel speed and wire feed controlled manually) or automatic (no manual control). The process selection depends on the material, the plate thickness, joint accessibility, and other factors. High-energy density processes or automated processes can often overcome material weldability problems but typically increase the cost of capital equipment and joint preparation. Electron-beam fusion welding is an example of such a process.

Welding is a decidedly nonequilibrium process and produces changes not only in the fused metal, but also in the adjacent base metal, called the heat affected zone (HAZ). The HAZ for a typical one-pass submerged arc weld in steel is depicted schematically in Figure 16.19. The rapid thermal cycle associated with the passage of the welding arc produces nine distinct regions in the vicinity of

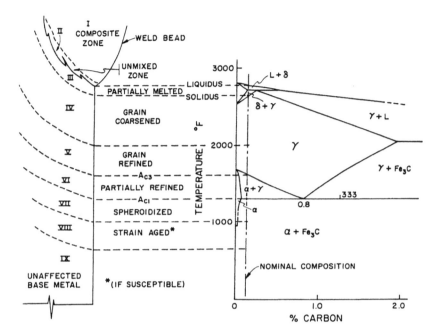

FIGURE 16.19 Schematic of the heat-affected zones (HAZ) in the fusion welding process. (Courtesy of D. Walsh, Cal. Poly.)

the weld, each with different microstructures and properties. Historically, in steels many problems have been associated with the partially melted and grain-coarsened regions, where the rapid thermal cycling may produce hard and brittle martensite.

Often, hydrogen present during the welding operation will diffuse into these areas and produce hydrogen-induced cracking. This may occur immediately postwelding or after a delay of as long as several weeks. This insidious problem obviously poses a challenge for nondestructive evaluation. Postweld heat treatment, preheat, and welding energy input limitations (to control cooling rates) are used in an attempt to circumvent all such problems.

Figure 16.20 highlights the grain-coarsened region of a submerged arc weld. Figure 16.21 shows the fusion-line region of a weld in 316L stainless steel. The dendritic structure of the fusion zone is evident. Ferrite is the dark etching vermicular phase. The partially melted region is also evident, as the complete grains in the HAZ are surrounded by liquated material. Full liquation of several alloy-rich inclusions is also evident. Such constitutional liquation is caused by rapid heating associated with the welding operation.

Hardfacing is the application of a hard, wear-resistant material to a metal surface by welding, spraying, or allied welding processes. This definition excludes nitriding, ion implantation, flame hardening, and similar processes. Hardfacing materials include a variety of cobalt-, nickel-, and iron-based alloys plus some ceramic materials such as carbides. Service performance requirements dictate

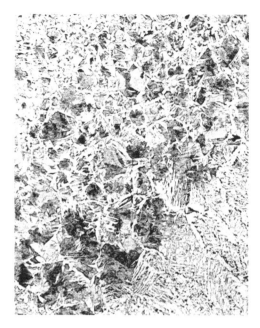

FIGURE 16.20 Microstructure showing the coarse-grained region in a submerged arc weld on a 1030 steel. x200. (Courtesy of D. Walsh, Cal. Poly.)

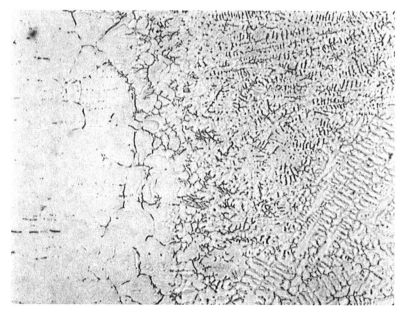

FIGURE 16.21 Fusion line and dendritic structure ina 316L stainless steel weld. x250. (Courtesy of D. Walsh, Cal. Poly.)

both the hardfacing material and the process selection. *Thermal spraying* is considered here as an allied welding process since the surface is fused. As a rule, welding is preferred for hardfaced coatings.

16.10.2 BRAZING AND SOLDERING

Brazing and soldering involve the use of low-melting-point filler metals that are placed between two higher-melting-point metals. The latter are not fused, whereas the filler metal is melted and chemically bonds to the metals being joined. It was once thought that the difference between brazing and soldering was that in brazing a chemical bond was formed, whereas in soldering it was a more mechanical-like bond, with the solder interlocking with irregularities on the metal surface. It is now realized that in both processes a chemical-like metallurgical bond is obtained, albeit to a miniscule depth. Thus, the American Welding Society has, somewhat arbitrarily perhaps, defined soldering as metal coalescence below 426.7°C (800°F). The term *coalescence* could also have a variety of meanings. Perhaps the best way to describe soldering is that the filler metal is distributed between the closely fitted surfaces by capillary action. To add to the confusion, brazing is defined in the same way, except that the temperature is above 800°F.

Brazing a large number of units is normally performed in a furnace, while torch brazing is used for smaller production runs or repair operations on a single unit. Filler metals are usually silver alloys or Cu–Zn alloys. The silver alloys have melting temperatures in the range of 593°C (1100°F) to 704°C (1300°F), while the Cu–Zn alloys are somewhat higher, in the neighborhood of 871°C (1600°F). Fluxes, usually consisting of boric acid, borates, or fluorides, are required to decompose oxides (except in vacuum brazing operations).

Formerly, soldering alloys consisted primarily of alloys of lead and tin, but because of the environmental hazards caused by lead, its use in industrial applications has been reduced dramatically, and it is being replaced by other alternatives.

DEFINITIONS

Age hardening: Heat-treating process that increases hardness via precipitation of some type of phase, often a compound.

Austempering: Isothermal heat treatment of steel below the nose of an IT curve but above the Ms temperature that results in an all-bainitic structure.

Bainite: Microstructure that consists of a mixture of α-iron with small Fe_3C particles that has a hardness in the Rc range of 40 to 45.

Binary system or binary phase diagram: Alloy or phase diagram in which more than 80% of the composition is made up of two elements.

Cementite: Fe_3C (iron carbide).

Dendrites: Grain structure in alloys in which growth is favored in certain directions, giving a Christmas tree-like appearance. Snowflakes also have a dendritic appearance.

Die casting: Process of forcing a liquid metal into a specially shaped die.

Dual-phase steels: Steels composed of the α-ferrite and martensite phases.

Eutectoid microstructure: Parallel platelets of two solid phases that result from the transformation that occurs on passing through the eutectoid point; this transformation is the *eutectoid reaction.*

Extrusion: Process of forcing a material through a die into a certain desired shape.

Forging: Method of shaping objects by forcing them into a die or into a shape dictated by the forging machinery and mold or die shape.

Gamma (γ'): Phase, usually an intermetallic compound, that exists in super-alloys and provides high-temperature strength.

Gray cast iron: Iron–carbon alloy of about 2 to 4 wt % carbon in which the carbon exists as flakes, and a fractured surface has a gray appearance.

Homogenization anneal: Heat treatment, generally at high temperatures and for long periods of time, used to diminish segregation of elements. It promotes the establishment of equilibrium conditions.

HSLA steels: High-strength low-alloy steels that are used primarily in large structures and in the machinery industry.

Malleable cast iron: Cast iron produced by annealing white cast iron.

Martensite: Metastable phase formed by quenching.

Matrix: In an alloy, the major and continuous phase.

Mechanical alloying: Process whereby alloying is achieved by repeated fragmentation and welding together of particles in a ball mill.

Microalloyed steels: Types of HSLS steel that result from a small dispersion of carbides combined with a fine grain size.

Monel: International Nickel Corp. trademark for a series of Cu–Ni alloys.

Nodular cast iron: Cast iron consisting of graphite nodules usually surrounded by ferrite and pearlite.

Overaging: Aging beyond the point of maximum strength in the precipitation-hardening process. The loss of coherency strains and coalescence of the precipitate occurs.

Phase diagram: *See* Equilibrium diagram.

Precipitation hardening: *See* Age hardening.

Proeutectic phase: Solid phase that forms first in solidification of a liquid solution of two elements that form a eutectic system. It appears at temperatures below the liquidus and above the eutectic temperature.

Proeutectoid phase: Solid phase that precipitates out of a single-phase solid solution as the solid solution phase enters a two-phase solid range (on the phase diagram) above the eutectoid temperature. Occurs only in eutectoid alloy systems.

Refractory metals: High-melting-point metals Re, W, Mo, Ta, and Nb.

Segregation: Nonequilibrium separation of elements that occurs during alloy solidification.

Solidus: Line or lines on a phase diagram below which all phases are in the solid state.

Spheroidite: Small Fe_3C particles in ferrite that result from a *spheroidizing anneal* for long periods at temperatures just below the eutectoid temperature.

Supersaturated solid solution: Solid solution that has solute in excess of that dictated by the phase diagram. It is a metastable phase usually formed by quenching from a temperature where the solute is not in excess of the solid solubility limit.

Tempering: Process of annealing martensite whereby it decomposes into α + Fe3C.

Terminal solid solubility: Solid solubility permitted by the equilibrium diagram near the pure base metal position on the diagram.

Ternary phase diagram: Phase diagram that shows the phases that exist as a function of temperature and composition for a three-base-element system.

QUESTIONS AND PROBLEMS

16.1   For the composition in Problem 6.1, sketch the resulting microstructure if this alloy were slowly cooled to room temperature.

16.2   List three nonferrous and one ferrous precipitation-hardening alloy along with their strengths and ductilities in the aged condition.

16.3   Using the Cu–Zn phase diagram of Figure 16.1, for the overall alloys composition, compute the relative quantities of α and β phases that exist at 600°C assuming equilibrium conditions.

16.4   Why must elements that have complete solid solubility in the solid state possess the same crystal structure?

16.5   For a 70% Cu–30% Ni alloy at equilibrium at a temperature of 1175°C, compute the amount of the solid phase. What is the composition for all phases within this alloy system at this equilibrium temperature?

16.6   If the alloy in Problem 16.6 became a solid via being poured into a mold, sketch or describe how the composition gradient might exist across the ingot. Sketch the grain structure that would exist in this casting if it solidified slowly but yet too fast for equilibrium to be attained, and show which regions have a nickel content greater than 30%.

16.7   Why is K Monel so much stronger than other Monel alloys?

16.8   Assume that a gold foil was placed on a silicon wafer (assume that foil is 1/10 the weight of the silicon) and heated to 365°C for a period of several hours and then suddenly cooled. Sketch the microstructure and identify the phases that would be present in the solidified compact.

16.9   For a 60 wt % Cu–40 wt % Ag alloy that was melted and cooled to 770°C, where equilibrium was established, compute the quantity of proeutectic α, eutectic α, and total α in the resulting microstructure. Sketch this microstructure.

16.10 Estimate the hardness that one would obtain on a 2024 Al alloy (a) quenched from the all-α region, (b) quenched and aged at 190°C for 24 h, and (c) furnace cooled from the α region.

16.11 Consider pure copper and a Cu–2 wt % Be alloy. Which has the higher strength and which has the higher electrical conductivity? Which would be the better material for a power line and which would be more suitable for electrical contacts where impact is involved?

16.12 The 2024 Al alloy is an age-hardened alloy, whereas most super-alloys are considered to be dispersion-hardened alloys. Explain the difference between these two alloy types.

16.13 Estimate the room temperature strength-to-weight ratio for a 2024 Al alloy aged to the T6 condition, an aged Ti–6Al–4V titanium alloy, and the highest-strength low-alloy steel (Tables 16.1 and 16.6). Assume that the densities of the alloys are the same as those for the pure base metals, that is, Al = 2.7, Ti = 4.5, and Fe = 7.87 g/cm³.

16.14 The refractory metals have a high density. Why are they of so much interest to the aerospace industry?

16.15 For a 1020 steel cooled to room temperature under near-equilibrium conditions, compute the total of (a) the proeutectoid α phase and (b) the total Fe_3C phase. Sketch the resulting microstructure and estimate the alloy's strength.

16.16 Describe two methods of obtaining a very desirable microstructure of about 40 to 45 Rc hardness in a 1080 steel.

16.17 What would be the benefits of adding small amounts (e.g., 0.5 wt %) of Cr and Mo to a 1080 steel? What would be the difference in the IT curves?

16.18 What heat treatment would you give to a [10100] steel slowly solidified alloy to make it useful? What strength would it have after your heat treatment?

16.19 Are there any advantages to oil quenching a 1080 steel vs. water quenching? List the phases and the microstructures that would result in each case.

16.20 Why are microalloyed steels suitable for welding? What types of steel should not be welded?

16.21 Construct a table listing the relative advantages and disadvantages of the three stainless steels: ferritic, martensitic, and austenitic.

16.22 How is malleable cast iron heat-treated to obtain its intended properties?

16.23   Where do the dual-phase steels fit into the overall HSLA category of steels in terms of strength and formability?

16.24   What is the highest-strength practical steel developed to date, and for what purpose might it be used?

16.25   Refer to the ternary Fe–Cr–Ni composition isotherms in Figure 16.4. In each of these isotherms locate the position of the alloy of composition 70% Fe–20% Cr–10% Ni and list the phases present at each temperature.

16.26   Oxygen-free high-conductivity copper (called OFHC) contains substantial quantities of oxygen in the form of oxides. Why is it labeled OFHC?

REFERENCE

1.   Guthrie, R. I. L., Steel processing technology, in *Metals Handbook*, Vol. 1, 10th ed., ASM International, Metals Park, OH, 1990, p. 107.

SUGGESTED READING

American Society for Metals, *Metals Handbook,* 9th ed., Vol. 6, *Welding, Brazing and Soldering,* ASM International, Metals Park, OH, 1983.

American Society for Metals, *Metals Handbook,* Desk ed., ASM International, Metals Park, OH, 1985.

American Society for Metals, *Metals Handbook,* 10th ed., Vols. 1 and 2, ASM International, Metals Park, OH, 1990.

Brandes, E. A., Ed., *Smithell's Metals Reference Book,* 6th ed., Butterworth, Stoneham, MA, 1983.

Dieter, G. E., *Mechanical Metallurgy,* 3rd ed., McGraw-Hill, New York, 1983.

Guy, A. G., *Elements of Physical Metallurgy,* 2nd ed., Addison-Wesley, Reading, MA, 1959.

Massalski, T. B., Ed., *Binary Alloy Phase Diagrams,* ASM International, Metals Park, OH, 1986.

Smith, W. F., *Principles of Materials Science and Engineering,* McGraw-Hill, New York, 1990.

17 Comparative Properties

17.1 INTRODUCTION

When design engineers are constructing their drawings, they must specify the materials for constructing the object or system being designed. Large corporations frequently have materials engineers who work closely with the design engineers throughout the design process. In smaller companies the design engineer often specifies the materials. In the latter case the design engineer will frequently rely on vendors' suggestions or use materials from existing designs of similar objects or systems. In some cases materials handbooks may even be used, and in a small percentage of cases a materials consulting engineer may be hired.

In selecting materials for designs, we can separate the products into two large categories:

- The mundane components, machinery, standard automobile parts, containers, and so on of the mature technologies: In most of these designs the person selecting materials selects those that have been used for years and have a proven record of satisfactory performance, but even for these established products, cost savings can often be made by switching to more recently developed materials whose properties have been well documented and certified by the producers. This is particularly true in the automobile industry, and perhaps these engineers are the more eager of those in the mature industries to risk using relatively new materials and processes after careful materials evaluation and testing in prototype systems.
- The high-tech systems of the aerospace, electronics, nuclear, medical, and some chemical industries: These designs frequently require new or recently developed materials, many of which are *engineered.* The aerospace industry leads the way in development of new materials, and some eventually filter down to the more established technologies. A good example of the latter are the boron and silicon carbide fibers and their composites that were first used in the space shuttle, other aircraft, and missiles that are now being widely used to manufacture sports equipment: skis, golf clubs, tennis rackets, fishing poles, boats, and the like.

In the materials selection process for either of these categories, assuming that we are considering materials that are available or can be made available on reasonably short notice, it helps to know in what ballpark the materials of interest for a particular application might be found. For example, if the materials are going to experience temperatures in excess of 500°C, should we waste our time and effort considering polymers and polymer matrix composites? Let us narrow our field of consideration as quickly as possible. After all, we have many thousands of materials from which to choose. Our fist objective is to reduce this number to a few hundred. This is analogous to the block diagram approach used for detecting failures in electronic equipment. We must first isolate the block of interest. It is the intent of this chapter to establish some general categories that should assist us in narrowing the field of possible materials for any given application.

In this chapter the cost of the materials is not included. These figures frequently change and for many of the newer materials are often unknown.

17.2 COMPARATIVE MECHANICAL PROPERTIES

The mechanical properties of paramount interest in most designs are those of strength, hardness, ductility, fatigue, and fracture toughness. For other than room-temperature applications, these properties must also be known in the temperature range of the intended application. For higher-temperature applications, we must add the creep and stress-rupture properties. In many room-temperature uses and in all elevated-temperature applications, the effect of the environment must be considered. Although in some electronic materials applications the mechanical properties of housings, substrates, and other components of electronic equipment designs are of some significance, usually the electronic and thermal properties are far more important. Consequently, the mechanical properties of electronic components are not considered in this section. Metals, polymers, ceramics, and composites are the materials that will be judged and compared on the basis of their mechanical properties in this chapter.

17.2.1 COMPARISON OF ROOM-TEMPERATURE MECHANICAL PROPERTIES OF METALS, POLYMERS, CERAMICS, AND COMPOSITES

In Figure 17.1 the ultimate tensile strengths of typical materials of the four categories are shown. To conserve space we have elected to use ksi (100 psi) instead of both systems of units. To obtain units of MPa, one can multiply the ksi figures by 6.9. Representative materials from each category have been selected. Space does not permit examination of a larger number of materials.

For other materials and for more detailed data, handbooks or computerized databases must be used. One could use average values of a large number of materials in each category, but many processing variables are involved, and such details affect the average values. By choosing specific materials, at least some of the processing variables, such as heat treatment, porosity, and fabrication methods, can be included. The ductilities and moduli for similar materials are depicted in Figures 17.2 and 17.3, respectively.

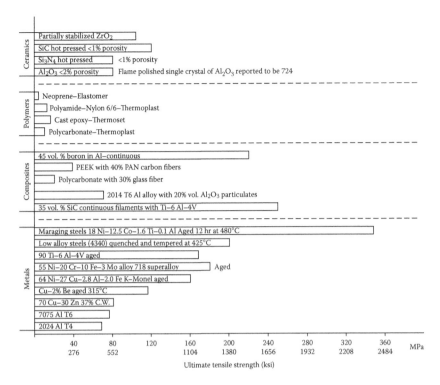

FIGURE 17.1 Ultimate tensile strengths of selected metals, composites, polymers, and ceramics.

17.2.1.1 Fracture Toughness

In Section 2.2.4 fracture toughness measurements by both the impact test, in which the toughness is reported in ft-lb of energy (or Joules) absorbed during fracture of a standard American Society for Testing and Materials (ASTM) Charpy V-notch test specimen, and by the more modern fracture mechanics measurement method were discussed. In the latter test method the results are reported as K_{1c} values, which are true material constants and are usually expressed in the SI units MPa/m or the English units of ksi/in. The conversion factor between the two units is 1.1, so for all practical purposes they are interchangeable. The K_{1c} value is a function of the applied stress and length of the crack at failure. The material must exceed a certain critical thickness as described in ASTM specification E399 in order for the measured K_{1c} to be a true material constant. In comparing the fracture toughness characteristics of the various materials, we must choose either the impact test or the fracture mechanics method. Unfortunately, most polymers have been measured by the impact test method, since the ductile ones, such as polycarbonate, require thickness on the order of 12.7 mm (0.5 in). Nevertheless, a sufficient number of K_{1c} values have been reported on polymers to permit us to choose this more up-to-date method for comparison of the fracture toughness of

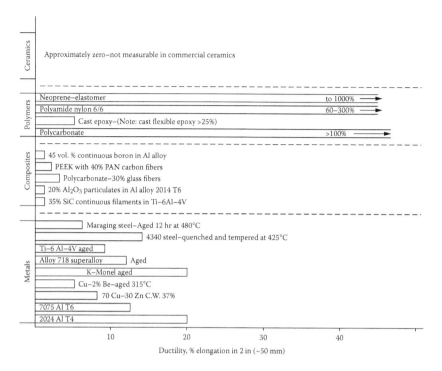

FIGURE 17.2 Ductilities of selected metals, composites, polymers, and ceramics.

the various groups of materials. Most polymers have rather low fracture toughness. The glass-reinforced thermosets, which are really composites, have some of the higher toughness values. In Figure 17.4 representative materials of each category were selected for toughness comparisons. It must be realized that there exists a wide variation in the fracture toughness of metals, depending on the heat treatment and other processing variables. Many designs will specify that K_{1c} exceed a certain value, for example, 25 MPa/m. Other designs may specify lower or higher values, depending on the application.

Some comments on the data in Figure 17.4 are in order here. The fracture toughness values for PMCs have not been reported in K_{1c} terms. Impact data show values ranging from 32 to 43 J/cm notch. In the ceramic–ceramic composites category the highest value of 25 MPa/m was reported for SiC fibers in a SiC matrix. The SiC–Al alloy particulate composites are usually in the range 10 to 30 MPa/m. The high value of 60 MPa/m was reported for an aluminum sheet composite, which probably had some directional variation in its fracture toughness. Perhaps this value should be treated with some caution until it has been well established and all of the conditions clearly understood. Many metals have fracture toughness values higher than those reported in Figure 17.4. One quenched and tempered steel of medium-carbon content (ASTM A533B) was reported to have a 200 MPa/m fracture toughness. Low-carbon steels also have very high

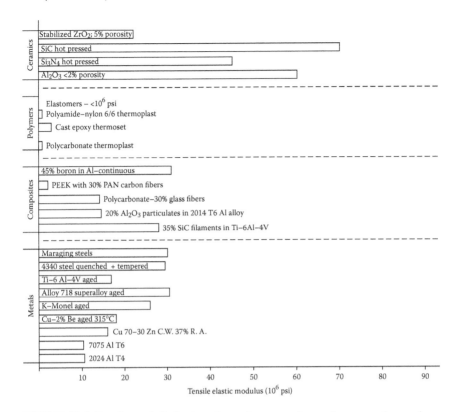

FIGURE 17.3 Tensile moduli of selected metals, composites, polymers, and ceramics.

values of fracture toughness, particularly in the annealed condition. Even the high-strength maraging steels have K_{1c} values on the order of 150 MPa/m.

Certain generalities are noticeable in Figure 17.4. First, the metals category, on the average, has the highest fracture toughness of the four classes of materials listed. There are many metals, including the copper and nickel alloys, that were not included. The nickel alloys were not shown since the Monels (Ni–Cu alloys) have fracture toughness values above those needed for structural purposes. Monel alloy K-500 has an impact strength of 37 ft-lb at room temperature, and the other Monels possess even higher fracture toughness. Many of the other nickel-based alloys consist of the superalloys used at elevated temperatures where fracture toughness is far more than adequate for these applications. Fracture toughness values for the copper alloys, including the high-strength copper–beryllium alloys, are not usually reported, even in the well-known handbooks. They are seldom, if ever, excluded for design purposes because of their fracture toughness numbers.

The continuous fiber SiC–metal composites have good fracture toughness, while the particulate SiC–Al composites have more than adequate toughness in most cases (the higher the SiC content, the lower the toughness). Polymers and polymer composites, together with ceramics and most ceramic composites, have

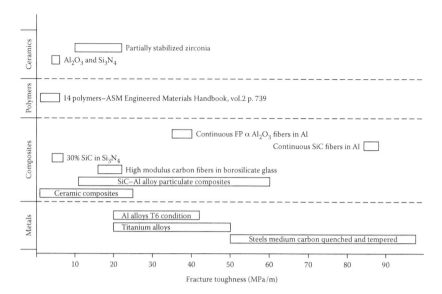

FIGURE 17.4 Fracture toughness of selected metals, composites, polymers, and ceramics.

very low fracture toughness, and their usage is currently limited by this factor. In the composites category, however, we will probably see higher fracture toughness values in the near future, especially in the SiC–SiC fiber composites.

17.2.2 COMPARISON OF ELEVATED-TEMPERATURE PROPERTIES OF METALS, CERAMICS, AND COMPOSITES

What is meant by *elevated* in terms of a temperature scale? Since atoms only recognize the Kelvin absolute temperature scale, the temperature of 77 K (liquid nitrogen) is an elevated temperature with respect to that of liquid helium (4.2 K). In our comparison we will be somewhat more practical and arbitrarily define temperatures above 500°C as elevated temperatures. This will rule out all polymers and most of the reinforced polymer matrix composites. Furthermore, the elevated temperature properties must include creep and stress-rupture data because lifetime is a very important design parameter in the high-temperature range. Just for comparison, Figure 17.5 shows the variation of the strength with temperature of representative members of all groups except for the polymers. Time was not a factor here; that is, the tests were conducted in a period of a few minutes at the usual tensile test strain rates of 0.05 to 0.5 per minute. In ceramic materials one finds a considerable amount of scatter in the data, being particularly large for silicon nitride. Also, ceramic strengths are most often determined in bend tests, which give flexural rather than tensile strengths. In Figure 17.5 the flexural strengths for the bend tests are so noted. The strength of the other materials reported are those obtained in tensile tests.

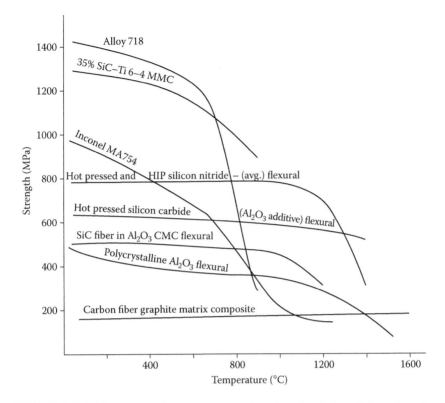

FIGURE 17.5 Ultimate strengths vs. temperature for selected materials. Unless otherwise noted (e.g., flexural), the values were determined in tensile tests.

The data in Figure 17.5 were obtained from a number of sources, most listed in the Suggested Reading, and must be considered as typical rather than precise values. Of particular note is the high strength of the superalloy 718 at room temperature and the sudden drop in strength at about 800°C. This is, of course, the reason behind the development efforts for nonmetallic materials for high-temperature applications. There is a considerable amount of high-temperature tensile strength, creep, and fatigue data on ceramics and composites now being generated at Oak Ridge National Laboratories (Figures 17.6 and 17.7). Returning to Figure 17.5, the ceramics, the MMCs, CMSs, and particularly the carbon–carbon composites, retain their strength to much higher temperatures, even though their room-temperature strengths are somewhat lower than the room-temperature strengths of the superalloys. The oxide-dispersioned superalloys (e.g., MA 754, Ma 6000, and MA 956) retain their strengths to higher temperatures than do the other superalloys. Their advantages will show more clearly in the stress-rupture tests.

The most important function of materials for elevated-temperature applications is their ability to withstand a particular stress and temperature for a reasonable length of time: that of the lifetime of the particular design of interest. In

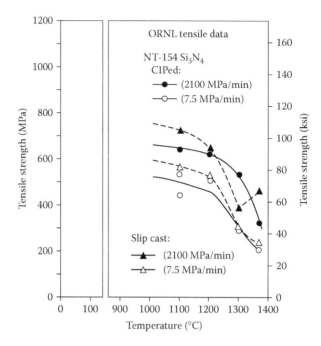

FIGURE 17.6 Tensile strength vs. temperature for CIP and slip cast Si3N4. (From K. C. Liu and C. R. Brinkman, *Proceedings of the Automotive Technology Department Contractors Meeting,* Paper 243, Society of Automotive Engineers, Warrendale, PA, October 1990, p. 238. With permission.)

many instances it is not necessary for the life of the component to be the same as that of the entire system of the design. Certain components are replaceable. The life of a turbine blade in an aircraft jet engine is far less than the life of the aircraft as a whole. The turbine blades can be and are replaced many times. We need to know with some degree of certainty the lifetimes of all components under the conditions of stress, temperature, and environment to which they will be exposed.

Following the argument above leads us to the subject of *creep* and *creep-rupture,* or as it is sometimes called, *stress-rupture.* The three stages of creep were discussed in Chapter 2. Creep and stress-rupture were discussed, but data were not presented. Now we must consider these properties. As a reminder, creep and stress-rupture are of concern only at temperatures above about half of the melting point in Kelvin (i.e., 0.5 Tm). Remember that atoms recognize only the absolute temperature scale. At about 0.5 Tm the atom vibrations are of such amplitude that atom-vacancy exchanges become very probable. Diffusion of atoms occurs. Dislocations can move readily with the assistance of very low stress, far below the short time yield stress at that particular temperature. Dislocation movement creates plastic deformation, albeit a very slow process compared to that in a short-time tensile test. Nevertheless, plastic deformation does occur,

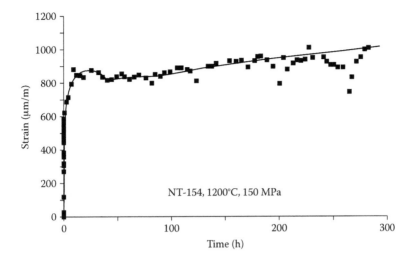

FIGURE 17.7 Creep of HIP Si,N4 at 1200°C and a stress of 150 MPa. (From K. C. Liu, H. Pih, C. O. Stevens, and C. R. Brinkman, *Proceedings of 27th Automotive Technology Department Contractors Meeting,* Paper 230, Society of Automotive Engineers, Warrendale, PA, April 1990, p. 217. With permission.)

and over a long period of time, substantial dimensional changes take place, which we call creep. Eventually, the creep strain will be sufficient to cause rupture, called creep-rupture. The term *stress-rupture,* which has the same meaning as *creep-rupture,* denotes the constant stress to which the material was subjected over this long period of time (from days to years, depending on the stress and temperature). Thus, for elevated-temperature usage, we are not particularly interested in the strength, as determined in tensile, compression, or bend tests, which may span a period of a few minutes. We are more interested in the creep rate, since perhaps the design specifies a maximum dimensional change during its lifetime, or, more often, the time to rupture at a specific constant stress. We will emphasize the latter (i.e., stress-rupture data) in our material comparison picture.

The time to rupture for a typical superalloy (Inconel 617) and two oxide dispersion-strengthened superalloys (Inconel MA 6000 and MA956) and a ceramic material (Si3N4) are shown in Figure 17.8. Unfortunately, there is only a limited amount of data available on the stress-rupture of ceramic materials and high-temperature composites, but the data that are available indicate that ceramics possess the ability to withstand stress at temperatures much higher than those for superalloys. Considerable experimental work is now being conducted on ceramics and composites for high-temperature applications.

In the above-mentioned figures, data are presented for certain stresses and temperatures. If one desires a different set of stress-temperature combinations, methods have been developed for extrapolating from one set of conditions to another over a wide range of temperatures and stresses by the use *master curves* for a given alloy. These master curves have been developed for numerous

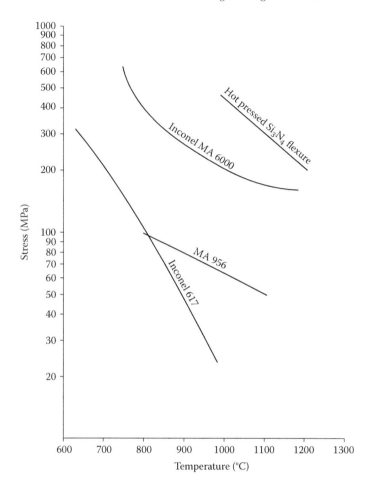

FIGURE 17.8 Thousand-hour stress-rupture curves for superalloys and hot-pressed Si3N4. (Superalloy data taken from Inco Alloys Int., Inc.; Si3N4 data from G. Quinn, *J. Am. Ceram. Soc.,* 73, 2374, 1990. With permission.)

superalloys and for some stainless steels and can be found in the published literature. Unfortunately they do not exist for composites and ceramics.

These curves are not difficult to use or interpret. Generally they involve subjecting the alloy to certain stresses at high temperatures, most often at higher temperatures than the intended application of the alloy, in order to enforce failure in a short period of time. These data are then used to extrapolate to lower temperatures of application where this particular alloy will be used for much longer periods than is practical to conduct stress-rupture tests. These methods are referred to as parameter methods, the best known and most widely used being the Larson–Miller (L–M) method developed in 1952. Other parametric methods have also been developed, but we will explore only the L–M method. The answers given by the other methods do not vary widely because they are all based on the

Arrehenius relationship, where a certain process, in the present case either creep strain rate or the time to rupture, is an exponential function of temperature. In the case of the time to rupture, t, it becomes

$$t = A \exp\left(\frac{Q}{RT}\right) \tag{17.1}$$

Taking logarithms of both sides, we have

$$\ln t = \ln A + \left(\frac{Q}{RT}\right) \tag{17.2}$$

If A and Q are functions of stress only, as assumed by Larson and Miller, this equation is linear in $\ln t$ and $1/T$ for any given stress. By studying a large amount of data, Larson and Miller concluded that the lines for different stresses representing Equation 17.2 converged at a common point on the t axis. This point then becomes the intercept in A, and the slope of the lines then becomes Q/R. The L–M parameter is then defined as

$$T(\ln A + \ln t) \tag{17.3}$$

where T is in degrees Kelvin and t is in hours.

The master curves are constructed by obtaining the time to rupture at various temperatures and stresses. This process can be best explained by referring to the data obtained on Inconel 718 illustrated in Figure 17.9. In Figure 17.9a the conventional log of stress vs. log of time to rupture is depicted for several temperatures. In Figure 17.9b constant time to rupture intercepts are determined, as denoted by the dashed lines, for various values of stress and temperature. These intercepts are used to construct the log t vs. $1/T$ curves of Figure 17.9c. The common intercept would be $\ln A$ in the linear equation 17.2 if the curves had been plotted in terms of degrees Kelvin. However, it is common practice in industry to use the Rankine temperature scale and log to the base 10 in the parameter methods of presenting data. Following this convention, Equation 17.2 now becomes

$$\log t = 2.3 \log A + 2.3 \left(\frac{Q}{RT}\right) \tag{17.4}$$

which can be rewritten as

$$\log t + C = m\frac{1}{T} \tag{17.5}$$

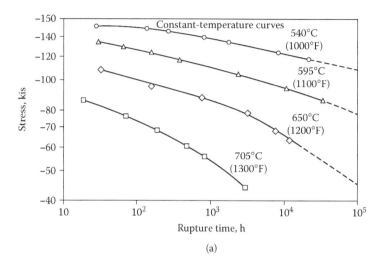

(a)

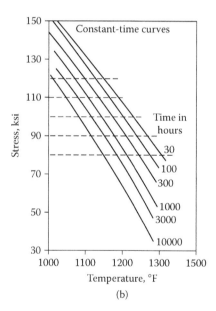

(b)

FIGURE 17.9 Method of constructing a master curve using data for Inconel 718. (From *Metals Handbook*, Vol. 3, 9th ed., ASM International, Metals Park, OH, 1980, p. 239. With permission.)

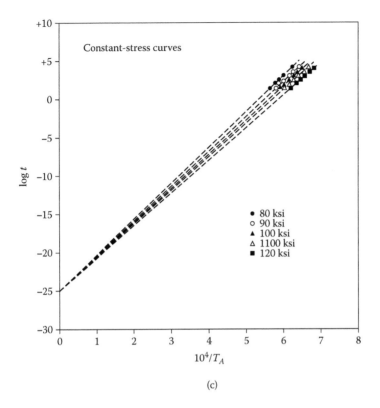

(c)

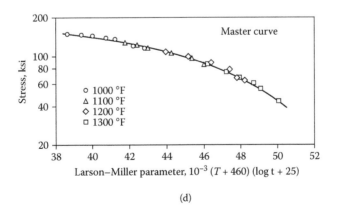

(d)

FIGURE 17.9 (Continued.)

where m = slope = 2 $3Q/R$, or as

$$T(\log t + C) = m = \text{L–M parameter } P \qquad (17.6)$$

where $C = 2.3 \log A$. The parameter P is usually defined as

$$P = T(\log + C) \times 10^{-3} \qquad (17.7)$$

where T is in degrees Rankine. C is now the constant determined by the intercept on the $\log t$ axis in Figure 17.9c. This constant C is found to fall in the range of 15 to 25, and many engineers use a value of 20 without doing the cross-plotting described above. Each point in the time, stress, and temperature family curves of Figure 17.9a can now be used to obtain the master curve of Figure 17.9d. Now any combination of time and temperature that equals the parameter P in Equation 17.7 will fail at the same stress.

Example 17.1 Using the L–M master curve of Figure 17.9d, determine the time to rupture of Inconel 718 for an applied stress of 80 ksi and a temperature of 600°C.

Solution The temperature of the stress application must be converted to the Fahrenheit scale, which is found to be 1112°F. Next a horizontal line is constructed from 80 ksi until it intersects the master curve. The abscissa at the point of intersection is the value of the parameter P, which for 80 ksi is 47. In this particular master curve it is not necessary to convert T to degrees Rankine since the parameter P is stated as

$$P_{\text{L–M}} = 10^{-3}(T + 460)(\log t + 25) = 10^{-3}(1572°\text{R})(\log t + 25)$$

where T is in degrees Fahrenheit.

$$\log t \; 47/1.572 - 25 = 29.9 - 25 = 4.9$$

$$t = 79{,}433 \text{ h} = 9.1 \text{ years}$$

If the L–M parameter is used in terms of the minimum creep rate instead of time to rupture, the parameter P becomes

$$P = T(C - \log \text{strain rate}) \qquad (17.8)$$

where T is in degrees Rankine.

Fatigue data for some representative materials are shown in Figure 17.10. Steels and the continuous fiber composites are the better materials with respect to fatigue life. The fatigue life of these composites will, however, vary considerably with fiber direction and loading direction. Because of their anisotropic characteristics, composite materials have very complex fatigue failure mechanisms.

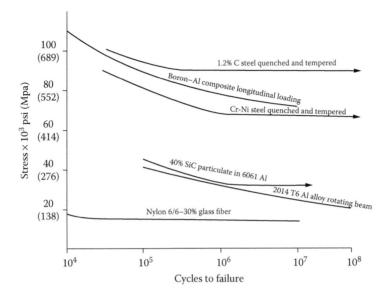

FIGURE 17.10 Fatigue behavior of selected composites and alloys.

Fatigue tends to occur by extensive damage throughout the specimen rather than by a single crack or a few cracks that propagate to failure as in the case of metal fatigue. The most common fatigue failures occur by delamination and interfacial debonding, although fiber fracture has also been observed, though the latter could be a result of the debonding and delamination. Cross-ply laminates tend to show a gradual reduction of strength until failure without a well-defined endurance limit, while the unidirectional laminates show little strength loss prior to failure. The aluminum alloys that contain particulates have a somewhat better fatigue behavior than do the nonreinforced aluminum alloys.

17.3 COMPARATIVE PHYSICAL PROPERTIES

We have already treated the electrical conductivity of metals, insulators, and semiconductors. Because there exists such a large variation among alloys, the electrical resistivity of metals is treated in a separate graph in Figure 17.11. The univalent and divalent metals and their alloys have lower resistivities than do the transition metals and their alloys. Probably the physical property of next most importance and interest is the thermal conductivity. These data for the various categories are summarized in Figure 17.12. Again, typical materials in each group have been selected. The values for composites vary considerably, as do all composite properties. For many composites physical property data are difficult to locate.

Most metals, particularly those with high electrical conductivities because of their large number of free electrons, also have the highest thermal conductivities,

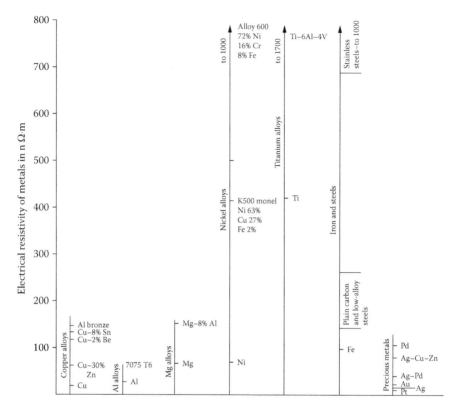

FIGURE 17.11 Electrical resistivities for selected metals. The variation is quite large, being smaller in the univalent, divalent, and precious metals than in the transition metals and their alloys.

while the covalently bonded polymers and ceramics have the lowest values of thermal conductivity. The thermal conductivities of ceramics tend to decrease with increasing temperature, while those for metals remain reasonably constant (see Figure 7.12). The thermoset fiber–epoxy composites, particularly the HMS graphite type, possess reasonable thermal conductivities. The metal matrix composites — particulates, discontinuous, and continuous fiber types — generally have good thermal conductivities. As a result, many of these, especially the boron–aluminum composites, are desirable for electronic packaging applications. Their low density and high moduli are additional benefits for this purpose.

The coefficient of thermal expansion must be considered in many designs in which the materials experience large variations in temperature. Just as in creep, only certain dimensional changes can be permitted. Furthermore, dimensional changes created by thermal expansion can cause localized stress gradients owing to nonuniform thermal conductivities. If large members are restricted in movement during changes in temperature, stresses will be set up since the change in

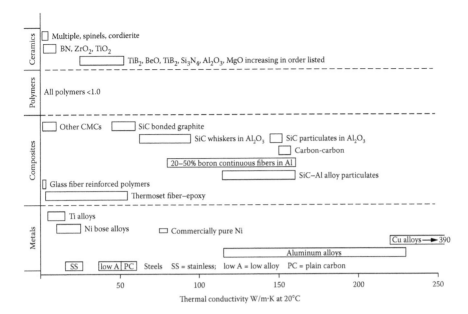

FIGURE 17.12 Thermal conductivities of selected materials. The polymers and ceramics have significantly lower values than do most metals and composites.

temperature and the corresponding desire to expand will place these members in compression. The coefficients of thermal expansion for a number of materials are listed in Table 17.1. The polymers appear to have the largest thermal expansion coefficients, but these materials do not usually experience wide temperature fluctuations, since they are already temperature limited in their applications. Some ceramics have surprisingly significantly high coefficients of thermal expansion, especially those with a high degree of ionic bonding, and along with their low thermal conductivity and high elastic moduli account for their susceptibility to thermal shock. Composites have a mixed bag in the way of thermal expansion coefficients, much as they behave with other properties. Some of those with low-thermal-expansion coefficients and high thermal conductivities are desirable electronic device packaging materials.

QUESTIONS AND PROBLEMS

17.1   If a steel rod of 10 in length with a coefficient of thermal expansion (CTE) of 10^{-6} K^{-1} is restricted in its motion by fixed nonexpandable members, how much stress is established if the rod is heated to 300°C above room temperature (20°C)? Assume a modulus of elasticity of 30×10^6 psi and assume that the CTE is constant over the temperature range of 20 to 300°C.

TABLE 17.1
Coefficient of Thermal Expansion at 20°C (10^{-6} K^{-1})

Metals	CTE	Composites	CTE	Polymers	CTE	Ceramics	CTE
Al alloys	20–24	SiC whiskers in Al alloy	16	ABS	53	BeO	7.6
Cu alloys	17–20	SiC particulates 30% in Al alloy	12	Nylon 6/6	40	MgO	12.8
Plain C steels	11–14	Continuous boron fibers in Al	6	PEEK	26	SiC_2	22.2
Low-alloy steels	12–15	Al_2O_3	6.7	PET	15	ZrO_2	2.1
Stainless steels	11–19	Si whiskers 30% in Al_2O_3	6.7	PMMA	34	SiC	4.6
Ti alloys	7–8	Carbon–carbon	0	Bakelite	16	Al_2O_3, Mullite	7.6
		Glass fibers in thermoplast	34–63	PPS	30		5.5
		Glass fibers in thermosets	9–32	Polysulfone	31		
		Graphite–epoxy thermosets	0.6				
		Boron–epoxy	4.5				

17.2 Locate, by the use of handbooks, vendors' literature, and data bases, materials in each category that have higher strength, higher ductility, and higher moduli than those possessed by the materials shown in Figures 17.1, 17.2, and 17.3.

17.3 Convert the value of 150 W/m · K thermal conductivity to the units cal/cm · s ·°C.

17.4 Convert the value of 200 ksi to MPa.

17.5 Using handbook data, find the ductility for the highest-strength maraging steel commercially available. Would this ductility limit its usage? What compromise materials would you select for optimum strength and ductility?

17.6 Why is Ce–Be much stronger than other copper alloys? For what applications is Cu–Be a suitable alloy?

17.7 Would a SiC continuous-fiber Al alloy composite be favored over a boron continuous-fiber Al alloy (a) for strength at 20°C and (b) for high-temperature strength at 500°C (assuming that 500°C is in the all-alpha range of the aluminum alloy)?

17.8 What are the advantages and disadvantages of the two composites above with respect to each other?

17.9 Assuming similar fatigue $S-N$ curves for both a steel and a boron–Al continuous-fiber composite, which would you choose (a) where strength-to-weight ratio is important and (b) where fatigue is the only consideration?

17.10 What advantage do the particulate composites have over the continuous-fiber composites in the SiC–metal composites?

17.11 Why do the titanium, steel, and nickel alloys have such high electrical resistivities compared to the other alloys in Figure 17.8?

SUGGESTED READING

American Society of Metals, *Advanced Materials and Processes: Guide to Selecting Engineering Materials*, ASM International, Metals Park, OH, 1990.

American Society of Metals, *Engineered Materials Handbooks*, Vols. 1–4, ASM International, Metals Park, OH, 1992.

American Society of Metals, *Engineered Materials Reference Book*, ASM International, Metals Park, OH, 1989.

American Society of Metals, *Metals Handbooks*, Vols. 1 and 2, 10th ed., ASM International, Metals Park, OH, 1990.

Richerson, D. W., *Modern Ceramic Engineering*, Marcel Dekker, New York, 1992.

See also data sheets from the following vendors:

Inco Alloys International
Textron Specialty Materials, Textron, Inc.
Duralcan USA
DuPuut Lanjide Composites
DuPont Kevlar
GTE Laboratories and GTE WESGO

Index

T - #0094 - 111024 - C544 - 229/152/25 - PB - 9780367388669 - Gloss Lamination